# SEMANTIC MULTIMEDIA ANALYSIS AND PROCESSING

# Digital Imaging and Computer Vision Series

**Series Editor**

Rastislav Lukac

*Foveon, Inc./Sigma Corporation*
*San Jose, California, U.S.A.*

# SEMANTIC MULTIMEDIA ANALYSIS AND PROCESSING

EDITED BY

**Evaggelos Spyrou**
Technological Educational Institute of Central Greece, Lamia, Greece

**Dimitris Iakovidis**
Technological Educational Institute of Central Greece, Lamia, Greece

**Phivos Mylonas**
National Technical University of Athens, Athens, Greece

CRC Press
Taylor & Francis Group
Boca Raton   London   New York

CRC Press is an imprint of the
Taylor & Francis Group, an informa  business

CRC Press
Taylor & Francis Group
6000 Broken Sound Parkway NW, Suite 300
Boca Raton, FL 33487-2742

First issued in paperback 2017

Version Date: 20140508

ISBN 13: 978-1-4665-7549-3 (hbk)
ISBN 13: 978-1-138-07538-2 (pbk)

---

**Library of Congress Cataloging-in-Publication Data**

---

Semantic multimedia analysis and processing / editors, Evaggelos Spyrou, Dimitris
  Iakovidis, Phivos Mylonas.
       pages cm. -- (Digital imaging and computer vision series)
  Includes bibliographical references and index.
  ISBN 978-1-4665-7549-3 (hardback)
  1. Semantic Web. 2. Semantic computing. 3. Semantic networks (Information
theory) 4. Multimedia systems. I. Spyrou, Evaggelos, editor of compilation. II. Iakovidis,
Dimitris, editor of compilation. III. Mylonas, Phivos, editor of compilation.

  TK5105.88815.S4264 2014
  006--dc23                                                                2014003911

---

Visit the Taylor & Francis Web site at
http://www.taylorandfrancis.com

and the CRC Press Web site at
http://www.crcpress.com

# Contents

# II   Semantic Knowledge Exploitation and Applications   133

## 6   Engineering Fuzzy Ontologies for Semantic Processing of Vague Knowledge   135

*Panos Alexopoulos*

## 7   Spatiotemporal Event Visualization on Top of Google Earth   163

*Chrisa Tsinaraki, Fotis Kazasis, Nikolaos Tarantilis, Nektarios Gioldasis, and Stavros Christodoulakis*

## 8   Enrich the Expressiveness of Multimedia Document Adaptation Processes    185

*Adel Alti, Sébastien Laborie, and Philippe Roose*

## III  Multimedia Personalization                           219

## 9  When TV Meets the Web: Toward Personalized Digital Media                                                          221

*Dorothea Tsatsou, Matei Mancas, Jaroslav Kuchař, Lyndon Nixon,*
*Miroslav Vacura, Julien Leroy, François Rocca, and Vasileios Mezaris*

## 10 Approach for Context-Aware Semantic Recommendations in FI 257

*Yannick Naudet, Valentin Groués, Sabrina Mignon, Gérald Arnould, Muriel Foulonneau, Younes Djaghloul, and Djamel Khadraoui*

## 11 Treating Collective Intelligence in Online Media 293

*Markos Avlonitis, Ioannis Karydis, Konstantinos Chorianopoulos, and Spyros Sioutas*

## IV Human–Computer Affective Multimedia Interactions 365

## 14 An Overview of Affective Computing from the Physiology and Biomedical Perspective 367

*Ilias Maglogiannis, Eirini Kalatha, and Efrosyni-Alkisti
Paraskevopoulou-Kollia*

# List of Figures

# List of Tables

# *Preface*

Over the last decade the massive creation of user-generated content, mostly in the form of digital images and video, together with the explosion of computational networking activities such as social networks, led to an unprecedented growth of raw digital multimedia content. Being part of this brave new digital world, online multimedia information is becoming increasingly dynamic, and more complex to process, track, and manipulate. Identification of meaningful patterns in the rapidly growing multimedia data repositories requires, now more than ever, efficient and scalable approaches.

Especially today, the need for efficient multimedia knowledge representation and processing methodologies seems more important than ever. One of the most interesting problems in the field of semi- or fully automatic interpretation of content is the detection of semantically meaningful concepts, actions, processes in it. The use of so-called mid-level information, such as knowledge obtained from the application of supervised or unsupervised machine learning methodologies on low-level image characteristics, and additional sources of peripheral knowledge such as geo-tags and user-generated tagging information, aids the semantic disambiguation of multimedia content.

This book provides an overview of the most recent advances in semantic multimedia analysis and processing. It has a broad scope, covering the most important aspects of semantic analysis and processing of multimedia, from algorithms to applications, and from trends to future challenges; it could also be used as a reference for the design and the implementation of contemporary multimedia systems. For consistency, and in order to further highlight authors' contributions, its contents are based on four (4) fundamental thematic pillars, namely: a) *information and content retrieval*, b) *semantic knowledge exploitation paradigms*, c) *multimedia personalization* and d) *human–computer affective multimedia interactions*.

In order for the book to increase global awareness of researchers and practitioners in this area, it includes a scattered distribution of fifteen (15) chapters, authored by researchers from fifteen (15) different research organizations and universities around the world, deriving from nine (9) different countries, namely: Cyprus, France, Germany, Greece, Hong Kong, Ireland, Luxembourg, Spain and the United Kingdom. Each of the accepted contributions provides an in-depth investigation of research and deployment issues, regarding state-of-the-art schemes and applications in the field of semantic multimedia analysis and processing. Moreover, it was among the main concerns of the editors that authors provide survey-based chapters, so potential readers can use it for catching up on recent and ongoing trends and applications with respect to the relevant literature. In the following, a short description of each pillar's chapters, as well as analytical prefaces of each chapter, is provided, summarizing the aims of each chapter and how the work described is related to each chapter topic.

# I. Information and Content Retrieval

The first chapter, entitled "Image Retrieval Using Keywords: The Machine Learning Perspective," prepared by Zenonas Theodosiou and Nicolas Tsapatsoulis, presents a machine learning–based methodology for automatic annotation of image content, and its application to image retrieval using keywords. A concise review of the recent literature and the advantages of this approach over the well-researched topic of content-based image retrieval (CBIR) are discussed. Implementation details are provided regarding low-level feature extraction and fusion, machine learning for image content classification and visual modeling of keywords, which are obtained through crowdsourcing.

The book continues with a chapter dealing with the visual search of objects in multimedia of complex visual context that may be effectively facilitated by saliency-based psycho-visual approaches, as the one described in "Visual Search for Objects in a Complex Visual Context: What We Wish to See" prepared by Hugo Boujut, Aurélie Bugeau and Jenny Benois-Pineau. The idea behind this approach is that the human visual system functions in a way that processes only relevant data as areas of interest. They consider visual saliency for catching spatio-temporal information related to the observer's attention on video frames, and propose an additional geometric saliency cue that models the anticipation phenomenon observed in subjects watching video content from a wearable camera.

Qianni Zhang and Ebroul Izquierdo conduct a survey on "Visual-Semantic Context Learning for Image Classification" and discuss the state-of-the-art research on image classification and retrieval, with specific focus on employment of context and interrelationships between semantic objects that are represented by classes, and using image regions as a representation for objects in such applications. They provide ideas on improving image classification performance by exploiting the semantic context in images, and they describe an innovative method for content-based image classification and semantic context learning, using a block-based representation scheme for semantic classes.

In "Restructuring Multimodal Interaction Data for Browsing and Searching," prepared by Saturnino Luz, the need to combine complementary information coming from different media and sources is addressed, in order to meet the challenges of modern multimedia information retrieval. A unifying architecture and a simple model of linked time-based events which seems general enough to accommodate the structures employed by current systems is proposed and applied for browsing and searching of interactive meetings.

Finally, the fifth chapter of this pillar, entitled "Semantic Enrichment for Automatic Image Retrieval" and prepared by Clement H. C. Leung and Yuanxi Li, provides a well-organized review of the current state-of-the-art of image retrieval and identifies current research challenges. The concepts of content-based and concept-based image retrieval are thoroughly investigated. Practices that have been proposed for injecting semantics in image retrieval are described, while pinpointing the value of ontological mechanisms.

## II. Semantic Knowledge Exploitation and Applications

Acting as a logical bridge between the first and second pillar, the sixth chapter of this book "Engineering Fuzzy Ontologies for Semantic Processing of Vague Knowledge" prepared by Panos Alexopoulos, discusses knowledge representation aspects of multimedia content, focusing on the challenge of vague knowledge treatment and utilization. Fuzzy ontologies that extend classical ontologies with principles of fuzzy set theory enable the assignment of truth degrees to vague ontological elements. Thus, vagueness can be quantified in a way that is exploitable for reasoning and knowledge inference from multimedia data. A step-by-step guide is provided for developing fuzzy ontologies, covering all the required stages from specification to validation and discussing, for each stage's execution, appropriate techniques, tools and best practices.

In "Spatiotemporal Event Visualization on Top of Google Earth," Chrisa Tsinaraki, Fotis Kazasis, Nikolaos Tarantilis, Nektarios Gioldasis and Stavros Christodoulakis address the popular problem of event modeling and visualization on top of spatial representations such as Geographic Information Systems (GIS). Event-related multimedia data that are annotated with their geographical context can be easily integrated in a GIS and associated with semantically represented spatial objects. Implementation aspects are described in the context of a Google Earth–based system called EVISUGE (Event VISUalization on Google Earth).

Considering the importance of multimedia portability in contemporary mobile and cloud-based environments, Adel Alti, Sébastien Laborie and Philippe Roose transfer the focus from pure semantics to applied adaptation problems within the eighth chapter entitled "Enrich the Expressiveness of Multimedia Document Adaptation Processes." In the process they focus on the issue of multimedia document adaptation, i.e., the transformation of any given multimedia documents to be displayable on any device. State-of-the-art approaches are reviewed and the role of semantics is highlighted and illustrated using specific examples.

## III. Multimedia Personalization

Chapter 9, entitled "When TV Meets the Web: Toward Personalized Digital Media" and prepared by Dorothea Tsatsou, Matei Mancas, Jaroslav Kuchař, Lyndon Nixon, Miroslav Vacura, Juliens Leroy, François Rocca, and Vasileios Mezaris, addresses the emerging topic of multimedia personalization and provides an overview of different aspects, knowledge bases and technologies involved in the process of personalizing and contextualizing user experience in networked media environments. A personalization approach within the LinkedTV platform, which enables browsing seamlessly interconnected TV and web content, is provided as a representative example.

In "Approach for Context-Aware Semantic Recommendations in FI," Yannick Naudet, Valentin Groués, Sabrina Mignon, Gérald Arnould, Muriel Foulonneau, Younes Djaghloul and Djamel Khadraoui present state-of-the-art research results toward providing context-aware recommendations in the Future Internet. The approach is based on semantic modeling of knowledge and recommending approaches mixing both semantic and fuzzy processing for better personalization. Methodological details on the use of ontologies and fuzzy sets in a semantic framework for content-based recommendations taking into account the context are provided.

Within the eleventh chapter "Treating Collective Intelligence in Online Media," Markos Avlonitis, Ioannis Karydis, Konstantinos Chorianopoulos and Spyros Sioutas provide a concise introduction to basic concepts for the study of collective intelligence behavior of Web users who share and watch video content. They also present a method that detects collective behavior of users' activity via the detection of characteristic patterns. The efficiency of collective activity detection is addressed by a real-time algorithm.

In the context of cloud-based multimedia exploitation for e-commerce applications, Yolanda Blanco-Fernandez, Martin Lopez-Nores and Jose J. Pazos-Arias explore the possibilities of the Semantic Web technologies for generating automatically interactive e-commerce services that provide users with personalized commercial functionalities related to the selected items. They propose a time-aware semantics-based recommendation strategy and provide practical application examples within the chapter entitled "Semantics-Based and Time-Aware Composition of Personalized e-Commerce Services on the Cloud."

In the thirteenth chapter, entitled "Authoring of Multimedia Content: A Survey of 20 Years of Research," Ansgar Scherp explains fundamental notions of multimedia, and provides a concise review of multimedia authoring and personalization methods. The review includes a classification of state-of-the-art approaches to authoring support for personalized multimedia content and a theoretical comparison between these approaches.

## IV. Human–Computer Affective Multimedia Interactions

Considering the fact that the role of human emotions for affective human–computer interaction has been receiving increasing interest by both the multimedia and semantics community, Ilias Maglogiannis, Eirini Kalatha and Efrosyni-Alkisti Paraskevopoulou-Kollia provide in Chapter 14 of this book, entitled "An Overview of Affective Computing from the Physiology and Biomedical Perspective," a detailed overview of the current state-of-the-art technologies for affective computing and multimedia applications with a focus on physiology and biomedical perspectives.

Finally, the book's last chapter, entitled "Affective Natural Interaction Using EEG: Technologies, Applications and Future Directions," prepared by Charline Hondrou, George Caridakis, Kostas Karpouzis and Stefanos Kollias, provides an overview of technologies that enable affective human–computer interaction through the use of electroencephalography (EEG). Such an approach is investigated under a framework of sensors for EEG signal acquisition and signal analysis methods mainly based on machine learning algorithms. Multimedia applications that would significantly benefit from an EEG-based user-friendly interface are described in a variety of contexts.

Following the division of this book into chapters, bibliographic links included within the latter provide a good basis for further exploration of the topics covered in this edited volume. The volume includes numerous examples and illustrations of semantic multimedia analysis and processing results, as well as tables summarizing the results of quantitative analysis studies. Given the above analytical descriptions covering a variety of research topics in the field of semantic multimedia, we, as editors, feel that this book provides an integrated approach and will act as future reference for the community.

As a last word, we would like to thank all the authors for their submitted inputs, which provided us with the opportunity to edit this book. We hope that all contributions will play a significant role toward a deeper understanding of the key problems in the popular research area of semantic multimedia analysis and processing, and that they will aid researchers, practitioners and developers in finding new solutions to existing problems, opening in parallel new research paths in uncharted waters.

We would also like to explicitly acknowledge the help of all referees involved during all review phases; we believe that their valuable comments and suggestions improved the overall quality of the final published outcome. Last but not least, we would like to express our gratitude to Dr. Rastislav Lukac, editor of the *Digital Imaging and Computer Vision* book series, as well as the Taylor & Francis Group, LLC/CRC Press group as a whole, and Nora

Konopka in particular, for all the support and guidance that were provided to us and the overall fruitful cooperation we had.

The Editors

Dr. Evaggelos Spyrou,[1] Dr. Dimitris K. Iakovidis,[2] Dr. Phivos Mylonas[3]

(1) Technological Educational Institute of Central Greece
Department of Computer Engineering
Lamia, Greece
e-mail: vspyrou@teilam.gr

(2) Technological Educational Institute of Central Greece
Department of Computer Engineering
Lamia, Greece
e-mail: iakovidis@ctr.teilam.gr

(3) National Technical University of Athens
Department of Electrical & Computer Engineering
Athens, Greece
e-mail: fmylonas@image.ntua.gr

# The Editors

**Dr. Evaggelos Spyrou** earned his diploma in electrical and computer engineering from the National Technical University of Athens (NTUA) in 2003, specializing in telecommunications. He was with the Image, Video and Multimedia Systems Laboratory (IVML) of NTUA from 2004 to 2010, working as a researcher and developer in many national and European R&D projects. In 2009 he completed his PhD thesis in the field of semantic image and video analysis. In 2011 he worked as an adjunct lecturer at the University of West Macedonia (UOWM). Since 2011, he has worked as a scientific associate at the Technological Educational Institute (T.E.I) of Central Greece. His current research interests lie in the areas of semantic multimedia analysis, indexing and retrieval, low-level feature extraction and modeling, visual context modeling, multimedia content representation, neural networks, and intelligent systems. He has published research articles in 8 international journals as well as book chapters and in 35 international conferences and workshops.

**Dr. Dimitris K. Iakovidis** earned his BSc degree in physics in 1997, his MSc degree in cybernetics with honors in 2001, and his PhD degree in informatics in 2004, all from the University of Athens, Greece. Since then he has been working in the broader field of image and video processing, pattern recognition and knowledge representation, with a special interest in biomedical applications. He has considerable experience from participation in several national and European research projects, and from adjunct academic positions at the University of Athens, and at the University of Central Greece. Currently he is a professor in the Department of Computer Engineering of the Technological Educational Institute of Central Greece. He is also head of the Institute of Information Technology, vice director of the Center for Technological Research of Central Greece, and serves as an independent expert for the European Commission. Dr. Iakovidis has been a member of IEEE and IAPR since 2005, and he has co-authored over 100 journals, conference papers, and book chapters. He is a book editor, editorial board member, and reviewer of several international scientific journals.

**Dr. Phivos Mylonas** earned his diploma in electrical and computer engineering from the National Technical University of Athens (NTUA) in 2001, his Master of Science (MSc) in advanced information systems from the National & Kapodestrian University of Athens (UoA) in 2003 and earned his PhD degree (with distinction) at the former University (NTUA) in 2008. He is currently a senior researcher at the Image, Video and Multimedia Laboratory, School of Electrical and Computer Engineering, Department of Computer Science of the National Technical University of Athens, Greece and an assistant professor in the Department of Informatics of the Ionian University, Greece. His research interests include content-based information retrieval, visual context representation and analysis, knowledge-assisted multimedia analysis, issues related to multimedia personalization, user adaptation, user modeling, and profiling. He has published articles in 28 international journals and in book chapters. He is the author of 55 papers in international conferences and workshops. He has edited 9 books and is a guest editor of 6 international journals, he is a reviewer for 19 international journals and has been actively involved in the organization of 40 international conferences and workshops. He is a member of the Technical Chamber of Greece since 2001, a member of the Hellenic Association of Mechanical & Electrical Engineers since 2002 and a member of W3C since 2009, as well as a past member of IEEE (1999–2010) and ACM (2001–2010).

# Contributors

**Panos Alexopoulos**
iSOCO S.A.
Madrid, Spain
palexopoulos@isoco.com

**Adel Alti**
Ferhat ABBAS University of Setif
Setif, Algeria
altiadel2002@yahoo.fr

**Gérald Arnould**
Henri Tudor Public Research Centre
Luxembourg
gerald.arnould@tudor.lu

**Markos Avlonitis**
Ionian University
Corfu, Greece
avlon@ionio.gr

**Jenny Benois-Pineau**
University of Bordeaux, LaBRI
Bordeaux, France
benois-p@labri.fr

**Yolanda Blanco-Fernandez**
University of Vigo
Vigo, Spain
yolanda@det.uvigo.es

**Hugo Boujut**
University of Bordeaux, LaBRI
Bordeaux, France
hugo.boujut@labri.fr

**Aurélie Bugeau**
University of Bordeaux, LaBRI
Bordeaux, France
aurelie.bugeau@labri.fr

**George Caridakis**
National Technical University of
   Athens
Athens, Greece
gcari@image.ntua.gr

**Konstantinos Chorianopoulos**
Ionian University
Corfu, Greece
choko@ionio.gr

**Stavros Christodoulakis**
Technical University of Crete
Chania, Greece
stavros@ced.tuc.gr

**Younes Djaghloul**
Henri Tudor Public Research Centre
Luxembourg
younes.djaghloul@tudor.lu

**Muriel Foulonneau**
Henri Tudor Public Research Centre
Luxembourg
muriel.foulonneau@tudor.lu

**Nektarios Gioldasis**
Technical University of Crete
Chania, Greece
nektarios@ced.tuc.gr

**Valentin Groués**
Henri Tudor Public Research Centre
Luxembourg
valentin.groues@tudor.lu

**Charline Hondrou**
National Technical University of
    Athens
Athens, Greece
charline@image.ece.ntua.gr

**Ebroul Izquierdo**
Queen Mary University of London
London, United Kingdom
ebroul.izquierdo@eecs.qmul.ac.uk

**Eirini Kalatha**
University of Central Greece
Lamia, Greece
ekalatha@ucg.gr

**Kostas Karpouzis**
National Technical University of
    Athens
Athens, Greece
kkarpou@image.ntua.gr

**Ioannis Karydis**
Ionian University
Corfu, Greece
karydis@ionio.gr

**Fotis Kazasis**
Technical University of Crete
Chania, Greece
fotis@ced.tuc.gr

**Djamel Khadraoui**
Henri Tudor Public Research Centre
Luxembourg
djamel.khadraoui@tudor.lu

**Stefanos Kollias**
National Technical University of
    Athens
Athens, Greece
stefanos@cs.ntua.gr

**Jaroslav Kuchař**
Czech Technical University in Prague
    and University of Economics
Prague, Czech Republic
jaroslav.kuchar@fit.cvut.cz

**Sébastien Laborie**
University of Pau, LIUPPA
Bayonne, France
sebastien.laborie@iutbayonne.univ-
    pau.fr

**Julien Leroy**
University of Mons
Mons, Belgium
Julien.LEROY@umons.ac.be

**Clement H. C. Leung**
Hong Kong Baptist University
Hong Kong
clement@comp.hkbu.edu.hk

**Yuanxi Li**
Hong Kong Baptist University
Hong Kong
yxli@comp.hkbu.edu.hk

**Martin Lopez-Nores**
University of Vigo
Vigo, Spain
mlnores@det.uvigo.es

**Saturnino Luz**
Trinity College Dublin
Dublin, Ireland
luzs@cs.tcd.ie

**Ilias Maglogiannis**
University of Piraeus
Athens, Greece
imaglo@gmail.com

**Matei Mancas**
University of Mons
Mons, Belgium
Matei.MANCAS@umons.ac.be

**Vasileios Mezaris**
Centre for Research and Technology
    Hellas
Thessaloniki, Greece
bmezaris@iti.gr

**Sabrina Mignon**
Henri Tudor Public Research Centre
Luxembourg
sabrina.mignon@tudor.lu

**Yannick Naudet**
Henri Tudor Public Research Centre
Luxembourg
yannick.naudet@tudor.lu

**Lyndon Nixon**
MODUL University
Vienna, Austria
lyndon.nixon@modul.ac.at

**Efrosyni-Alkisti Paraskevopoulou-
Kollia**
University of Central Greece
Lamia, Greece
ekolia@ucg.gr

**Jose J. Pazos-Arias**
University of Vigo
Vigo, Spain
jose@det.uvigo.es

**François Rocca**
University of Mons
Mons, Belgium
Francois.ROCCA@umons.ac.be

**Philippe Roose**
University of Pau, LIUPPA
Bayonne, France
Philippe.Roose@iutbayonne.univ-
    pau.fr

**Ansgar Scherp**
University of Mannheim
Mannheim, Germany
ansgar@informatik.uni-mannheim.de

**Spyros Sioutas**
Ionian University
Corfu, Greece
sioutas@ionio.gr

**Nikolaos Tarantilis**
Technical University of Crete
Chania, Greece
nicktaras@gmail.com

**Zenonas Theodosiou**
Cyprus University of Technology
Limassol, Cyprus
zenonas.theodosiou@cut.ac.cy

**Nicolas Tsapatsoulis**
Cyprus University of Technology
Limassol, Cyprus
nicolas.tsapatsoulis@cut.ac.cy

**Dorothea Tsatsou**
Centre for Research and Technology
    Hellas
Thessaloniki, Greece
dorothea@iti.gr

**Chrisa Tsinaraki**
Technical University of Crete
Chania, Greece
chrisa@ced.tuc.gr

**Miroslav Vacura**
University of Economics, Prague
Prague, Czech Republic
vacuram@vse.cz

**Qianni Zhang**
Queen Mary University of London
London, United Kingdom
qianni.zhang@eecs.qmul.ac.uk

# Part I

# Information and Content Retrieval

# 1

# Image Retrieval Using Keywords: The Machine Learning Perspective

**Zenonas Theodosiou**

*Cyprus University of Technology,* `zenonas.theodosiou@cut.ac.cy`

**Nicolas Tsapatsoulis**

*Cyprus University of Technology,* `nicolas.tsapatsoulis@cut.ac.cy`

## CONTENTS

In recent years, much effort has been expended on automatic image annotation in order to exploit the advantages of both the text-based and content-based image retrieval methods and compromise their drawbacks, having the ultimate goal to allow content-based keyword searching. This chapter focuses on image retrieval using keywords under the perspective of machine learning, by covering different aspects of the current research in this area. These include low-level feature extraction, creation of training sets and development of machine learning methodologies. Moreover, it presents the evaluation framework of automatic image annotation and discusses various methods and metrics utilized within it. Finally, it proposes the idea of addressing automatic image annotation by creating visual models, one for each available keyword, and presents an example of the proposed idea by comparing different features and machine learning algorithms in creating visual models for keywords referring to the athletics domain.

## 1.1 Introduction

Given the rapid growth of available digital images, image retrieval has attracted a lot of research interest during the last decades. Image retrieval research efforts fall into content-based and text-based methods. Content-based methods retrieve images by analyzing and comparing the content of a given image example as a starting point. Text-based methods are similar to document retrieval, and retrieve images using keywords. The latter is the approach of preference both for ordinary users and search engine engineers. Besides the fact that the majority of users are familiar with text-based queries, content-based image retrieval lacks semantic meaning. Furthermore, image examples that have to be given as a query are rarely available. From the search engine perspective, text-based image retrieval methods have the advantage of well established techniques for document indexing, and they are integrated into a unified document retrieval framework. However, for text-based image retrieval to be feasible, images must be somehow related with specific keywords or textual description. In contemporary search engines this kind of textual description is usually obtained from the web page, or the document, containing the corresponding images and includes HTML alternative text, the file names of the images, captions, surrounding text, metadata tags or the keywords of the whole web page [588]. Despite the fact that this type of information is not directly related to the content of the images it can be utilized only in web-page image retrieval. As a result, image retrieval from dedicated image collections can be done either by content-based methods or by explicitly annotating images by assigning tags to them to allow text-based search. The latter process is collectively known as "image annotation" or "image tagging."

Image annotation can be achieved using various approaches like free text descriptions, keywords chosen from controlled vocabularies, etc. Nevertheless, the annotation process remains a significant difficulty in image retrieval, since the manual annotation seems to be the only way guaranteeing success. This is partially a reason explaining why the content-based image retrieval is still considered an option for accessing the enormous amount of digital images. Despite the plethora of available tools, manual annotation is an extremely difficult and elaborate task, since the keyword assignment is performed on an image basis. Furthermore, manual annotations cannot always be considered as correct, due to the visual information that always allows the possibility for contradicting interpretation and ambiguity [395].

In recent years, much effort has been expended on automatic image annotation in order to exploit the advantages of both the text-based and content-based image retrieval methods and compromise their drawbacks mentioned above. In any case, the ultimate goal is to allow keyword searching based on the image content [1035]. Thus, automatic image annotation efforts try to mimic humans aiming to associate the visual features that describe the image content with semantic labels.

This chapter focuses on image retrieval using keywords under the perspective of machine learning. It covers different aspects of the current research in this area, including low-level feature extraction, creation of training sets and development of machine learning methodologies. It also presents the evaluation framework of automatic image annotation and discusses various methods and metrics utilized within it. Furthermore, it proposes the idea of addressing automatic image annotation by creating visual models, one for each available keyword, and presents an example of the proposed idea by comparing different features and machine learning algorithms in creating visual models for keywords referring to the athletics domain.

## 1.2   Background

Automatic image annotation has been a topic of ongoing research for more than a decade. Several interesting techniques have been proposed during this period [641]. Although it appears to be a particularly complex problem for researchers and despite the fact that annotation obtained automatically is not expected to reach the same level of detail as the one obtained by humans, it remains a research hot topic. The reason is obvious: Manual annotation of the enormous number of images created and uploaded to the web every day is not only impractical; it is simply impossible. Therefore, automatic assignment of keywords to images for retrieval purposes is highly desirable. The proposed methods attempted to address first, the difficulty of relating high-level human interpretations with low-level visual features and second, the lack of correspondence between the keywords and image regions in the (training) data.

Traditionally in content-based image retrieval, images are represented and retrieved using low-level features such as color, texture and shape regions. Similarly, in automatic image annotation, a manually annotated set of data is used to train a system for the identification of joint or conditional probability of an annotation together with a certain distribution of feature vectors corresponding to image content [83]. Different models and machine learning techniques were developed to learn the correlation between image features and textual words, based on the examples of annotated images. Learned models of this correlation are then applied to predict keywords for not yet seen images [1040]. Although the low-level features extracted from an image cannot be automatically translated reliably into high-level semantics [263], the selection of visual features that better describe the content of an image is an essential step for the automatic image annotation. The interpretation inconsistency between image descriptors and high-level semantics is known as a "semantic gap" [856] or "perceptual gap" [457]. Recent research focuses on new low-level feature extraction algorithms to bridge the gap between the simplicity of available visual features and the richness of the user semantics.

The co-occurrence model proposed by Mori et al. [686] can be assumed as the first automatic image annotation approach. This model tries to capture the correlations between images and keywords (assigned to them), and consists of two main stages. In the first stage, every image is divided into subregions and a global descriptor for each subregion is calculated. In the second stage, feature vectors extracted from sub-regions are clustered using vector quantization. Finally, the probability of a label related to a cluster is estimated by the co-occurrence of the label and the subregions within the cluster. Duygulu et al. [303] proposed a model of object recognition as a machine translator in order to annotate images automatically. Every image is segmented into object shape regions, called "blobs," and a visual vocabulary is created by feature quantization of the extracted feature vectors of the regions. Finally, the correspondence between blobs and words is found by utilizing the Expectation-Maximization algorithm. The Cross Media Relevance Model (CMRM) was introduced by Jeon et al. [468] in order to improve the machine translator model. They followed the same procedure for calculating the blob representation of images as Duygulu et al. and then utilized the CMRM to learn the joint distribution of blobs and words in a given image. The loss of useful information during the quantization from continuous features into discrete blobs that occurred on the translation model and CMRM, was treated by Lavrenko et al. [559]. The proposed Continuous Relevance Model (CRM) does not require an intermediate clustering stage and directly associates continuous features with words. Further improvement on annotation results was obtained by the Multiple Bernoulli Relevance Model (MBRM) [331], where the word probabilities are estimated using a multiple Bernoulli model, and the image feature probabilities are estimated using a non-parametric kernel density estimate. The computational cost of parameter estimation is probably one of the drawbacks of using statistical models in automatic image annotation approaches, since the learning of parameters lasts several hours. Nevertheless, object recognition based methods for image annotation are of limited scope because object recognition itself is a very hard problem and is solved only under strict constraints.

Automatic image annotation tries to learn the behavior of humans. Thus, utilization of machine learning methods is almost natural [548]. The objective is to learn a set of rules from training instances (pairs of images and keywords) and the creation of a classifier that can be used to generalize to new instances [537]. Several methods were developed within the machine learning framework. Classification and clustering based methods are among the most popular [548]. In classification approaches, image classifiers are constructed with the aid of training data, and are applied to classify a given image into one of several classes. Each class usually corresponds to a particular keyword. Several machine learning algorithms have been used for image classification into keywords classes. Support Vector Machine (SVM) [247], Hidden Markov Models [364], Decision Trees [469], are some of them. An extensive review of machine learning classifiers in image annotation is given in [537] while the general principal of machine learning utilization is revisited in Section 1.4.

Although classification based annotation methods give promising results, they are designed for small-scale datasets and they use a small number of keyword classes. As a result the trained classifiers do not generalize smoothly to allow accurate classification, of the large amount of images that are missing annotations, to the available classes. The limited number of manually annotated data (few positive and negative examples) that are used during training leads to ineffective keywords models without generalization ability. The limited number of classes, on the other the hand, is very restrictive to the number of text queries that will derive results. Users, in general, are reluctant to adopt search interfaces that are based on predefined sets of keywords because they are familiar with the free text searching paradigm used in web-search engines. Another problem of classification based annotation is that classifiers relate images with a single keyword while it is obvious that image content can be associated with many keywords. The multi-instance multi-label learning (MIML) proposed in [1046], [1047] where each training example is described by multiple instances and associated with multiple class labels tries to eliminate this problem. Although this method gives a fair solution to the problem of assigning more than one keyword to a given image, it has also several limitations. In order to model a valid probability of labels it is necessary to compute a summation over all possible label assignments leading to high computational cost. Furthermore, there is no provision to add new labels (keywords). The initial set of keywords remains unchanged while erroneous tagging is accumulated since the labels, that are assigned to a particular image, depend on its content similarity with other images that already have this label. Therefore, in cases where the initial label of an image is erroneous, the error is propagated to all images having similar content.

Creating independent keyword models, separately, appears to be a realistic solution to the drawbacks of the previous method. A given image could be associated with more than one keyword and a new keyword model can be trained irrespectively of the existing ones. This approach provides the required scalability for large scale text-based image retrieval. The idea of automatic image annotation through independent keyword visual models is illustrated in Fig. 1.1. The whole procedure is divided into two main parts: the training, and automatic image annotation. In the first part, visual models for all available keywords are created, using the one-against-all training paradigm, while in

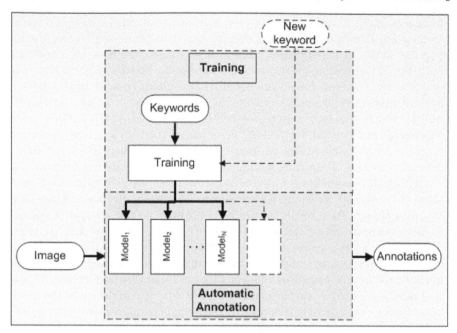

FIGURE 1.1: Automatic image annotation using visual models of keywords.

the second part, annotations are produced for a given image based on the output of these models, once they are fed by a feature vector extracted from the input image. On the other hand, whenever training data for new keyword are available, a new visual model is created for this keyword, and added into the unified framework. A detailed explanation of the use of keyword visual models for automatic image annotation is given in Section 1.4.

### 1.2.1    Key Issues in Automatic Image Annotation

There are some issues commonly encountered in automatic image annotation systems whether image retrieval is approached using the content-based or the text-based paradigm.

Initially, the problem of searching the enormous number of digital image collections that are available through the web or in personal repositories was tackled by efficient and intelligent schemes for content-based image retrieval (CBIR) [856]. CBIR computes relevance based on the visual similarity of image low-level features such as color, texture and shape [706]. Early CBIR systems were based on the query-by-example paradigm [391], which defines image retrieval as the search for the best database match to a user-provided query image. Under this framework, and in order to maximize the effectiveness of CBIR systems, it soon became necessary to mark specific regions in the images so as to model particular objects. During automatic annotation, labels corresponding to object classes are assigned every time a particular object instance is encountered in the input image. As already mentioned, this object instance almost always corresponds to an image region (part of the image). Therefore, region-based features must be computed and used for object modeling. Under this perspective, semantic labeling using object class labels is actually an object detection task. Unfortunately, object detection is a very hard task itself and is solvable only for limited cases and under strict constraints.

The low performance of CBIR systems along with their limited scope led researchers to investigate alternative schemes for image retrieval. It was quickly realized that the ultimate users were willing to search for images using their familiar text-based search interface. Therefore, the design of fully functional image retrieval systems would require support for semantic queries [745]. In such systems, images in the database are annotated with semantic labels, enabling the user to specify the query through a natural language description of the visual concepts of interest. This realization, combined with the cost of manual image labeling, generated significant interest in the problem of automatically extracting semantic descriptors from images.

As already mentioned in the previous section, automatic annotation of digital images with semantic labels is traditionally handled by utilizing machine learning emphasizing classification. In that case, semantic labels may refer to an abstract term, such as indoor, outdoor, athletics, or to an object class such as human, car, tree, foam mats, etc. The latter case reinforces the object detection problem encountered in CBIR systems. In order to learn a particular object class and create a classifier to recognize it automatically, you need accurate and region specific features of the training examples. Indeed, there exist some approaches [1013] that use some kind of object detection in order to assign semantic labels to images [192]. In contrast to object classes, abstract terms cannot be related to specific image regions. In the literature of automatic semantic image annotation, proposed approaches tend to classify images using only abstract terms or using holistic image features for both abstract terms and object classes. Nevertheless, extraction and selection of low-level features, either holistic or from particular image areas is of primary importance for automatic image annotation. This is true either for the content-based or for the text-based retrieval paradigm. In the former case the use of appropriate low-level features leads to accurate and effective object class models used in object detection, while in the latter case the better the low-level features are, the easier the learning of keyword models is. Low-level feature extraction is one of the key issues in automatic image annotation and is examined in detail in the next section.

The intent of the image classification is to categorize the content of the input image in one of several keyword classes. A proper image annotation may contain more than one keyword that is relevant to the image content, so a reclassification process is required in this case, as well as whenever a new keyword class is added to the classification scheme. The creation of separate visual models for all keyword classes adds a significant value in automatic image annotation since several keywords can be assigned to the input image. As the number of keyword classes increases the number of keywords assigned to the images also increases, and there is no need for reclassification. However, the keyword modeling incurred various issues such as the large amount of manual effort required in developing the training data, the differences in interpretation of image contents, and the inconsistency of the keyword assignments among different annotators. These key issues are also examined in detail in subsequent sections.

## 1.3    Low-Level Feature Extraction

Low-Level feature extraction is the first crucial step in the keyword modeling process; aims at capturing the important characteristics of the visual content of images. The low-level features are defined to be those basic features that can be extracted automatically from an image without any information about spatial relationships [712]. They can be broadly divided into two main types: (a) Local or domain-specific features, and (b) Global or holistic features. Selection of the most appropriate subset of features plays a significant role in efficient classification schemes as well as in visual modeling of keywords. Feature extraction and selection can be evaluated from three different perspectives: First, in terms of their ability to identify relevant features, second in terms of the performance of the created classifiers and third, in terms of the reduction of the number of features. The research in feature extraction is rich and dozens of methods have been proposed for keyword modeling.

### 1.3.1    Local Features

Local features are image patterns that differ from their immediate neighborhood. They are usually associated with a change of an image property or several properties simultaneously, although they are not necessarily localized exactly on this change. Image properties commonly considered for local feature derivation are intensity, color, and texture. Local invariant features not only allow finding correspondences in spite of large changes in viewing conditions, occlusions, and image clutter, but also yield an interesting description of the image content. Ideal local features should have repeatability, distinctiveness, locality, quantity, accuracy and efficiency [935]. Local features were first introduced by Schiele and Crowely [820], and Schmid and Mohr [821] and soon became very popular especially in machine learning frameworks.

The Scale Invariant Features Transform (SIFT) [619] and the Histogram of Gradients (HOG) [254] are two of the most successful local features categories. They are based on histograms of gradient orientations weighted by gradient magnitudes. The two methods differ slightly in the type of spatial bins that they use. The SIFT, proposed by Lowe [619], transforms image data into scale-invariant coordinates relative to local features and computes a set of features that are not affected by object scaling and rotation. Key points are detected as the maxima of an image pyramid built using difference-of-Gaussians. The multi-scale approach results in features that are detected across different scales of images. For each detected keypoint, a 128 dimensional feature vector is computed describing the gradient orientations around the keypoint. The strongest gradient orientation is selected as a reference, thus giving rotation invariance to SIFT features. On the other hand, HOG uses a more sophisticated way for binning. The image is divided into small connected regions and a histogram of gradient directions or edge orientations within each region is compiled. For the implementation of HOG, each pixel within the region casts a weighted vote for an orientation-based histogram channel.

Due to the large number of SIFT keypoints contained in an image, various approaches have been used to reduce the dimensionality or prune the number of detected keypoints before using them to train keyword models. Another difficulty in using the original SIFT features in machine learning is that the number of keypoints, and consequently the dimensionality of input vector, is image dependent. As a result they cannot be directly employed for creating and feeding keyword models. The PCA-SIFT is proposed in [503] by utilizing Principal Component Analysis (PCA) to normalized gradient patches to achieve fast matching and invariance to image deformations. Mikolajczyk and Schmid [676] presented an extension of the SIFT descriptor, the significance of the Gradient Location and Orientation Histogram (GLOH) which also applies the PCA for dimensionality reduction. Instead of PCA, the Linear discriminate Analysis (LDA) has also been applied to create a low-dimensional representation [89].

The effectiveness of SIFT and GLOH features led to several modifications that try to combine their advantages. Recently, the Speeded Up Robust Features (SURF) descriptor that approximates the SIFT and GLOH by using integral images to compute the histogram bins has been proposed [107]. This method is computationally efficient with respect to computing the descriptor values at every pixel and differs from SIFT's spatial weighting scheme. In particular, all gradients contribute equally to their respective bins, which results in damaging artifacts when used for dense keypoints' computation. The Daisy descriptor [915], on the other hand, retains the robustness of SIFT and GLOH and can be computed quickly at every single image pixel.

Other approaches use clustering techniques to manage the thousands of local descriptors produced by SIFT. Bag-Of-Features (BOF) methods represent an image as an orderless collection of local features [849], [1037], [592]. Usually, the k-means clustering algorithm groups visual patches into one cluster and creates a visual vocabulary. For each image, the number of occurrences of each word is counted to form a histogram representation. Besides the advantages of the BOF representation, these methods have important descriptive limitations because they disregard the spatial information of the local features. Lazebnik et al. [560] extended the BOF approach and proposed the Spatial Pyramid Matching method, which partitions the image into increasingly fine subregions and computes histograms of local features found inside each subregion.

SIFT features were originally proposed for object detection and recognition tasks. In these tasks a dedicated matching scheme is used to compare images or image regions. In keyword modeling this is not the case. The SIFT feature vector feeds the keyword visual models to produce an output indicating whether or not the corresponding keyword can be assigned to the image corresponding to this input vector. This difference, along with the dimensionality reduction, which is applied to produce SIFT based vectors of fixed dimensionality, leads to deteriorating performance in image retrieval compared to other types of features, like the MPEG-7 descriptors [910].

### 1.3.2 Global or Holistic Features

Global features provide different information than local ones, since they are extracted from the entire image. Statistical properties such as histograms, moments, contour representations, texture features and features derived from image transforms like Fourier, Cosine and Wavelets can be considered as global features. Global features cannot separate foreground from background information; they combine information from both parts together [935]. These features can be used when there is interest in the overall composition of the image, rather than a foreground object. However, in some cases, global features have been also applied for object recognition [933], [689]. The feature set in these approaches obtained from the projections to the eigenspace created by computing the prominent eigenvectors is based on the Principal Component Analysis of the image training sets.

Recently, the Compact Composite Descriptors (CCDs) [211] which capture more than one type of information at the same time in a very compact representation have been used for image retrieval applications [75], [207]. The Fuzzy Color and Texture Histogram (FCTH) [209] and the Color and Edge Directivity Descriptor (CEDD) [208] are determined for natural color images and combine color and texture information in a single histogram. The Brightness and Texture Directionality Histogram (BTDH) descriptor [210] describes grayscale images and captures both brightness and texture characteristics in a 1D histogram. Finally, the Spatial Color Distribution Descriptor (SpCD) [212] combines color and spatial color distribution information and can be used for artificial images. The performance of CCDs has been evaluated using several databases, and experimental results indicated high accuracy in image retrieval, achieving in some cases, better performance than other commonly used features for image retrieval such as the MPEG-7 descriptors.

The MPEG-7 visual descriptors [1] use a standardized description of image content and they were especially designed for image retrieval in the content-based retrieval paradigm. Their main property is the description of global image characteristics based on color, texture or shape distribution, among others. A total of 22 different kinds of features (known as descriptors) are included: nine for color, eight for texture and five for shape. The various feature types are shown in Table 1.1. The number of features, shown in the third column of the table, in most cases is not fixed and depends on user choice. The dominant color descriptor includes color value, percentage and variance, and requires specially designed metrics for similarity matching. Furthermore, the number of features included in this descriptor is not known a priori since they are image dependent (for example an image may be composed from a single color whereas others vary in color distribution). The previously mentioned difficulties cannot be easily handled in machine learning schemes and as a result the dominant color descriptor is rarely used in keyword modeling and classification schemes. The region shape descriptor features are computed only on specific image regions (and therefore they are not used in holistic image description). The number of peak values of the contour shape descriptor varies depending on the form of an input object. Furthermore, they require a specifically designed metric for similarity matching because they are computed based on the HighestPeak value. The remaining of the MPEG-7 descriptors shown in Table 1.1 can be easily employed in machine learning schemes and

**TABLE 1.1**

MPEG-7 Visual Descriptors.

| Descriptor | Type | #Features |
|---|---|---|
| Color | DC coefficient of DCT (Y channel) | 1 |
| | DC coefficient of DCT (Cb channel) | 1 |
| | DC coefficient of DCT (Cr channel) | 1 |
| | AC coefficients of DCT (Y channel) | 5 |
| | AC coefficients of DCT (Cb channel) | 2 |
| | AC coefficients of DCT (Cr channel) | 2 |
| | Dominant colors | Varies |
| | Scalable color | 16 |
| | Structure | 32 |
| Texture | Intensity average | 1 |
| | Intensity standard deviation | 1 |
| | Energy distribution | 30 |
| | Deviation of energy's distribution | 30 |
| | Regularity | 1 |
| | Direction | 1 or 2 |
| | Scale | 1 or 2 |
| | Edge histogram | 80 |
| Shape | Region shape | 35 |
| | Global curvature | 2 |
| | Prototype curvature | 2 |
| | Highest peak | 1 |
| | Curvature peaks | Varies |

since they are specially designed for image retrieval, they are an obvious choice
for keyword modeling.

Global features are a natural choice for image retrieval that is based on
machine learning. Since they are extracted from the image as a whole they
are also appropriate for creating visual models for keywords. This is because
training data can be created by defining the keywords that are related with
the images used for training and there is no need to define specific regions in
these images (which is by far more tedious). However, the choices of global
features from which one can select is unlimited and in some cases depend on
the type of features. Despite the fact that the MPEG-7 descriptors were initially proposed for CBIR systems they perform excellently within the machine
learning paradigm used either in classification based keyword extraction or in
keyword modeling. As a result they provide a good starting point in experimentation dealing with automatic image annotation and should be used as a
benchmark test before adopting different feature types.

### 1.3.3   Feature Fusion

Feature fusion is of primary importance where multiple features types are used
in training keyword models. Fusion can derive and gain the most effective
and least dimensional feature vectors that benefit final classification [1014].
Usually for each keyword group, various feature vectors are normalized and
combined into a feature union-vector whose dimension is equal to the sum
of the dimensions of the individual low-level feature vectors. Dimensionality
reduction methods are then applied to extract the linear features from the
integrated union vector and reduce the dimensionality. Principle Component
Analysis (PCA) and Linear Discriminant Analysis (LDA) are two widely used
approaches in this framework.

The PCA is a well-established technique for dimensionality reduction
which converts a number of correlated variables into several uncorrelated variables called principal components. For a set of observed $d$-dimensional data
vectors $X_i$, $i \in \{1, ..., N\}$, the $M$ principal components $p_j$, $j \in \{1, ..., M\}$ are
given by the $M$ eigenvectors with the largest associated eigenvalues $\lambda_j$ of the
covariance matrix:

$$S = \frac{1}{N} \sum_i (X_i - \overline{X})(X_i - \overline{X})^T \qquad (1.1)$$

where $\overline{X}$ is the data sample mean and $Sp_j = \lambda_j p_j$. The $M$ principal components of the observed vector $X_i$ are given by the vector:

$$c_i = P^T(X_i - \overline{X}) \qquad (1.2)$$

where $P = \{p_1, p_2, ..., p_M\}$. The variables $c_j$ are uncorrelated because the
covariance matrix $S$ is diagonal with elements $\lambda_j$. Usually cross-validation
is performed to estimate the minimum number of features required to yield

the highest classification accuracy. However, the computational cost of cross-validating is prohibitive so other approaches such as the maximum likelihood estimator (MLE) [581] are employed to estimate the intrinsic dimensionality of the fused feature vector by PCA.

LDA follows a supervised method to map a set of observed $d$-dimensional data vectors $X_i$, $i \in \{1, ..., N\}$ to a transformed space using a function $Y = wX$. The $w$ is given by the maximum eigenvector of the $S_w^{-1}S_b$ where $S_w$ is the average within-class scatter matrix and $S_b$ is the between-class covariance matrix of $X_i$.

The matrix $w$ is determined such that the Fisher criterion of between-class scatter over average within-class scatter is maximized [351]. The original Fisher criterion function applied in the LDA is,

$$J = \frac{wS_bw^T}{wS_ww^T} .$$ (1.3)

Obviously there are several fusion techniques that can be used to select the best feature set for training visual models for keywords. However, both PCA and LDA are based on a strong mathematical background and should be investigated before examining alternatives. Nonlinear fusion methods, on the other hand, might be proved more efficient in some cases.

## 1.4 Visual Models Creation

Creating accurate visual models for keywords depends not only on the low-level feature set that is used but also on the training data. Availability of training data and the use of specially designed earning algorithms are two important factors that must be also carefully investigate. This section summarizes the various approaches used to deal with these factors.

### 1.4.1 Dataset Creation

Training examples that are used for creating visual models for keywords are pairs of images and keywords. The low-level feature vector extracted from the image is considered as an example of the visual representation of keywords assigned to this image. Aggregating feature vectors across many images eliminates the case of having several keywords sharing exactly the same training examples. However, collection of manually annotated images to be used for creating the keyword visual models is a costly and tedious procedure. Furthermore, manual annotations are likely to contain human judgment errors, and subjectivity in interpreting the image due to differences in visual perception and prior knowledge. As presented in [908], there are several demographic factors that influence the way that people annotate images. As a result it is a common practice nowadays to use multiple annotations per image obtained from different people to alleviate this subjectivity as well as for detecting outliers or erroneous annotations. In the past, manually annotated datasets were obtained by experts. Since the majority of tomorrow's users of search engines are non experts, the idea of modeling the knowledge of several people rather than an expert can significantly improve the ultimate efficiency of image retrieval systems.

As already mentioned, multiple judgments per image, from several people, improve the annotation quality. The act of outsourcing work to a large crowd of workers is rapidly changing the way datasets are created [983]. The fact that differences between implicit and explicit relevance judgments are not so great [475] opened a new way, where implicit relevance judgments were considered as training data for various machine learning-based improvements to information retrieval [637], [925].

Crowdsourcing [430] is an attractive solution to the problem of cheaply and quickly acquiring annotations. It has the potential to improve evaluation of the keyword modeling by scaling up relevance assessments and creating collections with more representative judgments [501]. Amazon's Mechanical Turk [2] presents a practical framework to accomplish such annotation tasks by extending the interactivity of crowdsourcing using more comprehensive user interfaces and micro-payment mechanisms [307].

An alternative to manual annotation of training data is to explore the successful mechanisms of automatic keyword extraction in text-based documents adopted by contemporary search engines. The large amount of web images located in text documents and webpages can be used for that purpose. The text that surrounds theses images inside the web documents provides important semantic information that can be used for keyword extraction. Web Image Context Extraction (WICE) denotes the process of determining the textual contents of a web document that are semantically related to an image and associates them with that image. WICE uses the associated text as a source for deriving the content of images. In text-based image retrieval, the user provides keywords or key phrases, and text retrieval techniques are used for retrieval of the best ranked image. Successful web image search engines like the Google images[1] and Yahoo!Image Search[2] are well known WICE examples.

Image file names, anchor texts, surrounding paragraphs or even the whole text of the hosting web page are usually used as a textual content in WICE applications. Several approaches have been proposed to extract the text blocks as concept sources for images. A bootstrapping approach to automatically annotate web images based on a predefined list of concepts by fusing evidence from image contents and their associated HTML text in terms of a fixed size of sequence is presented in [329]. The main drawback of the proposed method is the low annotation performance because extracted text may be irrelevant

---

[1]http://images.google.com/
[2]http://images.search.yahoo.com/

to the corresponding image. Other applications use the DOM tree structure of the web page [326], [55] and the Vision based Page Segmentation (VIPS) algorithm [404] to extract image context using the surrounding text. Recently an interesting approach was presented in [920] where the context extraction is achieved by utilizing the VIPS algorithm and semantic representation of text blocks.

Usually, the datasets created either through crowdsourcing or based on the surrounding text principle suffer from the limited number of labeled images that correspond to the keyword classes of interest. A well structured framework can moderate the restrictions introduced by small labeled datasets and boost the performance of the learner using co-training algorithms [126]. In such a case, the labeled images are presented by two different views and two learning algorithms are trained separately on each view, and the prediction on unlabeled images of each algorithm is used to enlarge the training set of the other.

It is fair to mention here, however, that automatic extraction of keywords from webpages containing images so as to be used as training data is clearly inferior to using manually annotated images obtainedthrough crowdsourcing. The co-training mechanism mentioned above provides an attractive framework for combining crowdsourcing and automatic keyword extraction so as to get the advantage of both: accuracy and ease of training data collection. Initial training is performed using crowdsourced data while co-training is applied on training data collected automatically.

## 1.4.2   Learning Approaches

Machine learning methods play an important role in automatic image annotation schemes. Machine learning involves algorithms that build general hypotheses based on supplied instances, and then uses them to make predictions about future instances. Classification algorithms are based on the assumption that input data belong to one of several classes that may be specified, either by an analyst or because they are automatically clustered. Many analysts combine supervised and unsupervised classification processes to develop final output analyses and classified maps.

Supervised image classification organizes instances into classes by analyzing the properties of the supplied image visual features where each instance is represented by the same number of features. Training instances are split into training and test sets. Initially, the characteristic properties of the visual features of the training instances are isolated and class learning finds the description that is shared by all of the positive instances. The resulting classifier is then used to assign class labels to the testing instances where the values of the predictor features are known, but the value of the class label is unknown. Several supervised methods based on rules, neural networks, and statistical learning have been utilized for classifying images into class labels as well as for keyword model creation.

Decision trees are logic-based learning algorithms that sort instances according to feature values based on the divide-and-conquer approach. They are developed by algorithms that split the input set of visual features into branch-like segments (nodes). A decision tree consists of internal decision nodes where a test function is applied and the proper branch is taken based on the outcome. The process starts at the root node and it is repeated until a leaf node is achieved. There is a unique path for data to enter a class that is defined by each leaf node and this path is used to classify unseen data. A variety of decision tree methods have been used in classification tasks such as CART [163], ID3 decision tree [761], its extension C4.5 [762] that has shown a good balance between speed and error rate [600], and the newest, Random Forest [162].

Although decision trees offer a very fast processing and training phase compared to other machine learning approaches, they suffer from the problem of overfitting to the training data, resulting in some cases in excessively detailed trees and low predictive power for previously unseen data. Furthermore, decision trees were designed for classification tasks: Every input entering the tree's root is classified to only one of its leaves. Assuming that the leaves correspond to keywords and the input is a low-level vector extracted from an input image, the image is assigned at most one keyword during prediction. The keyword models, on the other hand, are based on the one-against-all paradigm. For each keyword there is a dedicated predictor which decides, based on the low-level feature vector it fed with, whether the corresponding image must be assigned the particular keyword or not.

The rules created for each path from the root to a leaf in a decision tree can also be used directly for classification. Rule based algorithms aim to construct the smallest rule-set that is consistent with the training data. In comparison with decision trees, rule-based learning evaluates the quality of the set of instances that is covered by the candidate rule while the former evaluates the average quality of a number of disjoint sets [537]. However, rule-based learning faces problems with noisy data. More efficient learners have been proposed such as the IREP (Incremental Reduced Error Pruning) [352] and Ripper (Repeated Incremental Pruning to Produce Error Reduction) [235] to overcome these drawbacks.

Assuming that every keyword is modeled with a rule, then rule-based learning is appropriate for creating visual models for keywords, since it provides the required scalability. That is, every time a new keyword must be modeled a new rule is constructed, based on available training data, without affecting the existing keyword models (rules). Unfortunately, the case is not so simple. Rule based systems perform well in cases where the dimensionality of input is limited. However, the low-level features that are used to capture the visual content of an image or image region are inherently of high dimensionality. Thus, despite their scalability, rule-based systems lack in performance compared to other keyword modeling schemes.

Neural Networks (NNs) have incredible generalization and good learning ability for classification tasks. NNs use the input data and train a network to learn complex mapping for classification. An NN is supplied by the input instances and actual outputs and then compares the predicted class with the actual class, and estimates the error to modify the weights. There are numerous NN based algorithms with significant research interest in Radial Basis Function (RBF) networks [785] since in comparison with the multilayer perceptrons, the RBFs train faster and their hidden layer has easier interpretation. Furthermore, in a comparative study on object classification [921] for keyword extraction purposes, RBFs proved to be the more robust and had the highest predicting performance among several state of the NN classifiers.

An RBF network consists of an input layer, a hidden layer with an RBF activation function and a linear output layer. Tuning the activation function to achieve the best performance is a little bit tricky, quite arbitrary and dependent on the training data. Thus, despite their significant abilities in prediction and generalization RBF networks are not as popular as the statistical learning approaches discussed next.

Support Vector Machines (SVMs), a machine learning scheme which is based on statistical learning theory, is one of the most popular approaches to data modeling and classification [952]. SVMs, with the aid of kernel mapping, transform input data into a high dimensional feature space and try to find the hyperplane that separates positive from negative samples. The kernel can be linear, polynomial, Gaussian, Radial Basis Function (RBF), etc. The hyperplane is chosen such that it keeps the distance between the nearest positive and negative examples as high as possible. The number of features encountered in the training data does not affect the model complexity of an SVM, so SVMs deal perfectly with learning tasks where the dimensionality of feature space is large with respect to the number of training instances. Training a classifier using SVMs has less probability of losing important information because SVMs construct the optimal hyperplane using dot products of the training feature vectors with the hyperplane. Sequential Minimal Optimization (SMO) [751], [504] and LibSVM [324] are two of the state-of-the-art implementations of the SVMs with high classification performance.

As far as the creation of visual models from keywords is concerned, the SVMs have many desirable properties. First, they are designed to deal with binary problems (they make decisions on whether or not an input instance belongs to a particular class) which provides the required scalability to train new keyword models independently of the existing ones. Second, they deal effectively with the large dimensionality of the input space created by low-level feature vectors extracted from images. Finally, they do not require as many training examples as other machine learning methods. As we have already mentioned, in automatic image annotation the availability of training data is a key issue. Therefore, methods that are conservative in this requirement are highly preferable.

## 1.5 Performance Evaluation

Keyword visual models are created using training data but their performance must be evaluated on unseen training examples. The experimental performance on unseen data tests the ability of the visual models to generalize. Using larger datasets leads to models that have better generalization ability. This is true, however, only when measures against overfitting are taken during training.

The performance of the created keyword visual models can be evaluated using a variety of measures. These measures are mainly taken from pattern recognition and information retrieval domains. The simplest measure used is the accuracy of correctly classifying instances in the corresponding keyword classes. Accuracy is defined as the percentage of correct classifications on test data while the error rate is the percentage of incorrect ones. The relation between accuracy and error rate is given below:

$$Accuracy = 1 - ErrorRate \qquad (1.4)$$

Although accuracy is representative enough to test the effectiveness of the created models, it suffers from several limitations. It assumes a relatively uniform distribution of training samples, as well as equal cost for misclassifications. The information about actual and predicted classification contained in the confusion matrix [533] can be used for deriving a variety of effectiveness measures. An example of a confusion matrix for two keyword classes (Positive, Negative) is shown in Fig. 1.2.

In the confusion matrix $a$ represents the number of samples in the test data that were correctly classified as negative (TN-True Negative) while $d$ indicates the number of data classified correctly as positive (TP-True Positive). In the same way, $b$ represents the number of samples that were negative but

| | | Predicted Values | |
|---|---|---|---|
| | | Negative | Positive |
| Actual Values | Negative | a | b |
| | Positive | c | d |

FIGURE 1.2: Confusion matrix.

incorrectly classified as positive (FP-False Positive) while **c** shows the number of positive samples that were incorrectly classified as negative (FN-False Negative). Based on the confusion matrix the accuracy can be calculated as:

$$Accuracy = \frac{a+d}{a+b+c+d} = \frac{TN+TP}{Total\ number\ of\ cases} \ . \quad (1.5)$$

The True Positive Rate (TPR) measure, which in the domain of information retrieval is known as Recall, indicates the proportion of positive samples that were correctly classified:

$$TPR = \frac{d}{c+d} = \frac{TP}{Actual\ positive\ cases} \ . \quad (1.6)$$

Similarly the True Negative Rate (TNR) indicates the proportion of negative samples that were correctly classified as negative:

$$TNR = \frac{a}{a+b} = \frac{TN}{Actual\ negative\ cases} \ . \quad (1.7)$$

The False Positive Rate (FPR) indicates the proportion of negative samples that were incorrectly classified as positive while the False Negative Rate (FNR) indicates the proportion of positive samples that were incorrectly classified as negative:

$$FPR = \frac{b}{a+b} \quad (1.8)$$

$$FNR = \frac{c}{c+d} \quad (1.9)$$

Finally, Precision, which is also a popular measure used in information retrieval, represents the proportion of the predicted positive samples that were correctly classified as such. It is given by the following formula:

$$Precision = \frac{d}{b+d} \quad (1.10)$$

The harmonic average of Precision and Recall gives the F-measure:

$$F - measure = \frac{2 \times Precision \times Recall}{Precision + Recall} \quad (1.11)$$

The precision-recall curve illustrates precision as a function of recall and it is a type of ROC curve. In several cases by tuning the parameters of the models you have the opportunity to decrease Precision in favor of Recall and vice-versa. By doing so you can create the corresponding curve and decide on an "operation point" (pair of precision and recall performances) that meets your requirements.

In cases where the ranking of results is important (i.e., some keywords are more related than others to a particular image) alternative measures are used.

The Mean Average Precision (MAP) is a popular measure used in this case. In order for MAP to be computed the precision in several queries is calculated and the mean score is obtained. However, in these queries the number of results is restricted. Examples of such queries are: "Find the best keyword of image X," "Find the two better keywords for image X," "Find the three better keywords for image X," and so on. The MAP is the average precision of these queries and is computed by:

$$MAP = \frac{\sum_{i=1}^{Q} P(i)}{Q} \qquad (1.12)$$

where $Q$ indicates the total number of queries and $P(i)$ is the precision obtained in the i-th query.

## 1.6 A Study on Creating Visual Models

In addition to the discussion included in the previous sections, we present an experimental study on the creation of visual models for crowdsourcing originated keywords within the athletics domain. Initially, 500 images[3] were manually annotated by 15 users using a predefined vocabulary of 33 keywords [910]. Manual annotation was performed with the aid of the MuLVAT annotation tool [909].

For our experiments we have selected eight representative keywords and for each keyword, 50 images that were annotated with this keyword were chosen. Twelve different visual models were created for each keyword class by combining three different feature types and four different machine learning algorithms. The keywords modeled are: "Discus," "Hammer," "High Jump," "Hurdles," "Javelin," "Long Jump," "Running," and "Triple Jump." The performance and effectiveness of the created models are evaluated utilizing the accuracy of correctly classified instances.

The unified framework of creating visual models for keywords obtained via crowdsourcing is illustrated in Fig. 1.3 while a detailed example is given in Fig. 1.4 and Fig. 1.5.

### 1.6.1 Feature Extraction

Three different low-level feature types, HOG, SIFT and MPEG-7 were extracted from each image group, and were used to create the visual models. In the case of HOG, the implementation proposed in [623] was used with the aid of 25 rectangular cells and 9 bins of histogram per cell. The 16 histograms with

---

[3]The images were randomly selected from a large dataset collected in the framework of FP6 BOEMIE project.

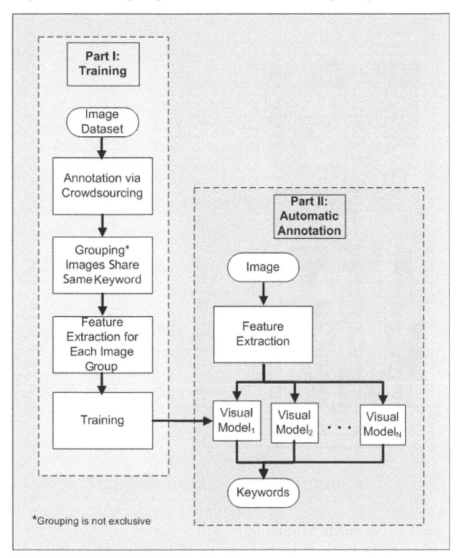

FIGURE 1.3: Automatic image annotation by modeling crowdsourcing originated keywords.

9 bins were then concatenated to make a 225-dimensional feature vector. In
the case of SIFT, the large number of extracted keypoints was quantized into a
100-dimensional feature vector, using k-means clustering. Finally, after an extensive experimentation on MPEG-7 descriptors (for details see also [921]) the
Color Layout(CL), Color Structure (CS), Edge Histogram (EH) and Homoge-
nous Texture (HT) descriptors were chosen. The combination of the selected
descriptors creates a 186-dimensional feature vector.

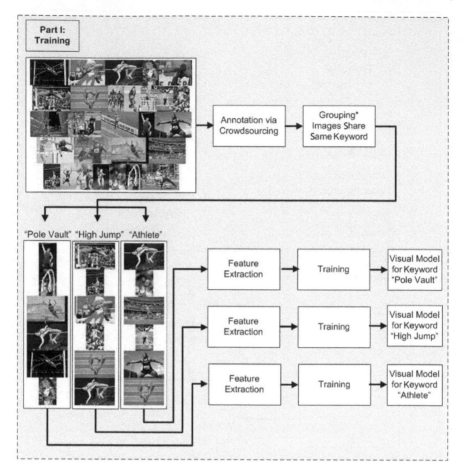

FIGURE 1.4: Example of creating visual models.

## 1.6.2 Keywords Modeling

As already mentioned in the previous sections, in order to ensure scalability and to fulfill the multiple keyword assignment per image, keyword models should be developed using the one-against-all training paradigm [904]. Thus, the creation of a visual model for each keyword was treated as a binary classification problem. The feature vectors of each keyword class were split into two groups: 80% were used for training models and the remaining 20% for testing the performance of these models. Positive examples were chosen from the corresponding keyword class while the negative ones were randomly taken from the seven remaining classes.

For the learning process we used four different algorithms: decision trees (in particular the Random Forest variation), induction rules (Ripper), Neural Networks (RBFNetwork) and Support Vector Machines (SMO). During train-

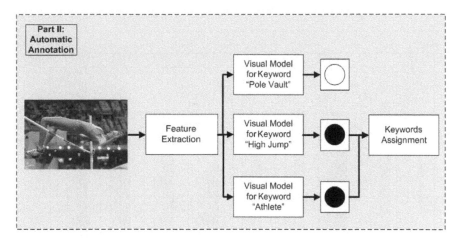

FIGURE 1.5: Example of automatic image annotation using visual models.

ing some parameters were optimized via experimentation in order to obtain the best performing model for each feature vector. The number of trees was optimally selected for the Random Forest models. The minimal weights of instances within a split and the number of rounds of optimization were examined for Ripper models. The number of clusters and ridge were tuned for each one of the feature vectors for the RBFNetwork. Finally, we experimented with the complexity constant and type of kernel for the SMO.

### 1.6.3 Experimental Results

Figures 1.6, 1.7, 1.8, 1.9 show the accuracy of correctly classified instances per keyword class using the four different machine learning algorithms mentioned earlier. The results shown in these figures can be examined under three perspectives: First, in terms of the efficiency and effectiveness of the various learning algorithms in modeling crowdsourced keywords; second, in terms of the appropriateness of the low-level features to accurately describe the visual content of images in distinctive manner; and third, in terms of the ability of the created models to classify the images into the corresponding classes and assign to them the right keywords.

The performance of the learning algorithms is examined through the time required to train the models (efficiency), the robustness to the variation of learning parameters and the effectiveness of the created models to identify the correct keywords for unseen input images.

The learning takes no more than a few seconds for the majority of the keyword models for all the machine learning algorithms examined. The fluctuation in classification performance during parameters' tuning is significantly

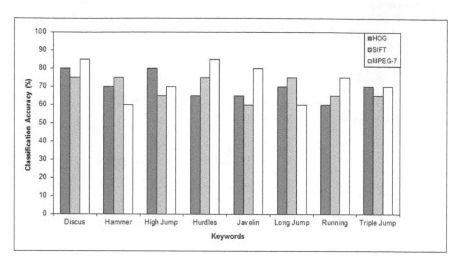

FIGURE 1.6: Performance of visual models using Random Forest decision tree.

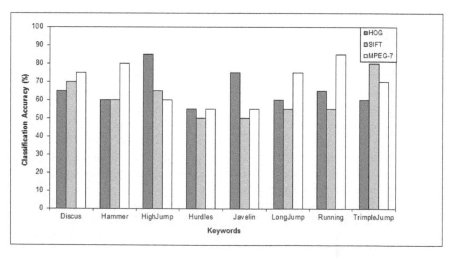

FIGURE 1.7: Performance of visual models using Ripper induction rule.

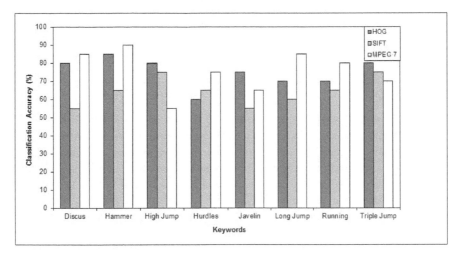

FIGURE 1.8: Performance of visual models using RBFNetwork.

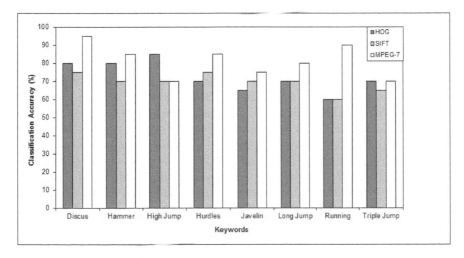

FIGURE 1.9: Performance of visual models using the SMO support vector machine.

**TABLE 1.2**

Average Classification Accuracy (%) Values.

| Classifier | HOG | SIFT | MPEG-7 | Overall |
|---|---|---|---|---|
| Random Forest | 70.0 | 69.38 | 73.13 | 70.83 |
| Ripper | 65.63 | 60.63 | 69.38 | 65.21 |
| RBFNetwork | 75.0 | 64.38 | 75.63 | 71.67 |
| SMO | 72.5 | 69.38 | 81.25 | 74.38 |

lower in Random Forest, Ripper and SMO than that of the RBFNetwork. This is something expected. As already discussed, tuning the activation function layer in RBF networks is a bit tricky and depends on the training data.

As far as the effectiveness is concerned, there is a significant difference on the performance of the models created using the individual learners. It is evident from Table 1.2 that SMO is the most reliable learner, with a total average classification accuracy equal to 74.38%. The RBFNetwork and Random Forest obtain nearly the same average classification accuracy, with both performing better when fed with MPEG-7 features. Random Forest performs well for the other type of features while the RBFNetwork performs fairly well when using HOG features and quite moderately when fed with SIFT type features. The Ripper inductive rule algorithm obtains the worst average classification accuracy score. The overall best classification accuracy occurs when combining the SMO classifier with MPEG-7 features while the worst performance occurs when combining the Ripper algorithm with the SIFT features. The majority, if not all, of the above results are in agreement with previous studies. SVM based algorithms perform well in learning tasks where the dimensionality of input space is high with respect to the number of training examples. Rule-based classifiers, on the other hand, face difficulties whenever the dimensionality of input space is high.

Concerning the effectiveness of the various low-level feature types, the experimental results indicate that the MPEG-7 features perform better than HOG and SIFT. The classification accuracy obtained using these features is better than the other two, independent of the training algorithm used. The second more reliable low-level feature type for modeling keywords is the HOG. HOG features when combined with an RBFNetwork classifier achieve an average classification accuracy of 75% which is quite high. The worst performance is obtained by the SIFT type features. It only achieves a maximum classification accuracy of approximately 75% for specific keyword classes. In particular, when combined with Random Forest decision tree it reaches an accuracy score 75% for the keyword classes "Discus" and "Hammer" (see also Fig. 1.6) while when combined with SMO it reaches a similar score for the "Hurdles" keyword class (see also Fig. 1.9). It is interesting to note that the best performance of the SIFT features is obtained for keyword classes corresponding to objects with well defined shape characteristics.

Once again, the results referring to the feature types are quite predictable. MPEG-7 descriptors are specially designed features to accommodate content based image retrieval. They were selected based on extended experimentation and comparative studies with other feature types. SIFT features, on the other hand, were primarily defined for object detection and object modeling tasks. Furthermore, in an attempt to fix the dimensionality of input space so as to be used in machine learning frameworks, SIFT keypoints are grouped together using either histograms or clustering methods. This grouping discards the information about the spatial distribution of keypoints and deteriorates significantly their visual content description power.

Nearly all models are able to assign the right keywords to unseen images. The overall accuracy scores are in the range 55%–95%. The best scores are obtained for keyword classes corresponding to objects with a well defined shape such as "Discus" and "Hurdles." In contrary, keyword classes corresponding to more abstract terms, such as "Running" and "Triple Jump" achieve relatively poor scores. Thus, keywords that are related with the actual content of the images can be more easily modeled, and as a result, automatic annotation of input images with such keywords is both feasible and realistic. On the other hand, modeling keyword classes which are not clearly related with the image content is by far more difficult. This is because the content of images corresponding to these keywords has many similarities with the content of images corresponding to other keywords.

## 1.7 Conclusion

Although Content Based Image Retrieval (CBIR) has attracted a large amount of research interest, the difficulties in querying by an example propel ultimate users towards text queries. Searching by text queries yields more effective and accurate results that meet the needs of the users while at the same time preserving their familiarity with the way traditional search engines operate. In recent years, much effort has been expended on automatic image annotation methods, since the manual assignment of keywords is a time consuming and labor intensive procedure. This chapter overviews the automatic image annotation under the perspective of machine learning and covers different aspects in this area. It discusses and presents several studies referring to: (a) low-level feature extraction and selection, (b) training algorithms that can be utilized for keyword modeling based on visual content, and (c) the creation of appropriate and reliable training data, to be used with the training scheme, using the least manual effort. Finally, we have proposed and illustrate a new idea for addressing the key issues in automatic keyword extraction by creating separate visual models for all available keywords using the one-against-all paradigm to account for the scalability and multiple keyword assignment problems.

We believe that the prospective reader of this chapter would be equipped with the ability to identify the key issues in automatic image annotation and would be triggered to think ahead to propose alternative solutions. Furthermore, the last section of the chapter can serve as a guide for researchers who want to experiment with automatic keyword assignment to digital images.

# 2

# Visual Search for Objects in a Complex Visual Context: What We Wish to See

**Hugo Boujut**

*University of Bordeaux, LaBRI,* `boujut@labri.fr`

**Aurélie Bugeau**

*University of Bordeaux, LaBRI,* `bugeau@labri.fr`

**Jenny Benois-Pineau**

*University of Bordeaux, LaBRI,* `benois-p@labri.fr`

## CONTENTS

This chapter addresses the problem of recognition/classification of objects in the so-called "egocentric video," i.e. one recorded by cameras worn by persons. It proposes the use of visual saliency for detecting *active* regions within video frames, and also proposes an improvement of the saliency model by adding a third saliency cue called *geometric*. Analysis of gaze fixations on egocentric videos and further studying of human anticipation from physiological studies is then taken into account toward a more efficient saliency based psycho-visual weighting of the BoVW for object recognition approach suitable for this kind of video, an approach that has been designed to identify objects related to videos recorded by a wearable camera.

## 2.1   Introduction

The problem of object recognition in the visual media data remains one of the most challenging tasks in the overall range of problems to be solved in order to build intelligent systems of multimedia data mining. Object recognition/classification is performed on several digital media such as pictures or videos. The recognition task is more or less obvious according to the visual scene complexity, and the object to find. It is indeed easier to find an object in a controlled environment than in a natural scene. Furthermore, in a real-life visual scene, objects can be numerous and located in the foreground, and the background as well. The object recognition task is often dependent on the global semantic interpretation task. When performing a visual understanding task, one does not merely seek to recognize all objects in a visual scene, but focuses in priority on those which are of interest. The examples of such a *selective* interest are numerous. For instance, when seeking to identify a person crossing the road, the observer will not focus on the surrounding buildings.

This chapter addresses the problem of recognition/classification of objects in the so-called "egocentric video," i.e. one recorded by cameras worn by persons. This kind of content has become a reality [568, 871] due to strong technological progress in production of video acquisition devices. It becomes common for sportsmen, but also is widely used for studies of human activities in research on human behavior, cognitive processes and motor control [494]. Especially we are willing to recognize manipulated objects of the *Instrumental Activities of Daily Living (IADL)*. For such videos, the wearable camera is either set on the subject's shoulder or tied on the chest. Both camera positions give an *egocentric* point-of-view of the visual scene. This point-of-view has the advantage to be the best to catch the action happening. However, nobody is behind the camera to center the object of interest. That is why the object of interest may be located in an unexpected area of the video frame. This issue is not usual in edited videos where objects of interest are almost always near the frame center. IADL video scenes are complex as well. Indeed several manipulated objects could be present in the frame, but only one or two of them could be *active*, that is, of interest for the observer. Hence additional information must be integrated in the recognition framework to catch the attention of the observer.

In this work, we propose to use the visual saliency for detecting *active* regions of the frame. The visual saliency represents the human visual attention within a visual scene. Therefore the saliency is well suited to distinguish active from inactive objects. Visual saliency modeling has captivated researchers since the early 80s with *Feature Integration Theory* [919] from A. Treisman, and G. Gelade. This research topic is still very active. In 2012, A. Borji, and L. Itti [144] took an inventory of 48 significant saliency models. Despite the fact that the visual saliency modeling is an old research topic, object recognition frameworks using such models is a new trend [325, 955]. Most of the visual saliency models only consider spatial information such as contrast. These models are called *spatial* and were designed at first for still pictures. There are also models called *spatio-temporal* based on the motion present in videos. The Human Visual System (HVS) is highly sensitive to the relative motion, which represents the difference between the global camera motion and specific motion of objects in the scene. This is why applying a *spatio-temporal* saliency model in object recognition framework in video is justified. Indeed, residual motion used in such models allows for integration of the dynamic aspect of video in the object recognition framework, while the majority of methods for object recognition nowadays perform on a frame-by-frame basis.

In this chapter, we also propose to improve the saliency model by adding a third saliency cue called *geometric*. Recent works [903] have shown that subjects tend to fixate on the screen center when watching a natural scene. In [294] the authors came to the same conclusion for natural edited videos. This is why the authors of [168] proposed a third cue modeling a 2D Gaussian centered in the middle of the frame. In our third cue, we considered this center hypothesis applied to egocentric videos. After analyzing gaze fixations on these videos we figured out that viewers anticipate the camera motion.

Furthermore, the cameras are worn by acting persons. They perform IADLs. And it is also a known fact, from physiological studies, that humans anticipate the action with the gaze and observers who watch the video recorded by an acting person also anticipate the action. Therefore, we take into account this phenomenon in the whole saliency extraction approach for egocentric videos.

Before going into further details on the use of saliency for object retrieval, we review in Section 9.2 the existing methods for object recognition in images or video frames considered as stills. Then Section 2.3 presents the state-of-the-art methods using visual saliency for object recognition as well as saliency models that are used in this work, and the proposed saliency *geometric* saliency cue. This section also describes our object extraction approach based on the Bag-of-Visual-Words (BoVW) weighted by saliency maps. Section 10.7 details the evaluation protocol, and the test video databases. Section 2.5 shows evaluation results. Finally, Section 4.5 concludes this chapter.

## 2.2    State-of-the-Art on Object Recognition

Object recognition and classification are very active research topics. Thousands of papers have been published on these subjects during the last ten years. Performing an exhaustive state-of-the-art survey is therefore unrealistic. Hence we focus on the works that have received the most attention and have given the most promising results. One common strategy for all these methods can be highlighted. First, the image or area of interest is described with the most possible pertinent information. The descriptors can either be local, global or semi-local. Next, a compact representation of the sets of all the descriptors is defined. Finally, distances or similarities between these representations are computed so that the current image can be classified or compared to a database in order to obtain the recognition result. In this section, all these steps are detailed.

### 2.2.1    Features Extraction

In order to analyze the content of images or videos, the first step consists in extracting some features which characterize the data. This step is useful for applications such as Content-Based Image Retrieval (CBIR), image classification, object recognition or scene understanding. The features can either be global, local or semi-local. They can all be applied for object recognition in the areas of interest detected in video frames. We here review some of the existing works on the topic.

### 2.2.1.1 Global Image Descriptors

Global image features are generally based on color cues. Indeed, color is an important part of the human visual perception. In images, the colors are encoded in color spaces. A color space is a mathematical model that enables the representation of colors, usually as a tuple of color components. There exist several models of this type, some motivated by the application background, some by the perceptual background of the human visual system. Among them we can cite the RGB (Red Green Blue) space, the HSV (Hue Saturation Value) or the luminance-chrominance spaces (YUV for instance).

Probably the most famous global color descriptor is the color histogram, which represents the distribution of colors within the image or a region of the image. Each bin of a histogram is the frequency of a color value within this area. It usually relies on a quantization of the color values, which may differ from one color channel to another. Histograms are invariant under geometrical transformations of the region.

Color moments are another way of representing the color distribution of an image or a region of an image. The first order moment is the mean which provides the average value of the pixels of the image. The standard deviation is the second order moment representing how far color values of the distribution are spread out from each other. The third order moment, named skewness, can capture the asymmetry degree of the distribution. It will be null if the distribution is centered on the mean. Using color moments, a color distribution can be represented in a very compact way [473, 615].

Other color descriptors that can be mentioned are the Dominant Color Descriptor (DCD) introduced in the MPEG-7 standard [645] or the Color Layout Descriptor (CLD).

### 2.2.1.2 Local Image Descriptors

The features that have received the most attention in the recent years are the local features. The main idea is to focus on the areas containing the most discriminative information. In particular the descriptors are generally computed around the interest points of the image and are therefore often associated to an interest point detector.

*SIFT*

Scale Invariant Feature Transform (SIFT) [619] has been designed to match different images or objects of a scene. The features are invariant to image scaling and rotation, and partially invariant to illumination changes and 3D camera viewpoint. They are well localized in both the spatial and frequency domains, reducing the probability of disruption by occlusion, clutter, or noise. In addition, the features are highly distinctive, which allows a single feature to be correctly matched with high probability against a large database of features, providing a basis for object and scene recognition. There are two main steps for extracting SIFT features: the key-point localization through scale-space extrema detection and the generation of key-point descriptors. First, a scale pyramid is built by convolving the image with variable-scale Gaussians and Difference of Gaussians (DoG) images are computed from the difference of adjacent blurred images. Interest points for SIFT features finally correspond to local extrema of these DoG images. To determine the key-point orientation, necessary for rotation invariance, a gradient orientation histogram is computed in the neighborhood of the key-point. The contribution of each neighboring pixel is weighted by the gradient magnitude. Peaks in the histogram indicate the dominant orientations. The feature descriptor finally corresponds to a set of orientation histograms, relative to the key-point orientation, on a 4×4 pixel neighborhood. As histograms contain 8 bins, a SIFT descriptor is a vector of 128 dimensions. This vector is normalized to ensure invariance to illumination changes.

*SURF*

SIFT has proven to be a powerful feature in many computer vision applications. Nevertheless, all the necessary convolutions make it computationally expensive. Speeded Up Robust Features (SURF) [108] have then been proposed as an alternative. These features describe a distribution of Haar-Wavelet responses within interest points neighborhood. They rely on integral images, which are the sums of all pixel values contained in the rectangle between the origin and the current position. SURF key-points are also extracted by scale-space analysis through the use of Hessian-matrices. Here again, the dominant orientation is extracted. It is estimated by computing the sum of Haar-Wavelet responses within a sliding orientation window. In an oriented square window centered at the key-point, which is split up into 4×4 sub-regions, each sub-region finally yields a 64 dimensional feature vector based on the Haar-Wavelet responses.

### 2.2.1.3    Semi-Local Image Descriptors

Most shape descriptors fall into this category. Shape description relies on the extraction of accurate contours of shapes within the image or region of interest. Image segmentation is usually fulfilled as a preprocessing stage. In order for the descriptor to be robust with regard to affine transformations of an object, quasi perfect segmentation of shapes of interest is supposed. Here, we just mention some shape descriptors but more can be found in literature. In particular, let us mention the Curvature Scale Space (CSS) descriptor [683] and the Angular Radial Transform (ART), descriptors in the MPEG-7 standard.

### 2.2.1.4    Bag-of-Visual-Words Approaches

The descriptors presented above, and in particular SIFT and SURF, have been widely used for retrieving objects in images. Local feature extraction leads to a set of unordered feature vectors. The main difficulty of the recognition, retrieval or classification steps consists in finding a compact representation of all these features and its associated (dis-)similarity measure. An efficient approach that has been widely used is the so-called Bag-Of-Visual-Words framework [850], that we now describe.

The Bag-of-Visual-Words (BoVW) approaches have four main stages: building a visual dictionary by clustering visual features extracted from a training set of images/objects, quantifying the features, choosing an image representation using the dictionary and comparing images according to this representation.

### Visual dictionary

In analogy with text retrieval, the features extracted in an image correspond to the words in a document. A visual dictionary must then be built. This is generally done by randomly selecting a sufficiently large set of features over a huge amount of images. A dictionary, $V = v_i, i = \{1, \ldots, K\}$, is then built by clustering these features into a certain number of $K$ classes or "visual words."

### Feature quantization

The second step consists in quantizing the features extracted in an image according to the visual dictionary. Each feature of an image is *quantized*. This quantization is generally achieved by assigning each feature to its closest word in the dictionary $V$.

### Pooling

Each image in the dataset can now be represented by a unique vector of $K$ dimensions. Each dimension represents the number of times a feature appears in the image. Therefore, this vector can be seen as a histogram representing the distribution of visual words in an image. This histogram is often normalized which enables comparing images containing a different number of features. These histograms were named *Bag-of-Visual-Words* [850] *(BoVW)*.

### Image comparison

All images being now represented by a histogram, the last step simply consists in comparing the histograms. Obviously, when the size of the database increases this step can become very computationally expensive. The computational time also depends on the size of the dictionary, which therefore needs to be chosen carefully. Several strategies have been proposed in the literature to improve the cost of this last step. In [850], this framework was applied with SIFT features. The vector quantization was carried out by k-means clustering, the number of clusters being chosen manually. In order to increase the discriminative power of the words produced by the clustering, a stop list was used. The idea of the stop list is to remove from the vocabulary the words which are very frequent, thus not discriminative enough, and those which are very rare, that can then be seen as noise. In [850], the authors removed from the list the 5% more frequent words and 10% less frequent.

*Limitations and improvements*

The method in [850] is considered to be at the origin of most recent works in the domain of image recognition and retrieval. Many improvements have been proposed since then.

### 2.2.1.5   Improvements of Bag-of-Visual-Words Approaches

*Feature quantization*

First, concerning the vector quantization, it is well known that k-means algorithm has no guarantee to converge to the global optimum and depends on the initialization of the centers of the clusters. An improved version of this algorithm, known as k-means++ has been proposed in [81]. In order to deal with an incremental amount of images in a dataset, a hierarchical quantization can be performed. For instance, a hierarchical k-means clustering, called vocabulary tree was proposed in [711]. The vocabulary tree gives both higher retrieval quality and efficiency compared to the initial BoVW framework of [850]. Until now, we have only been talking about Bag-of-Visual Words approaches in which only one type of feature is used. Note that if several different types of features are extracted from the images, the BoVW framework can also be applied. The set of all the feature vectors from one image is generally referred to as "Bag-of-Features."

*Soft and sparse coding*

In previous methods, the feature quantization, and thus the image representation, is obtained by assigning the feature vector to the closest word in the dictionary. This is called "hard coding." The coding step can be modeled by a function which assigns a weight $\alpha_{i,j}$ to the closest center $v_j$ of the feature vector $x_i$ :

$$\alpha_{i,j} = \begin{cases} 1 & \text{if } j = \arg\min_K \|x_i - v_K\| \\ 0 & \text{otherwise.} \end{cases} \tag{2.1}$$

Hence for each feature vector in the image, a code-vector can be computed by encoding the feature using the dictionary. The drawbacks of hard quantization are twofold: (i) word uncertainty: when a feature is close to several codewords, i.e., words of the dictionary, only the closest is considered; (ii) word plausibility: a codeword is assigned to the closest codeword no matter how far it can be. An illustration of these two drawbacks of hard coding are given in Figure 2.1 below. The square is an example of an "uncertain" word, the diamond-shape point is an example of an implausible word. A "consistent" word example is given by the triangle.

Instead of assigning a feature to a unique codeword, a soft assignment can

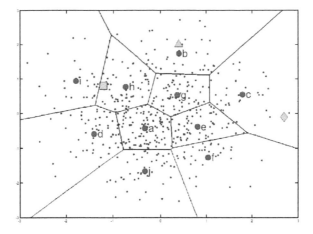

FIGURE 2.1: Soft quantization and sparse coding.

be used [946]. The weight $\alpha$ is not anymore a constant weight for all the feature vectors but will contain more information on the distribution of the feature vectors over the visual dictionary. For instance, the weight can be equal to the distance from the feature to the codeword. The resulting code is therefore not sparse contrary to what happens with hard coding. Some works also proposed to use kernel density estimators to estimate the weights [132]. Let us mention three models studied in [132], and relying on a Gaussian-shaped kernel: the kernel code-book (KCB), the codeword uncertainty (UNC), and the codeword plausibility (PLA). They all weight each word by the average kernel density estimation for each feature.

Sparse coding uses a linear combination of a small number of codewords to approximate a feature. The strength of sparse coding approaches is that one can learn the optimal dictionary while computing the weights $\alpha$ [640].

*Pooling in BoVW approaches*

The third step of BoVW approaches is pooling which consists in forming the final image representation. A good representation must be robust to different image transformations and to noise, and must be as compact as possible. A pooling operator aggregates the projections of all the input feature vectors onto the visual dictionary to get a single scalar value for each codeword. The standard BoVW [850] considers the traditional text retrieval sum pooling operator:

$$\forall j = 1 \ldots K \ , \ z_i = \sum_{i=1}^{N} \alpha_{i,j}.$$

Max pooling,

$$\forall j = 1 \ldots K \ , \ z_i = \max_{i=1,\ldots,N} \alpha_{i,j} \ ,$$

associated to sparse coding has also allowed more superior performance than sum pooling [1015]. Performance of max pooling and sum pooling has also been studied in [155]. Several extensions to these two traditional pooling operators have recently been proposed, some focusing on applying the pooling step on more local areas. The most powerful is probably the Spatial Pyramid Matching method (SPM) [561]. A fixed predetermined spatial image pyramid is first computed. The BoVW are then built on nested partitions of the image plane from coarse-to-fine resolutions. In other words, pooling is performed over the cells of a spatial pyramid rather than over the whole image. In [54], an approach called "Visual Phrases" is introduced to group visual words according to their proximity in the image plane as a sequence of features. The visual phrases are represented by a histogram containing the distribution of the visual words in the phrase. Spatial information has also been taken into account in [494]. Indeed, a spatial embedding of features with local Delaunay graphs is proposed. The advantage of Delaunay triangulation is that it is invariant to affine transformation of the image plane preserving the angles. Another BoVW improvement belonging to the aggregated coding class is the Fisher Kernel approach proposed in [738]. It is based on the use of the Fisher kernel with Gaussian Mixture Models (GMM) estimated over the whole set of images.

### 2.2.2  Classification and Recognition of Objects in Images

In previous section, we gave an overview of different types of features to represent the images and videos. The extraction of such descriptors is the preliminary step for any visual indexing and retrieval systems. The accuracy of all the methods presented in the following highly depends on the robustness of the chosen features to scaling, rotation and lightening.

Now that a good representation of each image or video has been extracted, the problem of classification or retrieval can be addressed. In case of image retrieval or indexing, the goal is to find, within a database, the image(s) that best matches a query image given by the user. In the context of classification, the purpose is to assign the image to the category to which it corresponds. The categories are defined beforehand by the user and a learning phase is necessary to learn the most important properties of each category. When relying to BoVW approaches, at the end of the pooling step, every object or image is represented by one histogram over the visual dictionary. In this section, we will see how these histograms can be used for image or object retrieval on one hand and for classification on the other hand.

## 2.2.2.1 Vector Distances

Many distances and strategies have been proposed to retrieve an image from its compact representation. Obviously, as BoVW approaches represent the distribution of visual words in an image by a histogram, the easiest way to perform retrieval is to compute (dis-)similarities between histograms. There exist two main categories of distances between histograms: the bin-to-bin distances and the cross-bin ones. Bin-to-bins measures require the histograms to have the same number of bins. Among the existing ones, let us mention the L2 and L1 metrics, the Kullback–Leibler divergence, the Chi-Square metric or the histogram intersection [891]. Among the cross-bins metrics, the most famous ones are the Mahalanobis distance and the Earth Mover's distance (EMD).

## 2.2.2.2 Feature Distribution Comparison

In previous sections, we considered a unique histogram per image. However, more recent approaches have shown that it can be more powerful to represent an image by several histograms, taking into account only a small part of the data or incorporating some local descriptions. When the data represented are illustrated by one or several histograms, kernel functions can be used to perform the matching. A kernel function allows evaluating the correlation between two data descriptions. In recent years, two principal types of kernels have been used in visual recognition system: the pyramid match kernel and the context dependent kernel. The pyramid match kernel was introduced in [377], for object recognition and document analysis. The principle is to map the features of some interest points using a multi-resolution histogram representation and to compute the similarity using a weighted histogram intersection. In [803], Sahbi et al. have introduced the context dependent kernel (CDK), which takes into account both the feature similarity "alignment quality" and the spatial alignment in a neighborhood criterion. The CDK is defined as the fixed-point of an energy function which balances a "fidelity" term, i.e., the alignment quality in terms of features similarity, a context criterion, i.e., the neighborhood spatial coherence of the alignment and an entropy term.

## 2.2.2.3 Image Classification

Image classification requires a pre-processing step of learning in order to find a decision rule (classifier) assigning Bag-of-Features representations of images to different classes. In general, a set of images belonging to the class (positive training examples) and a set of images not belonging to the class (negative training examples) are provided to the learning tool. Nowadays, the Support Vector Machine (SVM) is the most frequently used machine learning tool in a supervised context. Hence we find it necessary to briefly review its principles for object recognition purposes. SVM is a supervised statistical learning method that belongs to the class of kernel methods. The goal of SVM is to learn good separating hyper planes in a high dimensional feature space. The role of the kernel function $k$ is to map the training data into a higher dimensional space where the data is linearly separable. In other words, the feature vectors, $x_i$, are first mapped into feature vectors $phi(x_i)$ in an induced space. Next, a linear decision function $f(x_i) = wx_i + b$ is defined. The hyperplane, corresponding to $wx_i + b = 0$, separates the positive $y_i = +1$, from the negative training examples, $y_i = -1$:

$$\begin{cases} wx_i + b \geq 0 & \text{for } y_i = +1 \\ wx_i + b < 0 & \text{for } y_i = -1. \end{cases} \tag{2.2}$$

The SVM optimization problem can now be formulated as

$$min_{w,b} \frac{1}{2} \|w\|^2 \text{ subject to } \forall i, \ y_i f(x_i) \geq 1. \tag{2.3}$$

The decision function relies on the so-called "support vectors" which define the maximal margin between positive and negative subspaces in the target space. In its original formulation SVM is a binary classifier, but since the original framework it has been adapted to the multi-class problem.

### 2.2.3 Object Recognition in Videos

For recognition of objects in videos, a lot of work has been done using so-called spatio-temporal features [828] computed *around* spatio-temporal points [556]. Nevertheless, the key-framing and intra-frame object recognition still remains one of the most popular approaches [276]. Temporal dimension can be integrated in this case by fusion operators using multiple detections along the video [96] or by extraction of visually salient regions which are supposed to contain the objects of interest. In the following section we will introduce the notions of visual saliency for object extraction in videos.

## 2.3 Visual Saliency in Visual Object Extraction

### 2.3.1 State-of-the-Art in Visual Saliency for Object Extraction

Recently, the focus of attention in video content understanding, presentation, and assessment has moved toward incorporating of visual saliency information to drive local analysis process. The fundamental model by L. Itti and C. Koch [451] is that one the most frequently used for driving analysis process by visual saliency. If we simplify the concept of saliency to its very basic definition, we can reasonably say that visual saliency is what attracts human gaze. For visual object recognition in video, the first step which consists of extracting the potential area of object can be driven by extraction of saliency areas in video frames. Then features can be selected in these areas for object description. Numerous psycho-visual studies, which have been conducted since the last quarter of 20th century, uncovered some factors influencing it. Considering only signal features, the sensitivity to color contrasts, contours, orientation and motion observed in image plane has been stated by numerous authors [450, 562]. Nevertheless, these features alone are not sufficient to delimit the area in the image plane which is the strongest gaze attractor. In [903], the author states, for still images, that observers show a marked tendency to fixate on the center of the screen when viewing scenes on computer monitors. The authors of [294] come to the same conclusion for dynamic general video content such as movies and Hollywood trailers. This is why in [168] the geometrical saliency modeled by a 2D Gaussian located at the image center was proposed as the third cue. While signal based cues remain valuable saliency indicators, we claim that geometrical saliency depends on global motion and camera settings in the dynamic scene. camera motion, richness of the visual scene, especially if recorded in a home environment as it is done in [494] for studies of neurodegenerative diseases. Object recognition in such videos is difficult due to occlusions by hands, and the complexity of the environment. In egocentric motion as well as in all kinds of videos depicting human instrumental activities, the recognition of active objects, i.e., manipulated by persons, is difficult due to occlusions by hands and the cluttered natural environments. This is probably the most challenging content from the variety of user generated content from mobile devices. Hence the methods for visual object recognition in such content would add to the advances in a wide range of visual object recognition tasks in video. Some attempts to identify visual saliency for object recognition, mainly on the basis of the frequency of repetition of visual objects and regions in the egocentric video content, have recently been made [777]. We are specifically interested in the development and the application of visual saliency extraction methods for object recognition. We consider visual saliency as a combination of all cues in the pixel domain for the case of "egocentric" video content.

Recent research on object recognition extends the Bag-of-Features video representations [828] by making use of space-variant saliency mask [955]. In [828], space-time descriptors, capturing spatial appearance and motion properties, are sampled densely over the entire scene. The authors of [955] then propose to prune this set of densely extracted descriptors with the cumulative distribution function:

$$F(x; k; \lambda) = 1 - e^{-\left(\frac{x}{\lambda}\right)^k}, \tag{2.4}$$

where $k > 0$ is the shape parameter and $\lambda > 0$ is the scale parameter. The raw saliency value $x \in [0, 1]$ is derived from a saliency mask obtained with six different saliency models.

In the following section, we propose an automatic method for spatio-temporal saliency extraction on wearable camera videos with a specific accent on geometrical saliency dependent on strong wearable camera motion. We evaluate the proposed method with regard to subjective saliency maps obtained from gaze tracking. The saliency maps obtained will serve for weighting the visual features in the whole BoVW video object recognition scheme we use in this work. The advantage of our approach against [955] is that we propose a truly automatic way of building saliency maps. Specifically, for the "egocentric vision" the research on visual saliency is in its embryonic stage.

Let us now introduce the notion of visual saliences, the one *subjective* that we can obtain from human observers of the video content, and other *objective* saliency maps, that are automatically predicted from the video signal features.

## 2.3.2    Subjective vs. Objective Saliency Maps

Any *objective* human visual perception model expressed as an induced visual attention map in the image/video plan has to be validated and evaluated with regard to a *ground truth*. The *ground truth* is the *subjective* saliency. The subjective saliency map is built from eye fixation measurements with the help of eye-tracking. The eye positions are recoded with a device called an eye-tracker. The eye-tracker only collects eye positions at a regular rate, up to 1250 Hz for some models. Eye positions are first measured in the eye-tracker coordinates system. Then the measures are transposed to the experiment screen coordinates system, and recorded. The origin of the screen coordinates system is usually the screen center. Finally, the measures have to be transposed to the frame coordinates system. In this chapter, we consider that the eye measure coordinates $(x_0, y_0)$ are already transposed in the frame coordinates system. However, eye fixations cannot be directly used to represent the visual attention. First, the eye fixations are only spots on the frame, and do not represent the field of view. Second, to get accurate results, the saliency map is not built with the eye tracking data from one subject, but from many subjects. So the subjective saliency map should provide information about the density of eye positions.

The method proposed in [996] fulfills these two constraints. Moreover, this method was tested, using 5000 participants, on digitalized images of paintings from the National Gallery. In the case of video sequences, the method is applied on each frame $I$ of a sequence $M$. The process result is a subjective saliency map $S_{subj}(I)$ for each frame $I$. With this method, the saliency map is computed in three steps. In the first step, for each eye fixation measure $m$ of frame $I$, a two dimensional Gaussian is applied at the center of the eye measure $(x_0, y_0)_m$. The two dimensional Gaussian depicts the fovea projection on the screen. The fovea is the central retina part where the vision is the most accurate. The image falls on the fovea when an observer fixates. This region contains only cone photoreceptors, and has the highest cone density of the retina. The human eye contains two kinds of receptors, the rods and the cones. Rods are more sensitive at low light levels. However, rods do not allow color discrimination and provide poor information about detail. On the contrary, cones are efficient at high intensity lightning. Cones are responsible for color vision and for the fine detail detection. In [425], the authors stated

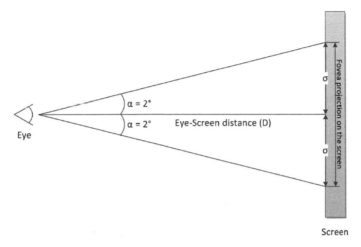

FIGURE 2.2: Fovea projection on the screen.

that the fovea covers an area from 1.5° to 2° in diameter at the retina center. It is also specified that the cone population falls sharply outside the fovea region, to reach a minimum at around 10° from the fovea center. The authors of [753] were the first to apply a two dimensional Gaussian to depict the fovea projection on the screen. In [996] the Gaussian spread $\sigma$ was set to an angle of 2° (Figure 2.2). Equation (2.5) is used to estimate the $\sigma_{mm}$ in $mm$ according to the fovea view angle $\alpha$ set to 2°. The distance eye-screen $D$ in $mm$ is also required. According to *ITU-R Rec. BT.500-11* [452], which stipulates the conditions for visual assessment of video quality and defines the parameter settings for psychovisual experiments for ground truthing, $D$ should be equal to three times the screen height $3H$.

$$\sigma - mm = D \times tan\,(\alpha) \qquad (2.5)$$

Nevertheless, measures in $mm$ are not convenient for processing video frames. So the $\sigma_{mm}$ value in $mm$ is converted in pixels ($\sigma$) with equation (2.6), where $R$ is the screen resolution in pixels per $mm$.

$$\sigma = R \times \sigma_{mm} \qquad (2.6)$$

For the eye measure $m$ of the frame $I$, a partial saliency map $S_{subj}(I, m)$ is computed (2.7).

$$S_{subj}(I, m) = Ae^{-\left(\frac{(x - x_{0_m})^2}{2\sigma_x^2} + \frac{(y - y_{0_m})^2}{2\sigma_y^2}\right)} \qquad (2.7)$$

with $\sigma_x = \sigma_y = \sigma$ and $A = 1$ .

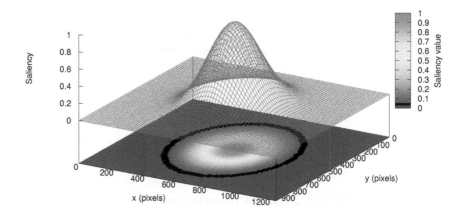

FIGURE 2.3: Psycho-visual 2D Gaussian (2° spread) depicting the fovea area related to one eye-tracker measure.

Then, at the second step, all the partial saliency maps $S_{subj}(I, m)$ of frame $S_i I$ are added into $S_{subj}{}'(I)$ (2.8).

$$S_{subj}{}'(I) = \sum_{m=0}^{N_I} S_{subj}(I, m),\qquad(2.8)$$

where $N_I$ is the number of eye measures recorded on all the subjects for the frame $I$. Finally, at the third step, the saliency map $S_{subj}{}'(I)$ is normalized by the highest value $argmax$ of $S_{subj}{}'(I)$ (Figure 2.4). The normalized subjective saliency map is stored in $S_{subj}(I)$ (2.9).

$$S_{subj}(I) = \frac{S_{subj}{}'(I)}{argmax(S_{subj}{}'(I))}\qquad(2.9)$$

Figure 2.5 shows an example of a subjective saliency map computed with the method from [996]. Why could not we use the subjective saliency maps for object extraction? The *subjective* saliency map processing requires eye-tracker measures from subjective experiments. Subjective experiments are time consuming and expensive to carry out. To get a *subjective* saliency map would require viewing of the video by several subjects. Thus, *subjective* saliency is not suited for real-life video analysis applications. To avoid this constraint we are interested in an automatic *objective* saliency map we proposed. The automatic spatio-temporal saliency map computation process is described in the

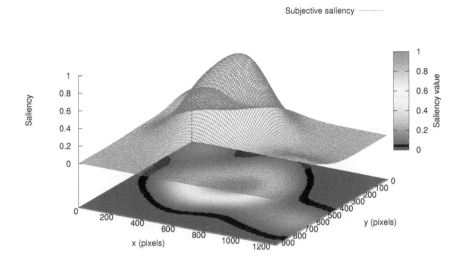

FIGURE 2.4: Normalized 2D Gaussian sum on an egocentric video frame.

next sections. However the *subjective* saliency map remains helpful for the objective saliency map accuracy evaluation. That is why the *subjective* saliency map is considered below as the reference saliency map. Hence, the *subjective* saliency maps will be used as the ground truth to assess the objective maps that are built automatically.

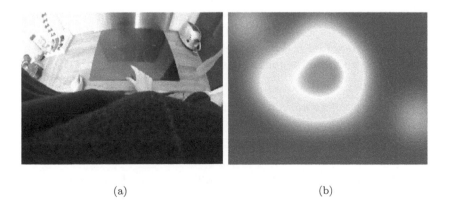

(a)            (b)

FIGURE 2.5: Subjective saliency map example computed with [996] from eye-tracker data. (a) Original frame. (b) Subjective saliency map.

### 2.3.3 Objective Saliency Map

To automatically delimit the area of video analysis in video frames to the regions which are potentially interesting to human observers we need to model visual saliency on the basis of video signal features. Here we follow the results of community research [447, 451, 450, 562, 168, 777] by proposing fusion of spatial, temporal, and geometric cues. We extend the state-of-the art approaches by a specific modeling of geometrical saliency and propose multiplicative fusion of all three cues.

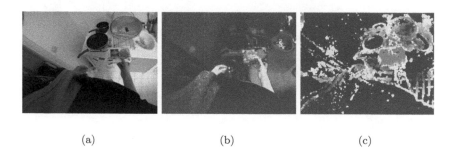

(a)                    (b)                    (c)

FIGURE 2.6: Objective saliency map example. (a) Original frame. (b) Spatial saliency map. (c) Temporal saliency map.

#### 2.3.3.1 Spatial Saliency Map

The spatial saliency map $S_{sp}$ is mainly based on color contrasts [90]. We used the method from [168]. The spatial saliency map extraction is based on seven color contrast descriptors. These descriptors are computed in the HSI color space [373]. Contrary to the RGB color system, the HSI color space is well suited to describe color interpretation by humans. The spatial saliency is defined according to the following seven local color contrasts $V$ in the HSI domain:

1. *Contrast of Saturation*: A contrast occurs when low and highly saturated color regions are close.

2. *Contrast of Intensity*: A contrast is visible when dark and bright colors coexist.

3. *Contrast of Hue*: A hue angle difference on the color wheel may generate a contrast.

4. *Contrast of Opponents*: Colors located at the hue wheel opposite sides create very high contrast.

5. *Contrast of Warm and Cold Colors*: Warm colors — red, orange and yellow — are visually attractive.

6. *Dominance of Warm Colors*: Warm colors are always visually attractive even if no contrast is present in the surroundings.

7. *Dominance of Brightness and Saturation*: Highly bright and saturated regions have more chances of attracting attention, regardless of the hue value.

The spatial saliency value $S'_{sp}(I, i)$ for a pixel $i$ from a frame $I$ is computed by a mean fusion operator from seven color contrast descriptors (2.10):

$$S'_{sp}(I, i) = \frac{1}{7} \sum_{\varsigma=1}^{7} V_\varsigma(I, i) .$$ (2.10)

Finally, $S'_{sp}(I, i)$ is normalized between 0 and 1 to $S_{sp}(I, i)$ according to its maximum value.

### 2.3.3.2 Temporal Saliency Map

The objective spatio-temporal saliency map model for video content requires a temporal saliency dimension. This section will describe how to build temporal saliency maps. The temporal saliency map $S_t$ models the attraction of attention to motion singularities in a scene. The visual attention is not grabbed by the motion itself. The gaze is attracted by the motion difference between the *absolute* motion scene and the global motion scene. The motion difference is called the residual motion. Many authors propose a temporal saliency map model that takes advantage of the residual motion [562, 168, 647]. In this work, we have implemented the model from [168].

The temporal saliency map is computed in three steps. The first one is the optical flow estimation. Then the global motion is estimated in order to get the residual motion. Finally a psycho-visual filter is applied on the residual motion.

To compute the optical flow, we have applied the Lucas Kanade method from OpenCV library [150]. The optical flow was sparsely computed on 4x4 blocks, as good results were reported in [152] when using $4 \times 4$ macro-block motion vectors from the H.264 AVC compressed stream. The next step in temporal saliency computation is the global motion estimation.

The goal here is to estimate a global motion model to differentiate the local motion from the camera motion. In this work, we follow the preliminary study from [152] and use a complete first order affine model (2.11):

$$\begin{aligned} dx_i &= a_1 + a_2 x + a_3 y \\ dy_i &= a_4 + a_5 x + a_6 y \end{aligned}$$ (2.11)

Here $\theta = (a_1, a_2, \ldots, a_6)^T$ is the parameter vector of the global model (2.11) and $(dx_i, dy_i)^T$ is the motion vector of a block. To estimate this model, we used the robust least square estimator presented in [539]. We denote this motion vector $\vec{V}_\theta(I, i)$. Our goal is now to extract the local motion in video frames i.e. residual motion with regard to model (2.11). We denote the macro-block optical flow motion vector $\vec{V}_c(I, i)$. The residual motion $\vec{V}_r(I, i)$ is computed as a difference between block motion vectors and estimated global motion vectors.

Finally, the temporal saliency map $S_t(I, i)$ is computed by filtering the amount of residual motion in the frame. The authors of [168] reported that the human eye cannot follow objects with a velocity higher than $80°/s$ [255]. In this case, the saliency is null. S. Daly has also demonstrated that the saliency reaches its maximum with motion values between $6°/s$ and $30°/s$. According to these psycho-visual constraints, the filter proposed in [168] is given by (2.12).

$$S_t(s_i) = \begin{cases} \frac{1}{6}\vec{V}_r(I, i), & \text{if } 0 \le \vec{V}_r(I, i) < \vec{v}_1 \\ 1, & \text{if } \vec{v}_1 \le \vec{V}_r(I, i) < \vec{v}_2 \\ -\frac{1}{50}\vec{V}_r(I, i) + \frac{8}{5}, & \text{if } \vec{v}_2 \le \vec{V}_r(I, i) < \vec{v}_{max} \\ 0, & \text{if } \vec{v}_{max} \le \vec{V}_r(I, i) \end{cases} \qquad (2.12)$$

with $\vec{v}_1 = 6°/s$, $\vec{v}_2 = 30°/s$ and $\vec{v}_{max} = 80°/s$. We follow this filtering scheme in temporal saliency map computation.

### 2.3.3.3   Geometric Saliency Map

Many studies have showed that the observers are attracted by the screen center. In [168], the geometrical saliency map is proposed as a 2D Gaussian located at the screen center with a spread $\sigma_x = \sigma_y = 5°$. In our work [153] we proposed to adapt the geometric saliency to the camera position estimated by a psycho-visual experiment with subjects watching recorded videos. We stated that in a shoulder-fixed wearable camera video the gaze is always located in the first upper third of video frames; see the scattered plot of subjective saliency peaks in Fig. 2.7. Therefore, we have set the 2D Gaussian center at $x_0 = \frac{width}{2}$ and $y_0 = \frac{height}{3}$. The geometrical saliency $S_g$ map equation is given by (2.13).

$$S_g(I) = e^{-\left(\frac{(x-x_0)^2}{2\sigma_x^2} + \frac{(y-y_0)^2}{2\sigma_y^2}\right)} . \qquad (2.13)$$

However, this attraction may change with the camera motion. This is explained by the anticipation phenomenon [553]. Indeed, the observer of video content produced by a wearable video camera tries to anticipate the actions of the actor. The action anticipation is performed according to the actor's body motion which is expressed by the camera motion. Hence we propose to simulate this phenomenon by moving the 2D Gaussian centered on initial *geometric saliency point* in the direction of the camera motion projected in the image plane. A rough approximation of this projection is the motion of the geometric saliency point computed with the global motion estimation model, equation (2.11), where $x = \frac{width}{2}$ and $y = \frac{height}{3}$.

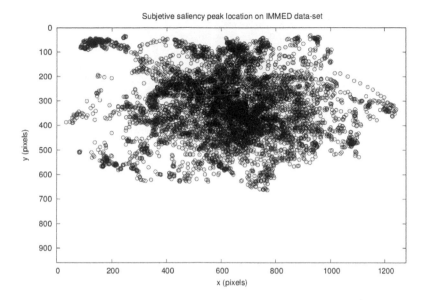

FIGURE 2.7: Scattered plot of subjective saliency peaks for all the frames from the IMMED database.

#### 2.3.3.4 Saliency Map Fusion

We now describe the method that merges these three saliency maps in the target *objective* saliency map. The fusion result is a spatio-temporal-geometric saliency map. In [152], several fusion methods for the spatio-temporal saliency without the geometric component were proposed. We have tested these fusion methods on an egocentric video database. The results show that the multiplicative fusion performs the best. Therefore for the full spatio-temporal-geometric saliency we compute multiplicative $S_{sp-t-g}^{mul}$ (2.14).

$$S_{sp-t-g}^{mul}(I) = S_{sp}(I) \times S_t(I) \times S_g(I) \qquad (2.14)$$

### 2.3.4 Object Recognition with Saliency Weighting

In the trending approach of Bag-of-Visual-Words, all the SURF descriptors from the learning dataset are quantized into a visual dictionary (or codebook) using *k-means* clustering. In this case, SURF points are detected by using the sparse SURF detector. Each image is then modeled by a distribution of its visual words. For this purpose, the descriptors computed on the image are matched with an $L_2$ norm with their closest representative in the codebook. In the traditional Bag-of-Visual-Words approach, the final image signature is the statistical distribution of the image descriptors according to the codebook.

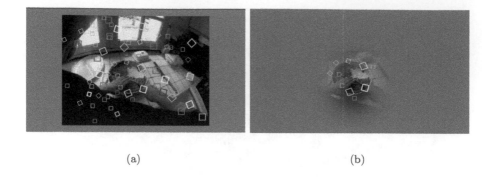

(a)                                                    (b)

FIGURE 2.8: Example of SURF points not weighted, and weighted by visual saliency. (a) Classical BoVW. (b) BoVW weighted by visual saliency.

We propose instead of a hard assignment for BoVW computation (see Section 2.2.1.4) to apply what we call *saliency weighting*. With saliency weighting, the contribution of each image descriptor is defined by the maximum saliency value found under the descriptor $\varphi_i$. In other words, descriptors over salient areas will get more weight in the image signature than descriptors over non-salient areas; see Figure 2.8. Therefore, the image signature $z_i$ is computed with equations (2.15), and (2.16):

$$\forall j = 1 \ldots N, z_i = \sum_{i=1}^{N} \alpha_{i,j} w_i \tag{2.15}$$

$$w_i = \underset{s \in \varphi_i}{argmax}(S(s)) \tag{2.16}$$

where $\alpha_{i,j} = 1$ if the descriptor $\varphi_i$ of feature point $p_i$ is quantized to class $K$. $w_i$ is the maximum saliency map value within the area covered by the descriptor $\varphi_i$ and $N$ is the number of classes.

## 2.4    Evaluation

This section presents the video databases and the methods used to evaluate the saliency model and the object recognition. Recognition methods were tested on the IMMED[1] (Indexing MultiMEdia data from wearable sensors for diagnostics and treatment of Dementia) and the ADL (Activities of Daily

---

[1]http://immed.labri.fr

Life) [748] video databases. Both are egocentric video corpora depicting Instrumental Activities of Daily Living (IADL). The evaluation of saliency models requires a subjective experiment to record eye fixations as presented in Section 2.3.2. The duration of these experiments is limited to 30 minutes per experiment because of the tiredness of participants [452]. For this reason, we only evaluate the automatic saliency models on the IMMED database. Object recognition is evaluated on both corpora. In the following subsection we describe these two datasets.

### 2.4.1 IMMED and ADL Video Databases

Egocentric video datasets are not largely available for this area of research. Hence in this work we used two datasets. Both of them were chosen since they were filmed by a GoPro wearable Camera which captures videos at the rate of 29.97 frames per second, with a resolution of 1280x960, and a 170° viewing angle.

The ADL is a publicly available academic dataset of 18 actions of daily life accomplished by 20 different people. All the 32662 frames extracted from these videos were annotated with an action label, object bounding boxes, object identity, and human-object interaction.

The IMMED corpus is composed of 53 videos of activities of daily living shot in a home environment. This corpus was recorded during the time life of a multidisciplinary project funded by French National Agency of Research (ANR) for studies of Alzheimer's disease. A psychologist was present during the shooting of the videos to suggest the activities. A total of 3641 frames extracted from the videos were annotated with temporal tasks, object locations and object categories.

### 2.4.2 Eye-Tracker Experiment

The subjective saliency maps expressing user attention are obtained on the basis of psycho-visual experiment consisting in measuring the gaze positions on egocentric videos from a wearable camera. The map of the visual attention has to be built on each frame of these videos. Videos from wearable cameras differ from traditional video scenes: the camera films the user point of view, including the user's hands. Unlike traditional videos, wearable camera videos have a very high temporal activity due to the strong ego-motion of the wearer. The gaze positions are recorded with an eye-tracker. We used the HS-VET 250Hz from Cambridge Research Systems Ltd. This device is able to record 250 eye positions per second. The videos we display in this experiment have a frame-rate of 29.97 frames per second. A total of 28 videos from IMMED database filming the IADL of patients and healthy volunteers are displayed to each participant in the experiment. This represents 17 minutes and 30 seconds of video. The resolution of the videos is 1280x960 pixels and the storage format is raw YUV 4:2:0. The experiment conditions and the experiment room is compliant to the recommendation ITU-R BT.500-11 [452]. Videos are displayed on a 23-inch LCD monitor with a native resolution of 1920x1080 pixels. To avoid image distortions, videos are not resized to screen resolution. A mid-gray frame is inserted around the displayed video. Twenty-five participants were gathered for this experiment, 10 women and 15 men. For 5 participants some problems occurred in the eye-tracking recording process. So we decided to exclude those 5 records. After looking at gaze position records on video frames, we stated that gazes anticipated camera motion and user actions. This phenomenon has been already reported in [553]. They state that visual fixation does precede motor manipulation, putting eye movements in the vanguard of each motor act, rather than as adjuncts to it. The observers somehow anticipate the motor act of camera wearer, in the same way as they are involved in the upcoming action. Nevertheless, gaze positions cannot directly be applied as ground truth to compare automatic saliency model we aim at. They must be processed in order to get the subjective saliency map as depicted in section 2.3.2.

### 2.4.3   Saliency Maps Evaluation

In this subsection, we compare the *objective* spatio-temporal saliency maps with *subjective* saliency maps obtained from gaze tracking $S_{subj}$.

Here, we use the Normalized Scanpath Saliency (NSS) metric that was proposed in [179, 647]. The $NSS$ is a Z-Score that expresses the divergence of the subjective saliency maps from the objective saliency maps. The $NSS$ computation for a frame $I$ is depicted by:

$$NSS = \frac{\overline{S_{subj} \times S_{obj}^{N}} - \overline{S_{obj}}}{\sigma(S_{obj})} \ . \tag{2.17}$$

Here, $S_{obj}^{N}$ denotes the objective saliency map $S_{obj}$ normalized to have a zero mean and a unit standard deviation, $\bar{X}$ means an average. When $\overline{S_{subj} \times S_{obj}^{N}}$ is higher than the average objective saliency, the $NSS$ is positive; it means that the gaze locations are inside the saliency depicted by the objective saliency map. In other words, the higher the $NSS$ is, the more objective and subjective saliency maps are similar. The $NSS$ score for a video sequence is obtained by computing the average of $NSS$ for all frames as in [647]. Then the overall $NSS$ score on each video database is the average $NSS$ of all video sequences. Saliency model evaluation results are presented in Section 2.5.1.

## 2.4.4 Object Recognition Evaluation

For the evaluation process we separate learning and testing images by a random selection. On each dataset, 50% of the images of each category are selected as learning images for building the visual dictionaries and for the retrieval task. Here we test the standard BoVW approach using 64 dimensional SURF descriptors. For the training, we applied the manually annotated masks on the object. For the testing, we compare the performance of queries by using the manually annotated mask, without mask, and the Spatio-Temporal-Geometric saliency map.

The performance is evaluated by the Mean Average Precision ($MAP$) measure using the *Trec Eval* [127] tool. For each test image, all images in the learning set are ranked from the closest (in terms of $L_1$ distance between visual signatures) to the farthest. The average precision $AP$ aims to evaluate how well the target images, i.e images of the same class as the query, are ranked among the $n$ retrieved images:

$$AP = \frac{\sum_{k=1}^{n} P(k) \cdot rel(k)}{c_p} , \qquad (2.18)$$

where $rel(k)$ equals 1 when the $k^{th}$ ranked image is a target image and 0 otherwise, $c_p$ is the number of target images and $n$ is the number of retrieved images. The average precision is evaluated for each test image of an object, and the $MAP$ is the mean of these values for all the images of an object in the test set. For the whole database we measure the performance by the average value of the $MAP$.

For the spatio-temporal-geometric saliency masking we use a 2D Gaussian function located at the screen center with a spread $\sigma_x = \sigma_y = 5°$. More details regarding the evaluation of the saliency-based masking are given in the following part of this section. For these preliminary results the dictionary size has been fixed to 500, 1000, and 5000.

## 2.5 Results

In this section, we first report the results of the correlation of the proposed saliency method with subjective saliency. Then we present the performance of object recognition by using the ideal mask, no mask, and the spatio-temporal-geometric saliency map for query images.

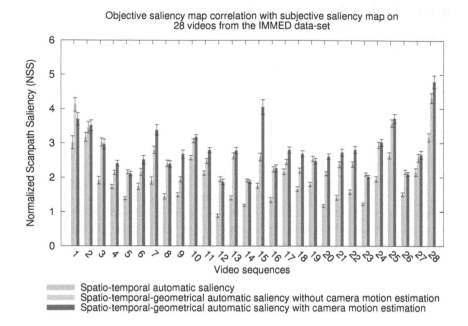

Objective saliency map correlation with subjective saliency map on 28 videos from the IMMED data-set

- Spatio-temporal automatic saliency
- Spatio-temporal-geometrical automatic saliency without camera motion estimation
- Spatio-temporal-geometrical automatic saliency with camera motion estimation

FIGURE 2.9: Objective saliency map correlation with subjective saliency maps.

### 2.5.1 Saliency Model Assessment

In this section, we compare the correlation of three automatic saliency maps with the subjective saliency. These three saliency maps are the spatio-temporal saliency map, the spatio-temporal-geometrical map without camera motion, and the proposed spatio-temporal-geometrical map with camera motion, expressing the anticipation phenomenon. The 28 video sequences described earlier from wearable cameras are all characterized by strong camera motion which is up to 50 pixels magnitude in the center of frames. As can be seen from the Figure 2.9 the proposed method with moving of geometrical Gaussian almost systematically outperforms the base-line spatio-temporal saliency model and the spatio-temporal-geometrical saliency with a fixed Gaussian. For few sequences (e.g. number 2), the performance is poorer than obtained by geometric saliency with a fixed Gaussian. In these visual scenes, the distractors appear in the field of view. The resulting subjective saliency map then contains multiple maxima due to the unequal perception of scenes by the subjects. This is more "semantic saliency" phenomenon (faces, etc) which can not be handled with the proposed model. The average NSS on the whole database also shows the interest of proposed moving geometrical saliency. The mean NSS scores are respectively 1.832 for patio-temporal, 2.607 for spatio-temporal with still geometrical Gaussian, and 2.791 with moving geometrical Gaussian. Which means 52.37% improvement of correspondence with subjective visual saliency map, which was our goal.

## 2.5.2 BoVW vs. Saliency-Based BoVW

For the ADL and IMMED datasets we present how the different masking approaches influence the results for the Bag-of-Visual-Words framework.

### 2.5.2.1 IMMED Corpus

The results for the different masking applied on the BoVW framework for these datasets are depicted in Figures 2.10, 2.11, 2.12, and 2.13. First of all, one can

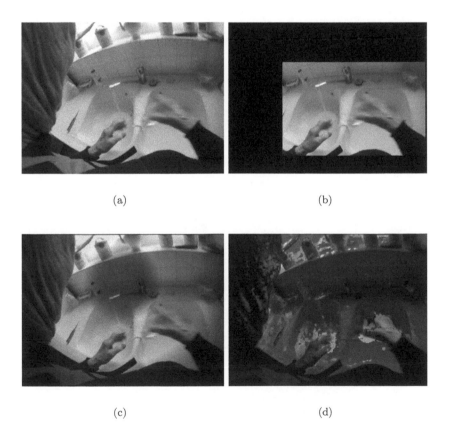

(a)  (b)

(c)  (d)

FIGURE 2.10: Visual representation of the 4 different masking approaches for an image from the IMMED corpus. (a) Original frame. (b) Manual masking. (c) Geometric saliency map. (d) Spatio-temporal-geometric saliency map.

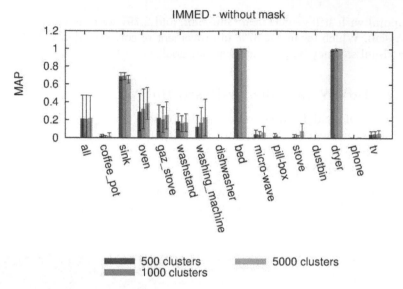

FIGURE 2.11: BoVW results on IMMED database without mask.

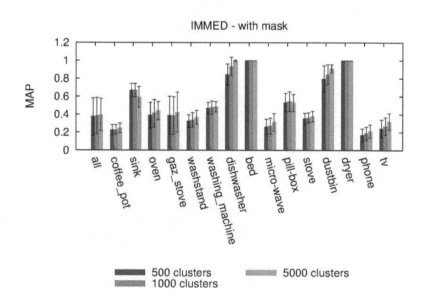

FIGURE 2.12: BoVW results on IMMED database with the ideal mask.

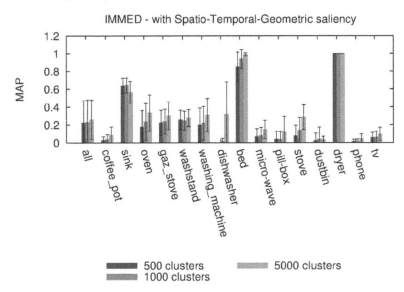

FIGURE 2.13: BoVW results on ADL database with the spatio-temporal-geometric saliency map.

notice the effect of manual masking on the performances. The overall performance is about 29.2%. The experimental result of spatio-temporal-geometric approaches did, as expected, improve in comparison to the baseline method (no mask). It is also important to emphasize that the results obtained from a simple geometric-based saliency map are better than those obtained from the analytical approach. We explain this phenomenon by pointing out that, similar to the Hollywood2 benchmark, the area of interest of the images extracted from the IMMED corpus have been designed to be at the center of the scenes.

### 2.5.2.2 ADL Dataset

The results for the different masking applied on the BoVW framework for the ADL dataset are depicted in Figures 2.14, 2.15, and 2.16. Similar to the IMMED dataset, the overall performance obtained by manual masking is the best. However, the overall experimental results of the spatio-temporal-geometric approach improved in comparison to the baseline method (no mask).

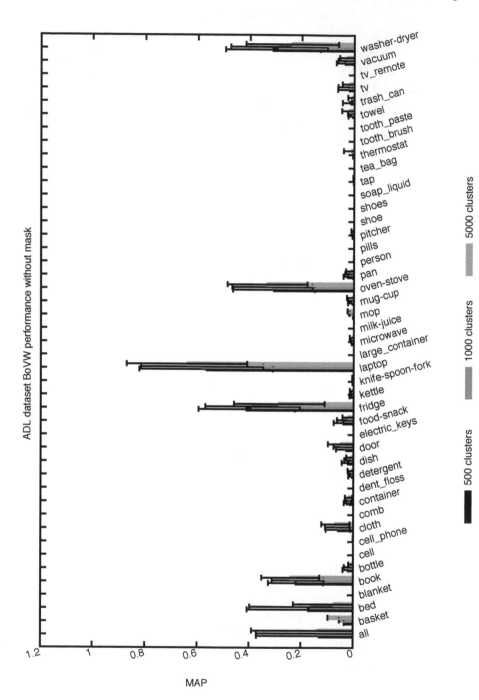

FIGURE 2.14: BoVW results on ADL database without mask.

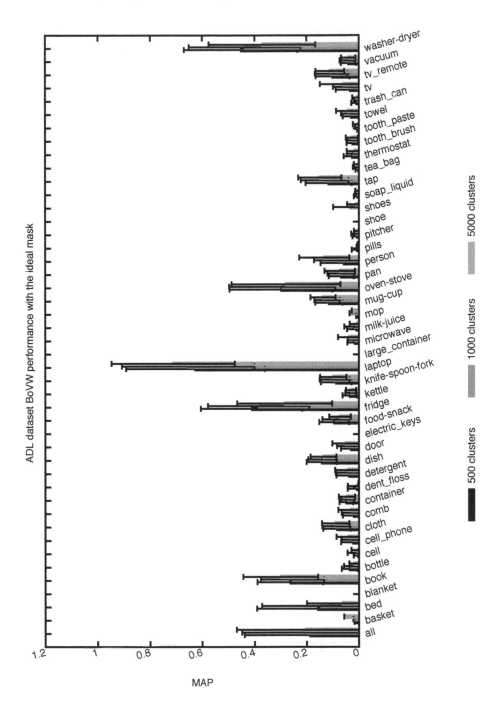

FIGURE 2.15: BoVW results on ADL database with the ideal mask.

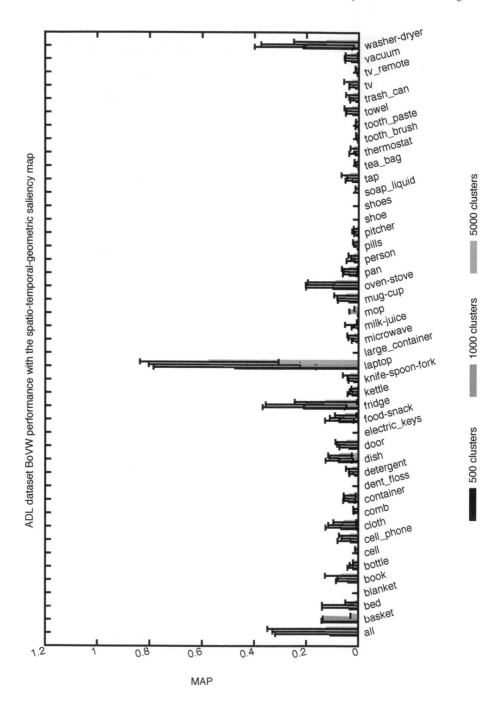

FIGURE 2.16: BoVW results on ADL database with the spatio-temporal-geometric saliency map.

## 2.6 Conclusion

In this chapter we proposed a saliency based psycho-visual weighting of the BoVW for object recognition. This approach has been designed to identify objects related to IADL in videos recorded by a wearable camera. These recordings give an egocentric point-of-view on the upcoming action. This point-of-view is also characterized by a complex visual scene with several objects on the frame plan.

The human visual system functions to process only the relevant data by considering areas of interest. Based on this idea, we proposed a new approach by introducing saliency models to discard irrelevant information in the video frames. Therefore we applied a visual saliency model to weight the image signature within the BoVW framework. Visual saliency is well suited for catching spatio-temporal information related to the observer's attention on the video frame. We also proposed an additional geometric saliency cue that models the anticipation phenomenon observed on subjects watching video content from the wearable camera. The findings show that discarding irrelevant features gives better performances when compared to the baseline method which considers the whole set of features in the images.

Thanks to these encouraging results, we believe that our propositions introduce a promising paradigm that can be used in future works to improve the quality of object recognition in complex user-generated videos.

## 2.7 Acknowledgment

This research has been supported by the European Commission under the 7th Framework Programme, research grant agreement 288199 Dem@care and French national grant ANR-09-BLAN-0165 IMMED.

# 3

# Visual-Semantic Context Learning for Image Classification

**Qianni Zhang**

*Queen Mary University of London,* `qianni.zhang@eecs.qmul.ac.uk`

**Ebroul Izquierdo**

*Queen Mary University of London,* `ebroul.izquierdo@eecs.qmul.ac.uk`

## CONTENTS

In realistic image classification scenarios, images are complex and they usually consist of many semantically meaningful objects interrelated to each other. The semantic context within the multimedia databases becomes useful in assisting semantic classification of images. This chapter aims to provide a forum for the state-of-the-art research in the emerging field of visual-semantic context analysis and exploitation, and to address the growing interest in automatic classification of multimedia content. Moreover, this chapter suggests some ideas on an image classification approach that jointly exploits the visual and semantic context embedded in images. A Bayesian learning procedure is first applied to derive a context model representing the relationships in semantic classes. This context model is later employed in inferring classes of images. The context learning and inference approach can work together with conventional content based image classification systems and provide improvements of classification performance on top of their results.

## 3.1 Introduction

The wide spreading of Internet services since the early nineties and the proliferation of digital content originated an acute need for effective search engines. Two fundamental reasons gave rise to this phenomenon: The exponential increase in the amount of available digital content, which in turn led to fast growing stocks of information; and the fact that such valuable information is rendered useless if it can not be found and used. Advances in conventional information retrieval technology over the last forty years led to very successful search-engine enterprises, such as Google, Yahoo and the like. However, this success has not been emulated yet in the context of generic multimedia data. The reason is simple: On the one hand conventional text based information retrieval profits from a fundamental coherence between queries and database content. Basically, both query and databases are made up of textual information. On the other hand multimedia and specifically visual information suffers from an irreconcilable disparity between queries usually expressed in semantic terms, i.e., free text or keywords, and the actual image or video data. This disparity between low-level features inherent to the actual content and the subjectivity of semantics in high-level descriptions of audiovisual media is called "the semantic gap." To bridge this gap has emerged as the holy grail of multimedia information retrieval. Solving the underlying problem hinges on developing generic, efficient and accurate technology for classification and recognition of semantically meaningful objects in images [856, 264].

In realistic multimedia search scenarios, high-level queries are usually used and search engines are expected to be able to understand underlying semantics in content and match it to the query. Researchers have naturally started thinking of exploiting context for retrieving semantics. Content is usually complex and may contain many semantically meaningful elements related to each other, and such interrelations together form a semantic context for the content. To understand the high-level semantic meanings of the content, these interrelations need to be learned and exploited to further improve the indexing and search process. In the latest content-based multimedia retrieval approaches, it is often proposed to build some forms of contextual representation schemes for exploiting the semantic context embedded in multimedia content. Such techniques form a key approach to supporting efficient multimedia content management and search in the Internet. In the literature, classification of visual content according to semantic meanings can incorporate domain knowledge to influence the classification decisions by context, grammar, semantics, and other high-level information. In this aspect, schemes for conceptual synthesis and reasoning at the semantic level have been developed, including statistical association, conditional probability distribution, and different kinds of monotonic and non-monotonic, and fuzzy relationships [272, 323, 535, 558, 1048]. Many object-based image classification frameworks relied on probabilistic models for semantic context modelling and inference [510, 52, 941, 1039, 586]. In contrast to object-based classification solutions, some works use context modeling in scene classification e.g., *indoor* vs. *outdoor* or *cityview* vs. *landscape* [941, 773]. However, these schemes are often limited as they have to rely on specifically defined domains of knowledge or require specific structure of semantic elements. In this chapter, we aim to provide a survey of the state-of-the-art (SoTA) in context exploitation for semantic image understanding, with specific focus on visual analysis based on block regions.

Following the outcome of the SoTA study, we introduce our ideas on how to enable systems to automatically construct a semantic context representation by learning from the content. Depending on the targeted source of content, which could be online databases, a model for its semantic context is directly learned from data and will not be restricted to the predefined semantic structures in specific application domains. This proposed method for context model learning and inference is referred to as the semantic context modeling (SCL) method. It is designed for jointly analyzing visual-semantic context information hidden in the content based on Bayesian models. This method can be applied based on the relevance ranking scores generated by any general image classification algorithms, and aims to improve the original classification performance.

## 3.2 Related Works

Much work in the area has been devoted to content-based image retrieval and classification under the well-known query-by-example paradigm [898, 203, 322, 736, 93, 859, 634, 1020]. At the end of the nineties, substantial research was conducted on semantic classification, semantic learning and inference for keyword annotation of images. Relevant works include [620, 242, 1043, 201]. The iFind system [620] approaches the problem by first constructing a semantic network on top of an image database and uses Relevance Feedback (RF) based on low-level features to further improve the semantic network and update the weights linking images with keywords. Chang et al. presented a content-based soft annotation scheme [201], which first trained an ensemble of classifiers for predicting label membership for images, then the ensemble was applied to each individual image to generate soft labels which were associated with label membership factors. In their work, both Support Vector Machines (SVMs) and Bayes point machines were used as the binary classifier, and their performances were compared. Zhao and Grosky utilized the latent semantic indexing technique in a content-based document retrieval system for multimedia web documents. This system employed both keywords and image features to represent the documents according to their semantic meanings. Cox et al. presented the PicHunter system [242], in which uncertainty about the user's goal is represented by a distribution across the potential goals and the representation is used to find the target image based on the Bayes' rule.

### 3.2.1   Methods in Image Region Analysis

In order to capture the extremely important semantic details about objects of interest, regional-based image retrieval is often considered. Some pioneering and more recent work directed toward region-based image retrieval include: [839, 978]. Region-based image retrieval systems were mostly based on individual region-to-region similarity matching such as the WALRUS system [698]. Some of these approaches also try to employ human assistance for accurately determining the proper segmentation of an image. In some systems such as Blobworld [196] and Netra [277], images are compared based on individual regions and the user is requested to select the region to be matched from the segmented regions of images, as well as the feature, e.g., color or texture, of the regions to be used for similarity estimation. Some approaches derive region weighting that is assumed to coincide with human perception through user RF [331]. In [913], similarly, the user is requested to choose a Region-of-Interest from regions of the query image. Among the approaches mentioned above, most of them analyze and exploit the relationship between semantic labels which is associated with image regions. Region-based techniques are very commonly applied in these approaches as it is easier to capture the extremely important semantic details about objects of interest in regions, compared to whole images. Most of the region-based approaches rely on the techniques that automatically segment images into regions which are expected to correspond to objects [839, 978]. By decomposing images to an object-level representation, these systems intend to imitate the perception of the human visual system. Many region-based visual information retrieval systems introduced in the previous section, such as [698, 331, 978, 84], obtain regions based on some sophisticate segmentation algorithms.

However, there is a disadvantage to these segmentation-based human-assisted approaches. Although the users are provided with more control over the querying process, at the same time they are confronted with more burdens, such as making the hard decision of choosing a query from various regions while none of these are clear representations of objects. Moreover, the segmentation algorithms, on which the above-mentioned approaches are based, often partition one object into several regions due to the natural complexity in the visual content, with only one of these regions being representative of the objects. This is because of the intrinsic limitation of segmentation algorithms – they only segment regions, but not objects. This problem also makes the task of choosing a representative region for an object more intractable. To relieve these problems, some approaches estimate the image similarities by calculating the total similarity of all regions in each image. All regions are either assumed to be equally important or are given some weighting. For instance, some approaches have utilized a learning process to obtain the weighting and have considered groups of regions as essential elements for retrieval [746]. The system internally generated many segmentations or groupings of each image's regions based on different combinations of features; then it learned which combination best represented the semantic categories given as exemplars by the user. The system requires supervised training of various parts of the image. In the SIMPLIcity system, the integrated region-matching is defined as a region-based similarity measure [590, 978]. It incorporates the properties of all the segmented regions so that information about an image can be fully used to gain robustness against inaccurate segmentation. This approach is assumed to be able to reduce the effect of inaccurate segmentation. In the querying system proposed in [861], an image was decomposed into regions with characterizations predefined in a finite pattern library, in which each pattern is labeled as a symbol. Each image was then represented as a string of pattern symbols of its regions. CRT descriptor matrices were obtained by converting the region strings of images and their distances were used as the similarity measures between images.

Although these approaches considered information from multiple regions, due to the uncontrollable nature of natural images, precise extraction of objects from images based on segments is beyond the reach of the state-of-the-art in computer vision. On the other hand, these segmentation algorithms often add heavy computation loads to the system. Recently, some newly developed systems have tried to solve the problem by utilizing some other scheme for image decomposition. For instance, for the framework proposed in [251], a grid-based approach was used which partitioned each image into blocks. A special query selection method was developed to remove the irrelevant background regions. The proposed search method was assumed to be free of spatial constraint and scale invariant. The proposed search method in [84] performs an initial region labeling based on matching a region's low-level descriptors with concepts stored in an ontological knowledge base and then utilizes contextual knowledge to readjust labeling results. Djordjevic and Izquierdo proposed an approach to adaptive structures of block regions for image classification based on non-linear Support Vector Machines (SVMs) [291]. Sivic and Zisserman proposed an efficient approach to search for objects in videos [851]. In this approach, each frame of the video is represented by a set of viewpoint invariant region descriptors. Then vector quantization is performed in these region descriptors, providing a visual analogy of a word, which is termed a "visual word." Efficient retrieval is then achieved by employing methods from statistical text retrieval.

### 3.2.2   Previous Works on Semantic Context Exploitation in Image Understanding

The work presented in the preview subsection focused on retrieving visual content based on interesting regions, which were assumed to represent the most important semantic clues to the visual content. However, in general the isolated regions of interest on their own are not able to describe the semantics. Consequently most recent research in this direction, such as that described in the last paragraph, has attempted to incorporate surrounding content and context with the regions of interest.

Different from machines, human beings have superior ability to interpret visual content into semantics. This ability comes from the knowledge stored in human brains which associates observations on different modalities in visual content and semantics. When a human watches a video (or an image), all the information is observed through a contextual sense in his/her knowledge. This knowledge in human brains is the basis that helps humans with reasoning and inference.

Therefore, to solve the problem of bridging the semantic gap, the retrieval of visual content according to their semantic meanings needs to incorporate the domain knowledge to influence the decisions on VIR by context, grammar, semantics and other high-level information. To tackle this aspect of the semantic gap problem, some schemes for conceptual synthesizing and reasoning on a semantic level have been developed. Especially, the exploitation of the surrounding context may greatly simplify the interpretation of regions as specific objects.

A number of models have been developed to explore the associations between terms and image regions, i.e., the visual context information, for image annotation and retrieval, such as the co-occurrence model [415], the translation model [303], the continuous relevance model [558] and the multiple Bernoulli relevance model [331]. Hidden associations among features of images are often explored according to the automatically extracted knowledge, and are well positioned to drive the multimedia content management and indexing more efficiently [289, 1004]. Context modeling and exploitation has also been widely applied in the field of video annotation and retrieval, based on techniques such as boosted conditional random field (CRF) [470] or domain adaptive semantic diffusion techniques [469].

In the literature many region or object-based image retrieval systems often utilize Bayesian networks for semantic-level modeling and inference [585, 332, 409, 419]. In such systems, the object's likelihood can be calculated from the conditional probability of feature vectors. These systems use probabilistic reasoning to exploit the relationship between objects and their features. Some other systems employ the Bayesian approach in scenario of scene classification e.g., indoor, outdoor, cityview, landscape [624, 940, 52]. However, this kind of classification has been restricted to mutually exclusive categories, and so is only suitable for images that have only one dominant concept. But in more realistic scenarios in image classification and retrieval, images are complex and they usually consist of many semantically meaningful objects. Therefore the relationships between semantically meaningful concepts of the objects cannot be ignored and need to be explored to a great degree. In [586], multi-categories are introduced with a simple "sub-class-of" relationship between some of the parent categories and their children categories. In above mentioned approaches, the useful semantics related to recognition and image retrieval, namely the conceptual features and their relationships, have barely been explored. Many approaches to modeling the relationships between concepts have been proposed, including statistical association (e.g. the co-occurrence of tiger and vegetation follows an observable proportion), conditional probability distributions (e.g. probability that vegetation is present given that tiger is present and water is present is 0.8), different kinds of monotonic and non-monotonic logics (e.g. if a dog doesn't bark, then it likes to bite) and fuzzy relationships (e.g. most Greek houses are white). Most real-life scenarios involve incomplete knowledge and can be more readily modeled by statistical relationships rather than logic. In [421, 443], a statistical analysis of the relationships among concepts is adopted and achieves image retrieving by using metadata (e.g. the RDF triples or the semantic web ontology model) that describe and organize the concepts in images.

As ontology becomes more and more widely used in multimedia management tasks, much research has been carried out employing ontologies as tools for knowledge based representations for semantic-based visual content retrieval and annotation. In [916], the author describes how an ontology, consisting of a ground truth schema and a set of annotated training sequences, can be used to train the structure and parameters of Bayesian networks for behavior recognition in video sequences. In [536], the proposed method exploits the structure of a multimedia ontology and existing interconcept relations for modeling semantic concepts in visual content for different semantic multimedia processing applications. In [323], a hierarchical concept learning algorithm is proposed by incorporating concept ontology and multitask learning to enhance the discrimination power of the concept models and reduce the computational complexity for learning the concept models for large amount of image concepts, which may have huge intraconcept variations and interconcept similarities on their visual properties. In addition to the abovementioned research which has much relation with visual descriptions, there are also other methods of addressing the multimedia search problem based purely on contextual information without using visual analysis techniques, such as [272].

In this section we have given an overview of the related works on two aspects, semantic context understanding and inference for image classification, and region based image analysis, which most semantic context inference systems are built upon. Later in this chapter, we introduce the SCL method for automatic context learning and inferences to boost the classification performance based on general region based image classifcation systems. In the SCL method, the aim is to explore and model the hidden interrelationships between different object classes in an image database. The context model consists of a set of semantic terms defining the classes to be associated to unclassified images. The modeling is conducted using a Bayesian network model, which can be learned from a small amount of training data. Semantic inference and reasoning is then carried out based on the learned model to decide whether an image can be assigned to an object class. The novelty and contributions of the SCL method is discussed in detail in the following section.

## 3.3    Semantic Context Learning and Inference

### 3.3.1    Background and Rationale

The SCL method is different from the existing research works targeting the task of context exploitation for image classification, in the sense that it can automatically learn a *visual-semantic context model* from a small training set without relying on dedicated model designs that are restricted to particular scenarios. Thus the definition of the model structure is a fully automatic learning process. There are two main advantages in such a method. First, the method is not restricted to a certain scenario or application domain but can be applied to any database with its own semantic context. Second, the user is not required to be an expert in the application scenario in order to be able to define the context model according to the domain knowledge. Rather, the context model is automatically learned using a small training set and the multi-feature similarities. The automatic learning process is conducted using a search-based algorithm — *K2*. The learned visual-semantic context model is then used to calculate the probabilities conveying the effects that the existing concepts have on each other.

Building on the initial concept-specific similarity scores of content, computed using a visual feature based image classification system [1038], Bayesian networks are used to exploit contextual information in the image database. These networks are constructed automatically by learning from manual annotations and similarities estimated over the multi-feature metric spaces in a small amount of training data. The aim is to model potential semantic descriptions of basic semantic concepts, the dependencies between them and the conditional probabilities involved in these dependencies. This information is finally used to calculate the probabilities of the effects that semantic concepts have on each other, in order to obtain precise labels to automatically annotate semantically meaningful objects in images. These semantic labels, i.e., terms or key-words, drive the subsequent image classification, annotation search and retrieval functionality of the proposed system. By using the initial visual feature based relevance scores as the input to Bayesian model learning, the target model encodes not only the semantic relationships between content items, but also their visual descriptions. This feature is very important as in most cases in image classification, there are no complete and accurate semantic descriptions on the whole dataset available. By jointly considering the visual and semantic evidence in the same model, these two kinds of descriptions, each with their own advantages and imperfections, can complement each other and derive more precise semantic information. The learned context model is then used to calculate the probabilities of the effects that those concepts have on each other in order to obtain more precise and meaningful semantic labels for the visual content. However, the proposed Bayesian context learning approach is not restricted to any specific method in calculating the initial similarity scores for each image.

An important aspect of the SCL method is that it can use regular image blocks as the basic units for visual representations of objects associated to semantic concepts. Feature extraction for visual representation, as well as the generation of training sets, are based on such image blocks. In the literature, most region-based approaches rely on image segmentation [84, 264, 474, 583, ?, 979]. However, there are several intrinsic difficulties in using image segmentation for semantic classification.

First, due to the complexity of natural images, segmentation algorithms based on visual features usually segment regions, not semantically meaningful objects. Indeed, precise extraction of objects from images using automated segmentation is an open problem in computer vision. Second, segmentation algorithms often add heavy computation loads to the system. Moreover, some human assisted segmentation algorithms may impose burdens on users and thus are unfeasible when large databases are processed. Considering these facts, alternative schemes for image decomposition have been exploited recently [251]. In this chapter, block regions are used as basic units without assuming precise segmentation or structures for representing objects. In general, the goal of block-region based object representation schemes is to reduce the influence of noise coming from the background and surrounding objects, and to identify suitable visual patterns for a given object without introducing errors or heavy computational cost associated with image segmentation. The work in [960] demonstrated that using block regions and semantic modeling for natural scene detection, good performance can be achieved. The concepts used in this work were relatively easy concepts for block regions, such as *sky*, *cloud*, etc. In the SCL method, we intend to test the block-based representation scheme on some more challenging object concepts, which have clear borders of their shapes, such as *lion* and *car*. An interesting observation on our experiments, is that this "simplistic" approach to object extraction does provide a reasonable performance when they are applied in the proposed image classification framework. However, without losing generality, the proposed framework has the full flexibility in supporting other representation schemes such as segmented regions, if effective and efficient segmentation algorithms are available [582]. In this chapter, it will be shown experimentally that by combining this approach with other appropriately designed methods, good performance can be achieved, while maintaining a low computational load.

Some works in the literature shared ideas similar in some aspects to the SCL approach. Despite these similar aspects, SCL has its unique and novel contribution to the research topic on semantic context modeling. For example, the work in [773] proposed a similar idea of context modeling using the output of some visual-feature based concept detection methods. However, this system was designed for a scene classification task; while in the SCL method, by using the block based representation, the proposed system tries to focus on local regions at object level in order to detect object concepts in natural complex images, which are often ignored at scene level. Moreover, their main approach used mixtures of Dirichlet distributions for learning the concept models, while the SCL uses *K2* algorithm context modelling. The work in [900] focused on detecting concepts from user generated content including images and their associated tags using a sparse graph-based semisupervised learning approach. However, it relied on textual data for learning and classification while the SCL method uses only intermediate semantic features which are directly inferred using independent content based image classification methods. [760] used a Correlative Multi-Label paradigm based on Gibbs Random Fields which simultaneously classifies concepts and models correlations between them. [767] used a CRF framework to maximise object label agreement according to contextual relevance. A commonality in these three works [760, 899, 767], was that they constructed correlative models from external resources, such as LSCOM and Google sets, in off-line settings. Compared to these works, SCL is a purely data-driven approach, which means it does not require any external guidance about the statistics of high-level contextual relationships, i.e., Google sets, LSCOM, etc. The inputs required by SCL are multi-feature similarities directly obtained from the MFL module and the ground-truth annotation of the class of concern.

### 3.3.2 Description of the Methodology

In the experiments presented in this chapter, the input to the SCL method is the image relevance scores of all considered object classes. These relevance scores are obtained using the multi-feature based image classifier as proposed in [1038], which shows that using block regions as basic units in object based image classification systems can generate good classification performance. However, there may be more potential for improvements beyond the intrinsic limitation of relying on low-level visual features to describe high-level semantics. Indeed, low-level visual features are generally considered as more ambiguous and semantically impoverished compared with sound patterns, phonemes, words or dialog.

This observation motivated us to exploit the semantic context existing in images and thus improve the system performance. To represent the underlying visual-semantic context of an image, we use the similarity values to represent

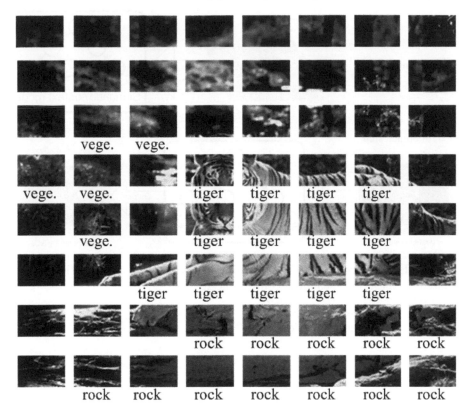

FIGURE 3.1: Image blocks used as regions for object-based image classification.

the probabilities of a block belonging to a particular class. Thus, the probability that an image belongs to class *tiger* should be increased when the same image also belongs to *vegetation* or *rock* classes, and should be decreased if the image belongs to *building* or *car* classes, as shown in Fig 3.1. We note that the semantic interrelations in images form a meaningful context, which can be potentially used to improve the image classification performance.

Semantic context learning is achieved by jointly considering the multi-feature similarities extracted from each block using the MFL module, and exploring the relationships between them. The proposed SCL module uses a Bayesian network to model these relationships. As shown in Fig. 3.2 (left), two steps are considered in the SCL module. In the *learning step*, a small set of images is used as a training set to obtain the *semantic context inference model*. In each one of these images, both visual and semantic evidence is taken into account as training data. Then, in the *inference step*, the learned context model is used to infer the interrelationships between multiple classes within

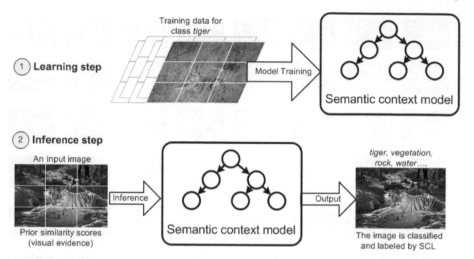

FIGURE 3.2: Illustration of the learning, inference, and classification process based on the visual–semantic context model.

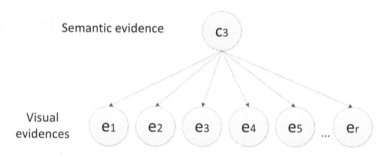

FIGURE 3.3: An example of a Bayesian network structure modeling a simple scenario of five classes.

an input image, synthesize visual-semantic evidences and generate improved semantic labels.

A key feature of the Bayesian Network is its capability to handle incomplete information gracefully. Thus, despite of the fact that the multifeature visual similarities extracted from the MFL module are not 100 percent correct, they can be used as a good evidence in building a meaningful Bayesian network model. Improved classification is then achieved by solving a joint inference problem relying on the integration of existing clues to disambiguate the visual-semantic evidence in a suitable context defined through an experimental scenario.

Assume a total number of $r$ semantic classes are considered in the experiment, denoted as $W = \{w_1, w_2, ...., w_r\}$. Multi-feature based classifications are performed on all the images for all these classes using the MFL module.

The corresponding input for obtaining a context model for class $w_i$ is a set of training images that come with two types of information. The first type of information is the ground-truth annotation on the class $w_i$. This data is referred to as the *semantic evidence*, denoted as $c_i$. The other type of input information is the multifeature similarities of the training images to all classes in $W$. This data is referred to as the *contextual visual evidence*, denoted as $E = \{e_i, i = 1, 2, ..., r\}$ where $e_i$ is the contextual visual evidence for class $w_i$ and $E$ is the collection of this evidence for all classes.

Taking the simple example as in Fig. 3.1 again,

if $W = \{cloud, rock, tiger, vegetation, water | r = 5\}$ and the class of concern is $w_3$ *tiger*, a typical naïve Bayesian network can be constructed as depicted in Fig. 3.2 (right). Since $w_3$ is the current concerned class, the node for semantic evidence $c_3$ is the parent node and the contextual visual evidence nodes $e_1, ..., e_5$ are the child nodes. For this example structure, the joint probability of $c_3$ and $E$ is denoted as $P(E, c_3)$. According to the Bayes' rule, $P(E, c_3)$ can be rewritten as:

$$P(E, c_3) = P(c_3) \cdot P(E|c_3) = P(c_3) \cdot P(e_1, e_2, ..., e_5|c_3). \tag{3.1}$$

Then if we apply the chain rule of probability theory on (3.1):

$$
\begin{aligned}
P(e_1, e_2, ..., e_5|c_3) &= P(e_1|c_3) \cdot P(e_2, e_3, e_4, e_5|e_1, c_3) \\
&= P(e_1|c_3) \cdot P(e_2|e_1, c_3) \cdot P(e_3, e_4, e_5|e_1, e_2, c_3) \\
&= P(e_1|c_3) \cdot P(e_2|e_1, c_3) \cdot P(e_3, e_4, e_5|e_1, e_2, c_3) \cdot P(e_4, e_5|e_1, e_2, e_3, c_3) \\
&= P(e_1|c_3) \cdot P(e_2|e_1, c_3) \cdot P(e_3|e_1, e_2, c_3) \cdot P(e_4|e_1, e_2, e_3, c_3) \\
&\quad P(e_5|e_1, e_2, e_3, e_4, c_3).
\end{aligned}
\tag{3.2}
$$

Intuitively, it is reasonable to assume that the multifeature similarity of one image to class $w_1$ is an independent value to its multifeature similarity to class $w_2$. Therefore, any two evidences in $\{e_1, e_2, ..., e_5\}$ should be independent of each other. Based on this assumption, the following terms can be rewritten:

$$
\begin{aligned}
P(e_2|e_1, c_3) &= P(e_2|c_3) \\
P(e_3|e_1, e_2, c_3) &= P(e_3|c_3) \\
P(e_4|e_1, e_2, e_3, c_3) &= P(e_4|c_3) \\
P(e_5|e_1, e_2, e_3, e_4, c_3) &= P(e_5|c_3).
\end{aligned}
\tag{3.3}
$$

Thus, the joint probability $P(E, c_3)$ can be rewritten as:

$$
\begin{aligned}
P(E, c_3) &= P(c_3) \cdot P(E|c_3) = P(c_3) \cdot P(e_1, e_2, ..., e_5|c_3) \\
&= P(c_3) \cdot P(e_1|c_3) \cdot P(e_2|c_3) \cdots P(e_5|c_3) \\
&= P(c_3) \cdot \prod_{i=1}^{5} P(e_i|c_3),
\end{aligned}
\tag{3.4}
$$

In a general case, assuming $r$ semantic classes are available and the class of concern is $c_k$, $k = 1, 2, ..., r$, instead of (3.4) we have:

$$P(E, c_k) = P(c_k) \cdot \prod_{i=1}^{r} P(e_i | c_k).$$

(3.5)

There are many methods for learning both the structure and parameters of Bayesian networks from the given training data. Learning the network structure and parameters is in fact a search through the space of all possible links and parameters of the set of nodes $n_i$, $i = 1, 2, ..., t$, where $t$ is the total number of nodes. In the proposed SCL method, given a set of predefined classes and the training data, a visual-semantic context model is constructed by applying the *K2* algorithm [239]. *K2* is a greedy search technique. It starts from an empty network with random initial settings and creates a Bayesian network by iteratively adding a directed arc to a given node $n_i$ from the parent node whose addition most increases the *K2* score of the resulting graph structure. The iterations terminate when no more possible additions could increase the *K2* score. The evaluation metric for calculating the *K2 score* of a network structure is described as follows.

Given a database $\Delta$, the *K2* algorithm searches for the Bayesian network structure $G^*$ with maximal $Pr(G^*|\Delta)$, where $Pr(G|\Delta)$ is the probability of network structure $G$ given the database $\Delta$. For two Bayesian network structures $G_1$ and $G_2$, we have

$$\frac{Pr(G_1|\Delta)}{Pr(G_2|\Delta)} = \frac{Pr(G_1, \Delta)}{Pr(G_2, \Delta)} .$$

(3.6)

Thus, the problem of calculating $Pr(G|\Delta)$ boils down to estimate $Pr(G, \Delta)$. Let $N(G) = \{n_i, i = 1, 2, ..., t\}$ be the set of nodes in $\Delta$, where each node $n_i$ has $p_i$ possible values $\{v_{ik} : k = 1, 2, ..., p_i\}$. Besides, each node $n_i$ has a set of parent nodes $\pi(n_i)$, with a total number of $q_i$ instantiations. Define $s_{ijk}$ to be the number of cases in $\Delta$ in which node $n_i$ has the value $v_{ik}$ and $\omega_{ij}$ to be a unique instantiation of $\pi(n_i)$, $j \in [1, q_i]$ and $s_{ij} = \sum_{k=1}^{p_i} s_{ijk}$. The set of conditional probability distributions associated to a directed acyclic graph $G$ is further denoted as $GPr$. Assuming that the cases occur independently and the conditional probability density function $pdf(GPr|G)$ is uniform, then the *K2 score* can be computed as:

$$Pr(G, I) = Pr(G) \prod_{i=1}^{t} \prod_{j=1}^{q_i} \frac{(p_i - 1)!}{(s_{ij} + p_i - 1)!} \prod_{k=1}^{p_i} s_{ijk}!,$$

(3.7)

where $Pr(G)$ is the prior on the network structure that equals to a constant, thus it can be ignored. Therefore, the evaluation metric for computing the *K2 score* of a network structure $G$ is given by:

$$K2score = \prod_{i=1}^{t}\prod_{j=1}^{q_i} \frac{(p_i - 1)!}{(s_{ij} + p_i - 1)!} \prod_{k=1}^{p_i} s_{ijk}!. \qquad (3.8)$$

Assume the nodes are in a given order and $n_i$ cannot be a parent of $n_j$ if $i > j$. The algorithm starts an iterative process for each node $n_i$, including the following steps: 1) calculate the score for the case where $n_i$ has no parents; 2) calculate the score for the case where $n_i$ has a parent among all nodes that have smaller indices than $i$. If any of these are greater than the case with no parents, select the node $n_j$ which gives the maximum and add an arrow from $n_i$ to $n_j$. This process is iterated by adding more parents and continuing until no further nodes increasing the score can be found.

After the structure $G^*$ of the Bayesian context model is learned, an inference process is carried out to measure joint probability distributions between the *contextual visual evidences* and the semantic class in each input image, as in (3.5). The multi-visual-feature similarities of images are replaced by the *posterior probabilities* inferred using the visual-semantic context model learned for a class. By applying the Bayes classifier using the context model of a class, images in the database are classified and labeled accordingly.

## 3.4 Experiments

In this section, we present our experimental results and analyze the performance of the multi-feature fusion model and semantic context model obtained using MFL and SCL approaches. In Section 3.4.1, the results are evaluated based on a dataset called DB-COMB10000. In Section 3.4.2, we compare the performance of the proposed approaches with some other state-of-the-art approaches using a benchmarking dataset in the MIRFlickr database.

Without losing generality, seven visual primitives have been used to assess the performance of our experiments: Color Layout Descriptor (CLD), Color Structure Descriptor (CSD), Dominant Color Descriptor (DCD), Edge Histogram Descriptor (EHD), Texture based on Gabor Filter (TGF), Grey-Level Co-Occurrence (GLC) and Color Histograms in HSV Space (HSV). Observe that the first four primitives are MPEG-7 descriptors [202], while the other three are well established descriptors from the literature [891, 929]. It is important to stress that the proposed approaches are not tailored to a given number of low-level descriptors; instead, any descriptor bank can be used.

Training sets of SCL modules were randomly selected subsets from the experimental image collections. In our experiments each training set contained five hundred images. The representative groups for the MFL module were also selected from these training sets. Positive representative samples were collected directly based on existing ground-truth of the training sets. In case negative samples were needed, they were selected by using the first a few false positive samples in the result of $R^+$ based MFL classification.

### 3.4.1 Experiments in DB-COMB10000 Dataset

The proposed image classification approaches were first evaluated using an image collection named *DB-COMB10000*. DB-COMB10000 contains 10000 images selected from the *Corel* database and the *LabelMe* database [795]. The aim is to produce a dataset that is close to realistic image repositories in which images are complex and contain multiple semantic objects in both background and foreground. The context formed by multiple concept classes is an essential ground for inference and classification in the SCL method. For that reason this research did not use popular object classification databases such as PASCAL VOC [316], in which each image only contains a single object class. The DB-COMB10000 dataset has pictures depicting a variety of topics, taken in different time, places and conditions. As a consequence, only a small portion of images belong to some of the considered classes used in this experiment, especially for the classes representing uncommon objects such as *elephant* and *tiger*. Thus, the classification task in this dataset was very difficult and represents a realistic image classification scenario. Notice that the DB-COMB10000 dataset is fully annotated and of reasonable size compared to some of the state-of-the-art image databases in the literature that also have full annotations.

The SCL method mainly targets object-based image classification. Thus, a set of object classes were selected here for experiments, covering most object-like semantic elements of image content. For DB-COMB10000, ten classes have been selected for testing based on subjective observation throughout the experimental dataset. These classes are: *building, car, cloud, elephant, flower, lion, rock, tiger, vegetation,* and *water*.

The first experiment was performed based on the multi-feature fusion models generated by the MFL approach. To assess the performance of the multi-feature fusion models, the obtained results were compared to similar classifications but employing only single features. Thus for each class, eight experiments were performed, including one Multi-Feature Based Classification (MFBC) using the obtained class-specific multi-feature fusion models, and seven Single Feature Based Classifications (SFBCs) based on each one of the seven visual features. Fig. 3.4 (left) shows the mean precision–recall curves across ten classes for the MFBC result and seven single feature based classification results. It can be observed in this figure, that the proposed multi-feature fusion models significantly improved the classification performance against any of the single visual features used.

A second experiment was performed to assess the performance of the SCL approach. Here, the multifeature similarities were used as the *contextual vi-*

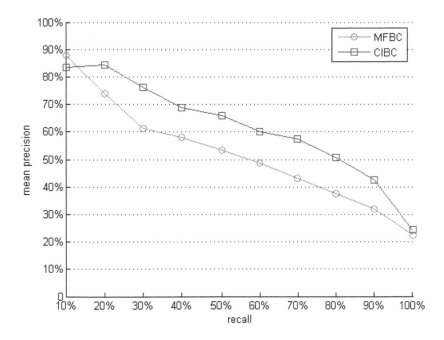

FIGURE 3.4: Mean precision–recall curves of ten classes using the initial similarity basic retrieval and the proposed SCL approach.

*sual evidence* for SCL based inference and classification. A Bayesian network was constructed to model this semantic context by carrying out the learning process as depicted in Fig. 3.2. In the inference process, a posterior probability can be calculated through this Bayesian model, indicating how likely it is that each test image belongs to a target class given both its multi-feature similarity and the visual-semantic context model. This round of experiments is referred to as the Context Inference Based Classification (CIBC). To show the classification performance of CIBC, the same mean precision–recall curve is depicted in Fig. 3.4 (right). For comparison, the curve for MFBC is also depicted in the same figure. A clear observation from this picture is that the CIBC results have shown clear advantage over the MFBC results.

The results of MFBC and CIBC experiments show that the classes of an image can be effectively determined using the two proposed approaches: MFL and SCL. MFBC performed much better than any single feature based classification, which means that the MFL approach was able to build a suitable multi-feature fusion model for a particular semantic class. The obtained multi-feature fusion model synthesized the useful aspects from different features, and thus provided better discrimination power compared to the single features. Furthermore, the CIBC produced even better results compared to MFBC. This means that the SCL approach was able to build a useful visual-semantic context model using the imperfect visual evidence, which in our case comprised the multifeature similarities generated by the MFL approach. The inference process based on this model can boost the classification performance significantly.

### 3.4.2 Image Classification in MIRFlickr25000 Dataset

MIRFlickr25000 dataset [439] is an image collection consists of 25000 images that were downloaded from the social photography site Flickr.com through its public API. These images are representative of a generic domain. MIR-Flickr mainly targets concrete visual concepts in high-quality color images, which suits the purpose of the object-based image classification system. This image collection has been used in image annotation tasks of the ImageCLEF benchmarking forum. In this task, usually both visual features and seman-tic features, i.e., user tags associated to the images, are used. However, the proposed image classification system is purely visual based.

MIRFlickr provides full annotation of 24 potential labels and 14 regular (subjective) annotations. These concept classes were considered in our exper-iments except for two types of concepts. First, concepts about scenes were not considered in this experiment. The task for the proposed framework is object-centered image classification. Scene classification is out of the scope of this research. Second, concepts about humans, i.e., female or baby, were not used in our experiments. Human detection and recognition has been a famous and well-studied research topic. Many specialized systems are available in the literature dedicated to this task. Our proposed system aims at general image classification, and is not suitable in detecting human-related concepts. There-fore, the object classes being considered in our experiments include 14 labels: *sky, plant life, structure, clouds, sea, river, tree, flower, dog, bird, car, water, animals, lake*; and 8 regular annotations: *clouds(r), sea(r), river(r), tree(r), flower(r), dog(r), bird(r), car(r)*.

The same seven visual features were used in the MFBC and CIBC experi-ments. Classifiers of MFBC and CIBC were trained for each of the classes and the mean precision–recall curves across the 22 classes are presented in Figure 3.5.

For benchmarking, the resulting average precisions (APs) of each of the 22 classes were compared with the output of two visual feature based classi-fication methods: a linear discriminant classifier (LDA) and a support vector machine classifier (SVM) with an RBF kernel, provided in [440]. In Fig. 3.6, we present AP scores per concept class, for MFBC, CIBC, LDA, and SVM. For reference, a rate of relevant images per concept in the ground-truth database is also included.

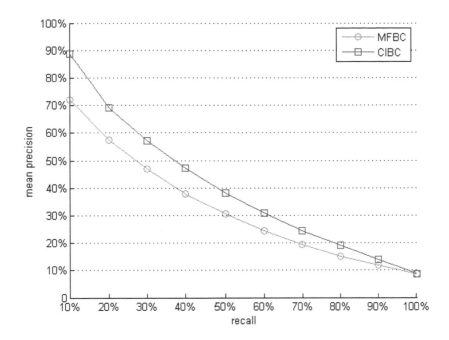

FIGURE 3.5: Mean precision–recall curves of 14 classes using the two proposed approaches.

For 17 classes out of 22, the CIBC approach yielded higher AP scores compared to the MFBC approach. On average, CIBC had a higher mean AP score compared to LDA, while MFBC had slightly lower AP than LDA (less than 1%). SVM performed better than the other three approaches in most classes. It should be noted that the better performance of the SVM approach came at a much higher cost in calculation, compared with the two other proposed approaches.

In terms of performance, there were mainly two real-time steps to be considered in our experiment: 1) distance calculation using the multi-feature fusion model in MFL and 2) semantic context inference in SCL. A computer with a Windows XP operating system and Intel Core-Duo E5200 CPU was used in all experiments. For one classification process in about 20 thousand images on concept *car*, step 1) used 13.37 seconds and step 2) used less than 1 second. As these online steps are based on straightforward arithmetic, the overall system has potential to be extended for application in large-scale image databases. Some of the offline processes, such as feature extraction, may be time consuming but because they take place offline, they will not affect the performance of the system.

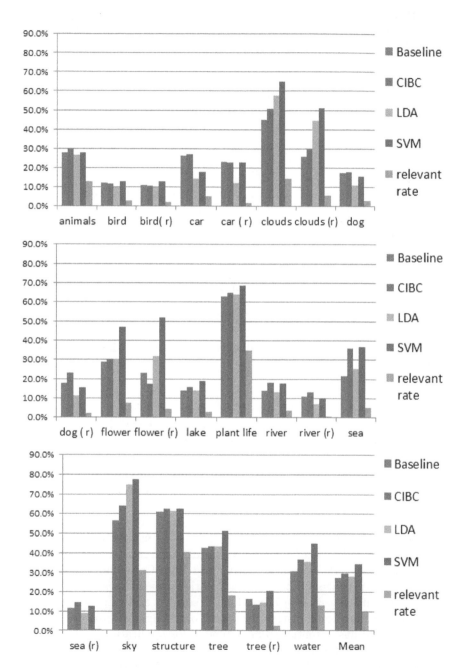

FIGURE 3.6: Comparison in terms of AP of MFBC, CIBC, LDA, and SVMs.

## 3.5 Conclusions

In this chapter, we have described an innovative method for content based image classification, semantic context learning based on a block-based representation scheme for semantic classes. The block-based representation scheme is an intuitive but effective way to extract object classes from complex images. The semantic context inference method has then been employed to enhance the classification performance by jointly exploring the context among object classes represented by blocks of an image. The visual-semantic context, which can be represented by the interrelationships among object classes, is modeled using a Bayesian network. This Bayesian model has been generated through a learning process based on a search based algorithm *K2* in a fully automatic manner.

We have shown that, with extensive experiments, the proposed framework for image classification was able to effectively classify visual content that belongs to a class. We have also shown that the proposed approaches were able to handle a variety of problems in image classification without requiring extensive human expertise and efforts for constructing the Bayesian model. We first demonstrated the adequacy of the multifeature learning approach in semantic classification based on block regions. The multi-feature based classification employing the learned fusion model outperformed single visual feature based classifications, and other similar multi-feature based classification methods described in the literature. On top of that, we have shown that the proposed semantic context learning method can further boost the classification performance. The advantage of the semantic context learning method is that it exploits the very important, but usually ignored, visual and semantic context information to assist in the classification. Thus, it can often improve the classification accuracy compared to the classifiers relying on purely visual features. The improvement introduced by the semantic context inference method has demonstrated the effectiveness of exploring the resourceful context information in addition to the isolated visual content and vocabularies.

# 4

## Restructuring Multimodal Interaction Data for Browsing and Searching

**Saturnino Luz**

*Trinity College Dublin, luzs@cs.tcd.ie*

## CONTENTS

This chapter presents recent work on techniques for indexing and structuring recordings of interactive activities, such as collaborative editing, computer-mediated and computer-assisted meetings, and presentations. A unifying architecture is presented which encompasses modality translation (e.g., speech recognition and image analysis) as well as the underlying recording under a single linked time-based model. This model is illustrated through case studies and implemented prototypes, which support remote multiparty collaboration and collocated recorded meetings. Requirements and the issues they pose for the design of this kind of system are discussed.

## 4.1   Introduction

Consider a remote collaboration between two people who were given the task of organizing a final-year weekend trip for staff and students. The trip planners communicated through a dedicated audio conferencing system which included a real-time audio channel, a text chat channel, and a shared editor on which the participants could write as well as highlight and point to objects such as sentences and paragraphs, while talking [147]. During the meeting, the participants used the shared surface as a common notebook where they could jot down their thoughts and options considered (or to be considered), record decisions and "action points", share references to external data (such as URLs containing useful resources), and so on. The excerpt in Table 4.1 is an example of a typical text that remained on the shared editor once the meeting was concluded.

This excerpt, by itself, can hardly be regarded as a complete record of the discussion, or even the part of it that the text spans. However, while producing a structured meeting minutes was probably not the intention of the participants, a certain amount of structure remains and can be identified in the text. It seems clear, for instance, that the text up to line $t_{12}$ was used during the decision making phase of the meeting, where options were considered, discussed, and in some cases rejected. From line $t_{13}$ on[1], the text seems to record the final decisions made, costs etc.

In meetings that are recorded for review at a later time, such as the meeting from which the above text has been extracted, there is a possibility of enhancing the textual structure by listening to the recorded audio and watching the video or alternatively, if video is not available, some form of visual playback of relevant actions such as pointing and adding text to the shared surface. Since these complementary modalities are essentially rendered through time-based media, the text can be easily linked to them if the actions that produce and modify it are time stamped. Line $t_8$ of the text fragment in Table 4.1, for instance, was modified and pointed at while the participants discussed entertainment options for Saturday afternoon. The audio recorded during (or around) these typing and pointing actions contained the dialog shown in Table 4.2.

Having listened to the audio, a person reviewing the meeting record would arguably gain an understanding of why greyhound racing made it into the program, how the program was decided, how much it will cost, what other options were considered (the abbey, which did not appear in the final textual

---

[1] Note, however, that the actual text log for this meeting contains a considerable amount of text between what is identified here as lines $t_{12}$ and $t_{13}$. This intervening text has been omitted here for clarity, as indicated by the use of the vertical ellipsis symbol. This convention is used throughout this chapter for both text and transcribed speech. Textual content is denoted in this chapter by $t_1, t_2, \ldots$ and quoted speech is denoted $s_1, s_2, \ldots$

**TABLE 4.1**
Traces of Textual Interaction on a Shared (Synchronous) Text Editor.

 

$t_1$. Organization of Final-Year/Staff Weekend

$t_2$. How much will it cost per person... 124

$t_3$. budget of 3000 from the student union

$t_4$. maybe charge people more?

$t_5$. travel 1500

$t_6$. hotel 6120

$t_7$. Sat night :1375

$t_8$. Saturday afternoon (Greyhound racing) - 825

$t_9$. total 9820 - 3000 = 6820 / 55 = 124

$t_{10}$. Booking Hotel...

$t_{11}$. The [X] clube hotel - 240 for double

$t_{12}$. 4400 for students

$\vdots$

$t_{13}$. **Information about the weekend...** Weekend in Kilkenny City

$t_{14}$. Travel by hired bus

$t_{15}$. Staying in [Y] Hotel.

$t_{16}$. Saturday afternoon's activities: Visit Castle and then go Greyhound racing!

$t_{17}$. Saturday night: [Z]'s Steak & Ale House, traditional cuisine and music!

$t_{18}$. Total cost- 124 per person (subsidized by the student union)

---

record but was later referred to by B, who stated that a visit to it would have been her prefered option), and so on.

This fairly obvious combination of media for recording and reviewing meetings has been the basis for most approaches to automatic meeting record creation and access found in the literature [755, 973]. In fact, the idea of "anchoring" the content of transient conversations to permanent textual objects has also been suggested by qualitative observational studies in human-computer interaction [685, 986] and exploited in early applications that attempted to structure conversations temporally around documents, from text chat [231] to real-time audio [972, 630]. The timeline has proven a powerful metaphor for access to multimedia records [792], and enhanced methods of meeting brows-

**TABLE 4.2**

Dialogue Produced while Line $t_8$ of Table 4.1 Was Being Referred to through Deictic Gestures or Modified.

---

$s_1$. A: we have hm two options for the Saturday afternoon... so we've got, like, the greyhound racing... or there is kind of this abbey...

$s_2$. B: hmm

$s_3$. A [reading]: "It is regarded as one of the most interesting Cistercian ruins... offers a unique insight into the lives of the monks because many of its domestic arrangements are still recognizable"...

$s_4$. A: it is 10 per person to wander around there

$s_5$. B: and how much is the greyhound racing?

$s_6$. A: 15 per person

$s_7$. B: I wonder, could we give people a choice? or would that complicate matters.

$s_8$. A and B: [discuss the options]

$\vdots$

$s_9$. B [returning to the text after a few minutes]: OK

$s_{10}$. A: that's gonna cost 825, is that ok?

$s_{11}$. B: ok. So Saturday is going to be... 1375

---

ing usually take the timeline as a basis on which more sophisticated indexing methods based on modality translation can build [335, 778].

However, for all its intuitive appeal, the timeline may lead to inefficient access to content. To illustrate this point, let us take a closer look at the decision to go to greyhound racing on Saturday, and the statement that the total cost for the night would be 1375 (line $s_{11}$, in the excerpts above). It is not immediately clear from the text and speech transcripts why it is that, in line $s_{11}$, B concludes that the cost will be 1375, rather than 825. A closer look at the timestamps (not shown in the transcripts) reveals that changes in the text referring to the greyhound racing overlapped with a segment of the dialog which in turn overlapped with text modification and pointing events in line $t_{17}$. These text actions partially overlapped with a discussion about dinner arrangements that took place several minutes earlier. One can hence conclude that the cost of dinner accounts for the difference.

Although there is a clear relationship between the text and the audio streams, sequentiality only reveals this relationship up to a point. Thus, a browsing interface consisting of a timeline representing speech turns annotated with text, whether extracted from segments of the text record that overlap temporally with the speech or text generated through automatic speech recognition would be of limited use in identifying the sort of relationship described in the previous example.

This chapter argues that a model that can minimally account for such recursive inter-media relations based on different levels of text and speech (and optionally video) segmentation is needed. It then goes on to propose a model based on temporal links that induce a graph structure on multimedia records of multiparty communication and collaboration. This model implies a restructuring of such records and often results in the clustering together of segments located far apart in the original linear structure of the data. Instantiations of this model are illustrated through two case studies on information access in multimedia meeting records.

## 4.2 Recording and Analysis of Multimodal Communication in Collaborative Tasks

Once restricted to denoting face-to-face interaction, the word "meeting" has, with the spread of information and communication technologies, taken on a much broader meaning. It now generally refers to any form of multiparty interaction mediated through video, audio, synchronous text (including collaborative editing, text chat etc) or, as is most often the case, a combination of these communication media. Technological developments have also encouraged recording of meetings for analysis and information retrieval [973, 149]. In both face-to-face (co-located) and remote meetings, it is now possible to capture a wide variety of information exchanged through different communication modalities. This abundance of data, however, needs to be properly indexed if it is to be useful in review and retrieval contexts.

Initial research on automatic indexing of time-based media focused on *modality translation* [858], that is, on translating content originally encoded in transient and sequential form (i.e., audio and video) into a parallel and persistent presentation [502, 855]. Text, key frames, and other static content generated through modality translation may then be used in conjunction with existing retrieval techniques to create a form of content-based indexing. The role time naturally plays in structuring data recorded during meetings, lectures and other such presentations is crucial to these approaches. Thus, content extracted from the audio track through speech recognition, for instance, may provide valuable links to video content. The temporal structure of these data also suggests visualization and retrieval interfaces that emphasize sequential access, enriching the basic playback metaphor with media and domain specific improvements, such as skimming [79], parsing with compressed data [1036], generation of visual summaries [855], text tagging [148], dialog act annotation [879], and other techniques [755].

### 4.2.1   Requirements and Fieldwork

Observational and ethnographic work has been an influential line of research into the functional requirements for systems to support access to recorded meeting content. Moran et al. [685] looked at how people used audio recordings to review and report on the meetings they attended. The researchers defined a notion of *salvaging* content as "an active process of sense-making" which may encompass browsing and retrieval but goes beyond these activities. Salvaging content involves sorting through the artefacts manipulated and produced during the meeting, and the activities performed by the participants, in order to reassemble them so as to make the meeting record more accessible to "potential consumers" of its content [685].

It is assumed that this salvaging activity is targeted at particular kinds of consumers and therefore may vary considerably in terms of its goals, from producing short summaries (minutes) to explaining the reasons behind the decision making as illustrated in our earlier example. It is also assumed that salvagers would have been aided by tools designed specifically for the purpose of recording and relating activities and artefacts. While the study reported in [685] mainly aimed at studying this salvaging activity as done manually by humans, the prospect of automating the process of identification of key activities, events and artefacts in the multimedia record of a meeting, and allowing the content consumers themselves to specify the goals of the content (re)structuring process has motivated much work on meeting analysis and browsing [149].

Following this line of work, Whittaker et al. [986] conducted ethnographic studies to investigate how meeting participants make records. They categorized such records as *public* or *private*, and identified several shortcomings of existing meeting browsers in supporting the production of both types of records. Shortcomings in support for public records relate to the focus on individual meetings (i.e., the lack of ways to link a meeting record to another) and the mismatch between the output produced by automatic speech recognition and the formality requirements commonly associated with public records such as minutes, which have archival and sometimes legal implications. From the perspective of private records, the authors note that existing browsers fall short in terms of support for extraction of personal actions, focusing instead on low level annotation such as speech turns and key events such as slide changes.

Other qualitative studies confirmed the importance of multimedia records for sense making. Jaimes et al. [456] conducted a survey with people who regularly participate in meetings and grouped their goals when using multimedia (video) records into the following broad categories: verifying what was said by a particular participant, understanding parts that were missed or not understood during the meeting, reexamining contents under a different context, record keeping, and recalling ideas not explicitly discussed. They then conducted a study with 15 participants to determine which facts relating to meetings that had been attended by the participants were more easily remembered and which were more easily forgotten. Perhaps unsurprisingly, items more easily forgotten were dates and times, participants' dress, posture and emotional expressions. On the other hand, items such as seat positions, table layout, participant roles, and major topics discussed were easily remembered even three weeks after the meeting took place. Based on these goal categories, the authors proposed an interesting framework for retrieval anchored on visual cues anchored on those items they found to be more easily recalled.

## 4.2.2   Corpora and Meeting Browsers

Popescu-Belis et al. [755] summarize the findings of a number of studies (observational, questionnaire- or interview-based, and laboratory-based) aimed at eliciting requirements. They highlight the importance of topic lists [100], summaries, and observations of interest which formed part of a proposed *browser evaluation test* [984].

These studies aimed at understanding the goals, tasks and needs of potential users of meeting browsers and happened in connection with research projects focused on meeting technologies. These efforts include the ICSI Meeting Recorder project [463], the European funded AMI/AMIDA projects [779], the ISL Meeting Room project [180], the M4 project [656], VACE-II [218] and the ECOMMET project [488, 627]. These projects have produced a wealth of recorded meeting data and, together with the NIST Meeting Corpus [357], have helped lay the foundations for automatic analysis of meeting contents. In addition to producing corpora, these projects contributed to advancing the state of the art in a number of technologies deemed necessary to satisfy some of the needs identified through fieldwork and user studies.

In order to be able to index recorded audio content according to speaker, for instance, a necessary first step is *speech diarization* (who said what). Speech diarization is usually performed through change detection with the Bayesian information criterion [219] followed by clustering of audio feature vectors. The best performing techniques employ Gaussian mixture models as emission probabilities for continuous density Hidden Markov Models [51]. While diarization can still be rather error prone, specially in noisy environments, great advances have been made in this area [335, 918],

Following the segmentation of the audio stream according to speakers and speech activity performed through diarization, it is necessary to segment the dialogs into topics. This task has been approached in different ways, and no dominant approach seems to have emerged in the literature. Most approaches to topic segmentation employ a combination of features. Commonly used features are lexical features (or "bags of words") obtained from the output of a speech recognizer, conversational features (lexical cohesion statistics as well as dialog structure, vocalization and silence statistics) [355], prosodic features [842], video features [284, 433], and other contextual features such as dialog type and speaker role [433]. The generation of such features usually poses its own challenges in terms of machine learning and signal processing techniques, and can be performed reliably only to a certain extent. High word error rates for automatic speech recognition, for instance, would hinder the use of lexical features and lexical cohesion statistics, in spite of the fact that topic modeling is resilient to moderate word error rates [433]. An approach to topic segmentation that avoids speech recognition input altogether is presented in [627] and tested in the ECOMMET and in the AMI corpora [626].

Large vocabulary continuous speech recognition (LVCSR) is nevertheless generally regarded as an essential component of any meeting record indexing system, as a source of input features to topic segmentation as described above, as a source of keywords for annotation of audio segments and video sequences [502, 148, 755], and as general transcripts for browsing and information retrieval [973, 779].

In some cases, summaries of the transcription may be more effective as an aid to meeting browsing [1031] than simple annotation by keywords. In addition, other high level functions can arguably only be performed if LVCSR attains a minimum level of accuracy. Stolcke et al. [879], for instance, propose a system that attempts to recognize *dialog acts* (i.e., to group utterances into classes such as *statements, yes-no-question, wh-question, quotation*, etc.) in spontaneous speech. However, the best accuracy achieved by their system was only 65% for automatically recognized words, compared to a chance baseline accuracy of 35%.

Other techniques for recognizing high-level events identifiable in meetings, and regarded as useful in the creation of personal records [986] have also been proposed. Hsueh and Moore propose a method for detecting decision points in a discussion [432]. In a similar vein, Dielmann and Renals employ dynamic Bayesian networks to segment and categorize dialog acts [285], according to the annotation scheme used by the AMI project. Despite relatively good performance in segmentation, classification error for dialog acts is still quite high for this challenging task.

As regards the visual modality, techniques developed for browsing of video data such as highlight extraction, static key frame selection, shot boundary detection, and other methods that combine various source modalities (e.g. high motion and pitch for detection of discussion activities) [864]. Such combinations have been employed with some success in identifying significant meeting actions. In [656], for instance, a technique that uses low-level audio and visual features (speech activity, energy, pitch and speech rate, face and hand blob) is employed to characterize a meeting as a continuous sequence of high-level meeting group actions (monologue, presentation, discussion, etc.) using Hidden Markov Models.

## 4.2.3 Issues and Open Questions

Despite the fact that recording a vast range of information from several communication modalities is well within current technological capabilities, that the analytic methods reviewed in this section are improving and that in some cases they have reached a level of maturity which permits their use in practical multimedia browsing systems, these technologies still have a long way to go before all communication modalities can be properly accounted for.

The issues that may arise are exemplified by consideration of the process of *grounding*, a core phenomenon in human dialog. Grounding is the process of updating the shared information needed for dialog participants to coordinate and maintain their communication activity [233]. Sometimes grounding is explicit, as in utterance $s_{11}$ where "ok" signals agreement and confirms that both participants know the cost, which then becomes part of their *common ground*. Most often, however, the process is managed either (a) through backchannels, as in when B utters "hmm," in $s_2$ to signal to A that she is listening and has understood his utterance $s_1$, or (b) implicitly, as in when B simply initiates a relevant next turn in $s_5$ by asking a question. While these are examples of positive feedback (evidence that one has been heard and understood), negative feedback is also common. The example shown in Table 4.3, which was extracted from the meeting quoted above, illustrates how explicit negative evidence can be presented as part of the grounding process.

**TABLE 4.3**
Dialog Illustrative of the Use of Negative Evidence in Grounding.

---

$s_{12}$. A: should we work out how much that costs, then?

$s_{13}$. B: —so far?

$s_{14}$. A: yes. [...]

$s_{15}$. A: ... so... 24 rooms, that would be... like... 48 people? [pause]

$s_{16}$. B: 55?

$s_{17}$. A: yeah, 55.

---

B utters $s_{13}$ in order to indicate that she has not quite understood A's suggestion, and later utters $s_{16}$ to indicate disagreement with A's figure ($s_{15}$), which is uttered as a question (a "try marker," followed by a short pause) which invites A to confirm or correct the information just given.

A proper treatment of grounding is important for indexing of multimedia records of dialogs and meetings because grounding activities are good indicators of segments of a dialog where indexable terms (keywords), such as references and verbatim descriptions [233, pp. 227–230] are uttered and resolved. The shortcomings of automatic speech transcription for this kind of task are evidenced by the fact that transcription alone will fail to identify patterns such as the try marker-correction exchange in $s_{15}$-$s_{16}$ or positive confirmation evidence provided by verbal backchannels such as "hmm," "m," etc. While prosody extracted from acoustic features might help restore question marks and other relevant information [614] the best performing approaches are still quite inaccurate. In addition, grounding does not necessarily take the form of vocalizations. Eye contact, gestures (e.g., pointing as a way of identifying an object) and monitoring of facial expressions are all techniques unconsciously employed by interlocutors for maintaining common ground.

## 4.3 Modeling Interaction

The requirements and issues discussed suggest that, in addition to technologies capable of capturing and analyzing interaction data such as speech, gestures, facial expressions, text editing etc., one needs to be able to integrate these capabilities under a unified information structure.

The timeline provides a natural structure for interaction data. However, as we have seen, additional structure is needed in order to link activities that are discontinuous on the sequential record but are semantically connected. These connections often have a hierarchical structure, as illustrated by the following fragment of a remote meeting of three participants who are planning an HCI course (see Tables 4.5 and 4.4). The participants collaboratively write a course syllabus on a shared whiteboard. The shared surface of this whiteboard is also the focus of deictic gestures that are visible to all participants and thus serve to identify parts of the text as referents which form part the participants' common ground. The text produced (also part of the participants' common ground) includes the excerpt shown in Table 4.4.

Line $t_{26}$ was edited, highlighted, or pointed at 14 distinct and discontinuous time time intervals. Presumably, the speech exchanged at each of these intervals is potentially related to $t_{26}$. The first interval, for instance, contains the exchange $s_{18}$-$s_{19}$, between participants C and D (Table 4.5). Utterance $s_{18}$ overlaps in time with the pointing gestures that target $t_{21}$, $t_{22}$ and $t_{23}$. The first of these text objects, for instance, is active at another 12 time intervals scattered over the duration of the meeting. They will therefore overlap with other speech segments that will contain, for instance, information about high-level course organization which is not shown in the text, such as the speech turn $s_{20}$, in Table 4.5.

These kinds of linkage structures are common in multimodal interaction and can be exploited for information browsing and retrieval. In [629] a basic model was introduced which related chunks of text to speech segments at two elementary levels: temporal proximity and co-occurrence of key words. This model, however, does not extend beyond time-stamping, and basic LVCSR

**TABLE 4.4**

Text Written on a Shared Editor as Part of a Course Design Meeting.

---

$t_{19}$. course assessment. Project to mirror flow of lectures: – 1) model users/tasks/ human context... paper based interaction design... 2) working prototype... 3) evaluation ... user manual...3 chunks 20 30 15 marks??? group or individual??

$t_{20}$. web site resources

$\vdots$

$t_{21}$. 3- Designing for people

$t_{22}$. 4- Modeling users and their tasks

$t_{23}$. 5- Designing user interfaces

$\vdots$

$t_{24}$. 8- Experimental evaluation

$t_{25}$. 9- User support materials

$t_{26}$. 10- Advanced topic: e.g. Computer supported Cooperative Work [...]

---

**TABLE 4.5**

Dialog from the Course Design Meeting.

---

$s_{18}$. C: well, but I mean... specially for someone like me who has done CSCW work it would be nice to have CSCW there, but we also have to consider how much we will be teaching in these parts [points to $t_{21}$, $t_{22}$ and $t_{23}$].

$s_{19}$. D: I think the for this type of introduction to HCI course it is good to have a pointer to advanced topics [writes on $t_{20}$]

$\vdots$

$s_{20}$. D: this group here [highlights $t_{22}$ and $t_{23}$] deals with a kind of contextual and human side of things. These three lectures are hmm... kind of a chunk as are these two [points at two other lines of text].

---

output, and therefore cannot accommodate richer data sources and automatically extracted features. In the following section, an alternative is presented which arguably captures the above discussed requirements in a more comprehensive manner.

### 4.3.1    A Linked Time-Based Model

The elementary entities in the framework proposed here are (a) interaction *events*, (b) content *segments* and (c) time *intervals*.

Interaction events are discernible occurrences that alter the recorded content. The most common events of interest for multimedia indexing are actions performed by the people participating in the recorded interaction. Therefore one might refer to writing events, speech events, gesture events, etc. These are the only types of events to be distinguished in this framework. Of course not all events are (or result from) actions performed by the participants. A fire alarm may go off during the meeting, for example. Such types of events will only be addressed indirectly via actions performed by the participants in response to the event (e.g., a participant shouts "Fire!").

Segments are sets of results (traces) of interaction events which contain indexable information. At the most basic level, the minimally indexable content of an interaction event is the record of the very fact that it occurred (e.g., participant A was speaking at a certain point in time). Segments may consist of permanent and predominantly static content or transient and predominantly dynamic content. The first group includes textual and graphic segments such as words, sentences, paragraphs, icons, images, drawings etc. The second group includes vocalizations, sounds, gestures, video sequences etc. Super-segments can often be identified by means of clustering and classification methods. Thus in text and speech one can attempt to segment according to topics [842, 433, 125, 405], tasks [627], activities [284], etc. Speech can also be segmented according vocalizations to turns, speech act sequence patterns [1026] an so on.

Time intervals can be regarded as anchor points to which information extracted from different media sources are linked through the interaction events that produced that information. Of particular interest are the following temporal logic relations [63] between the time intervals associated with two segments: *equal* ($e$), *meets* ($m$), *overlaps* ($o$), *during* ($d$), *starts* ($s$), and *finishes* ($f$). These relations are depicted in Figure 4.1. They are the subset of the possible 13 basic interval relations (and $2^{13}$ possible indefinite interval relations [591]) that encompass continuity and concurrency. We take these to be the basis for relating the contents of two segments. Note that the segments in question may come from any of the media sources.

Content segments can be linked to time intervals by means of *time stamps*. A time stamp is an annotation of a time interval $[a, b]$ on a content segment. We will represent the set of time stamps of a segment by means of a function $\tau : \mathcal{S} \to \wp(\mathbb{R}^2)$, so that $\tau(S_i)$ is the set of all intervals in which events related to $S_i \in \mathcal{S}$ took place (e.g., the intervals in which any participant points at a paragraph of text). A segment made up exclusively of transient events such as speech or gestures has a single timestamp corresponding to the duration of the event(s) that produced the segment. Temporal links between segments can then be defined as follows:

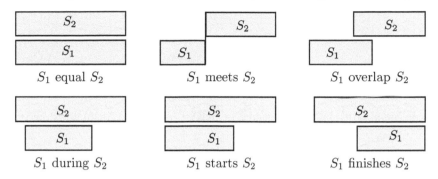

FIGURE 4.1: Time relations of co-occurrence between segments $S_1$ and $S_2$ spanning the time intervals represented by the respective rectangles.

**Definition 1** Temporal link: *a segment $S_1$ is temporally linked to a segment $S_2$, written $S_1 L S_2$, if there is at least one pair of intervals $i_1 \in \tau(S_i)$ and $i_2 \in \tau(S_2)$ such that $i_1 R i_2$ where $R$ is one of the temporal relations listed above (e, m, o, d, s, or f) or $m_1$, the inverse relation of m. R is symmetric and irreflexive.*

Given a strategy for individuating segments and the above defined link relation one can proceed to define a meta-structure for a multimedia record of an interactive activity as a graph.

**Definition 2** *A temporal interaction graph is a tuple $(\mathcal{S}, \mathcal{L})$ where $\mathcal{S}$ is a set of segments extracted from a multimedia record and $\mathcal{L}$ is a set of pairs $(S_i, S_j)$ such that $S_i, S_j \in \mathcal{S}$ and $S_i L S_j$.*

The definitions above do not specify what constitutes a segment or how specifically those segments are time-stamped. Our aim here is simply to set out a general framework into which different segmentation and timing techniques can be incorporated and tested. A simple example should help illustrate this point. The setting is the one in which the course planning example was recorded. Participants were remotely located and communicated through a networked audio tool and a shared (virtual) editor/whiteboard. Speech segments were individuated through simple adaptive average-energy analysis of separate streams of the speech signal (each participant wore a lapel microphone, which generated individual speech streams and obviated the need for speaker diarization [147]). Straightforward synchronization ensured that each segment was properly time-stamped with respect to text events. Text segments were defined as chunks of text enclosed by paragraph-marking boundaries (similar to "boxes" in TEX [526]) and time-stamped through monitoring of the participants' activities on the shared editor. Figure 4.2 shows the a temporal interaction graph constructed from the recorded interaction data for text chunk $t_{20}$ of the course syllabus discussion fragment presented above

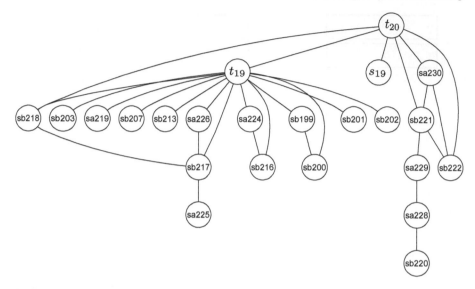

FIGURE 4.2: Temporal interaction graph for segment $t_{20}$, from the example given in Table 4.4, showing its connections to $t_{19}$ and $s_{19}$, as well as other (arbitrarily labeled) segments.

(Table 4.4). Note the connection with the text labeled $t_{19}$ in Table 4.4 as well as the connection to speech turn $s_{19}$, from Table 4.5.

One can promptly observe that $t_{19}$ is linked to a number of speech segments both directly (adjacently in the graph) and indirectly (connected by a path of more than one edge). Since it extends beyond the node neighborhood of the segment under consideration (in this case text segment $t_{19}$), the temporal interaction graph can potentially capture semantic relationships between segments that are both spatially (as in the placement of text and graphics on a page, slide or whiteboard) and temporally discontiguous, and thus would appear to an observer who scanned the recording sequentially (e.g., by playing back a video) to be semantically unrelated. This underlying structure may therefore help to produce more useful summaries than the current transcription-summarization approaches adopted for instance by "meeting browsers structured around main topics [...] or action items" [755]. Summarization techniques that gather facts scattered in the time-based media in order to reassemble the relevant discussions, rationales, counter arguments behind a decision or action item would come closer to automating the activity of content salvaging which, as we have seen, meets important requirements of the tasks of reviewing and reporting based on multimedia meeting records [685]. In addition to browsing, the graph structure may also be useful for retrieval. Algorithms for extraction of keywords related to a target segment [148, 931] could, for instance, extend the set of candidate words to include words in each of the segments reachable from the target in its temporal interaction subgraph and use link analysis to rank these words [505]. The temporal link structure may also be naturally complemented by spatial context analysis for records that include still images or video sequences [690, 691].

In the following section we briefly examine two examples of meeting recording and browsing activities and the technologies that support these activities in relation to the above described model.

## 4.4   Case Studies

Two different cases of meeting activities are investigated: one based on meetings recorded in our laboratory (including both normal work meetings among members of the research group and scenario-based meetings) using tools specifically implemented for this purpose [148, 147], and another based on "real-world" meetings held regularly at a busy teaching hospital [488].

### 4.4.1   Recording and Browsing Artefact-Based Interaction

The goal of this study was to investigate the semantic relationships between the contents of interaction events on orthogonal modalities (visual and auditory) as delivered by different media. A dedicated software environment was implemented to support remote meetings and enable multiparty real-time audio communication, text interaction through a shared editor, and gestures (pointing, highlighting, circling) performed on the text. The effect of a participant's gestures on his remote collaborator's screen is represented in the system through a color-coded overlay that remained visible long enough to attract the collaborator's attention [147]. Thus, the multimedia record comprises three types of data: permanent visual data (text), transient auditory data (speech, audio), and transient visual data (gestures).

A total of 31 meetings were recorded. The data contain a basic form of segmentation provided by the recording environment: audio was automatically segmented into talk spurts and text was segmented into paragraph chunks, as mentioned in the preceding section.

Meeting browsers were then built whose design drew on two different approaches to structuring the browsing activity. The first was based on the idea of identifying segments of the record which exhibit the greatest levels of *interleaving* between text and speech streams and within speech streams [625], and highlighting these segments on the timeline [628]. This approach provided higher-level segmentation by clustering together talk spurts and text chunks in the regions of high interleaving but such groupings were restricted to contiguous regions of the recording. User trials of the browsing prototype suggested that users explored the record non-sequentially, alternating between reading and listening to the recorded speech. This observation motivated the

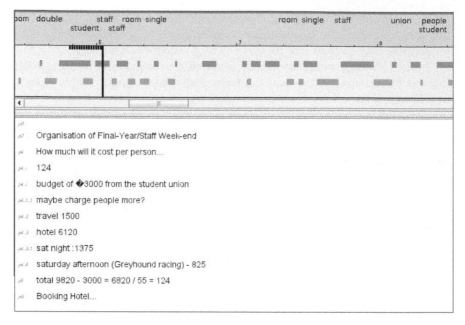

FIGURE 4.3: Main user interface components of a speech-centric meeting browser.

development of the second meeting browser [148], which was based on the "neighborhoods" model presented in [629] for structuring and presenting information.

This segment neighborhood browser followed the typical design of speech-centric browsers, according to the categorization of [755], displaying a text component coupled with a timeline on its main window, as shown in Figure 4.3. This coupling of the two main components meant that while the audio was played back (middle panel in Figure 4.3), the actions performed on the text segment as the participants spoke were displayed on the text component (bottom panel in Figure 4.3). The browser structured these data as a tree which encoded text-speech segment links implicitly defined according to action time stamps. Automatic speech recognition was also performed but the output was deemed too inaccurate to be read as running text. Instead, these imperfect transcripts were time-stamped and used to augment the text segments temporally linked to them for the purpose of keyword extraction. The keywords extracted in this manner were used to annotate the timeline (top panel in Figure 4.3) thus providing an index into the audio record. Keyword-based search was also available on the interface, as a separate component.

As can be gleaned from this brief description, the information structuring approach employed by this browser can be readily accommodated by the model described in Section 4.3.1. Definition 1 is general enough to cover both text and speech segments. In the specific case of the neighborhood browser, the definition could be further specialized to reflect the text-speech linking patterns by making the relation $L = L_t \cup L_s$ a union of a relation $L_t \subseteq \mathcal{T} \times \mathcal{S}$ linking text segments (in $\mathcal{T}$) to speech segments to a relation $L_s \subseteq \mathcal{S} \times \mathcal{T}$ where $L_s$ and $L_t$ are not required to be reflexive. The temporal interaction graph could then be constrained to having a tree structure by defining a partial order on $L$.

Furthermore, the model could be used to extend that approach in a number of useful ways. First of all, the temporal graph could be used to produce dynamic summaries adapted to different contexts. Selection of a text segment, for instance, would cause the entire subgraph for this segment to be produced and displayed on the user interface. This could take the form of highlights on the text component and on the timeline panel which the user would be able to select for closer inspection, thus updating the subgraph. Keyword extraction and the search function could also be improved by link analysis techniques as mentioned above. Finally, static textual topic summaries could be generated by combining information from the different segments in the temporal interaction graph. Unlike the topic summaries generated by most meeting browsers, summaries based on the model presented here would not be constrained to encompassing only data from adjacent segments.

Similarly, other meeting browsers, both speech- and document-centric [755], can also be modeled in terms of this linked time-based model. There is reason to suppose that the techniques described above can also benefit those browsers in terms of improved adaptation to context, mitigation of the negative effects of speech recognition errors on the browsing task, and production of more effective summaries. Improved adaptation to context would be obtained by providing the user with better support for content salvaging, since there is evidence that switching between text and audio modalities appears to better reflect what users actually do [685, 628] than sequential reading of transcripts or audio listening. Similarly, poor speech recognition output can be compensated for by giving the user prompt access to related (temporally linked) text. Finally, new forms of content summarization can be implemented by employing the temporal relations as a segment clustering criterion.

## 4.4.2   Structuring Medical Team Meetings

Nearly all meeting corpora available to researchers, including the corpus described in Section 4.4.1, have been gathered in laboratory, either under controlled conditions with meeting participants recruited as experiment participants [779, 180] or under naturalistic conditions but in meetings among researchers [463]. While data gathered under such conditions are necessary and useful to researchers since they allow for standardized comparisons among different research efforts on multimedia indexing, machine learning algorithms, natural language processing techniques and other quantitative methods, these data offer little information to guide the use of these methods in real application situations.

In order to investigate the production and use of meeting data in a real-world situation we carried out extensive naturalistic observation of a multi-disciplinary medical team in their meetings over a period of two years and collected video recordings of their interactions for analysis [488]. Due to the stringent constraints of medical work (time pressures, confidentiality, assurances given to the ethics committee that the research would not interfere with the work of the staff in any way, and so on) the recordings took place under less than ideal conditions. The data were collected from two media sources. The first was an S-VHS recording facility available through the teleconferencing system used by the team, which recorded audio and the screen display being broadcast to the meeting. The majority of participants were collocated at the main hospital but the teleconferencing equipment was always used for image presentation and display of patient data even when there were no remote participants connected to the system. When remote participants joined the meeting, outgoing video data were captured through the picture-in-a-picture view that was displayed on a TV monitor during the conferences. A second high-end camcorder mounted on a tripod was placed at the back of the room and captured participants' gestures, direction of gaze and activities (e.g., note taking) in addition to audio, through a directional microphone. Two wall-mounted cardioid condenser boundary microphones were also used to capture the speech stream. These media streams were synchronized and annotated with a dedicated media annotation tool.

As in the previous examples of meeting data, interaction and interleave between and within modalities was found to be ubiquitous in medical team meetings. Participants of different specialities (radiologists, pathologists, oncologists, surgeons, nurses) gather in these meetings to present evidence and discuss patient cases. Therefore it is common for some specialities (notably radiologists and pathologists) to augment the description of their findings by images [487, 348]. Others interact with text by taking personal notes or updating the patient sheet, which is sometimes visible to the entire group on the teleconferencing system's screen. There is empirical evidence (from studies using laboratory collected corpora) that personal notes can be an effective aid to the creation of meeting summaries [145]. These facts suggest that it makes good sense to gather synchronized data from different modalities also in medical meetings.

Due to the high levels of noise in the speech record, the basic segmentation unit adopted in the analysis of these meetings was not talk spurts as in the previous case, but rather *dialog states* defined in terms of *vocalization* events. These states are defined in [627] as follows:

- *Vocalization*: the event that a speaker "has the floor." A speaker takes the floor when he begins speaking to the exclusion of everyone else and speak uninterruptedly; without pause for at least 1 second. The vocalization ends when a silence, another individual vocalization or a group vocalization begins. Talk spurts shorter than 1 second are incorporated into the main speaker's vocalization.

- *Group vocalization*: the event that occurs when an individual has fallen silent and two or more individuals are speaking together. The group vocalization ends when any individual is again speaking alone, or a period of silence begins. Individual speaker identities are lost when a group vocalization state is entered.

- *Silence*: periods when no speech is produced for over 0.9 seconds between vocalizations (including group vocalizations). A silence ends when an individual or group vocalization begins. A silence can be further classified as:

  - a *pause*: a silence between two vocalizations by the same participant,

  - a *switching pause*: a silence between two vocalizations by different participants,

  - a *group pause*: a silence between two group vocalisations, or

  - a *group switching pause*: a silence between a group vocalization and an individual vocalization.

The temporal interaction graph in this case also included the speaker's (medical) *role*, as symbolic information. The structure of these vocalizations augmented with role information has proved quite effective at segmenting medical team meetings into high-level topics. A Bayesian approach based on these structures alone, without transcribed speech, has been shown [627] to match the levels of accuracy obtained by other state-of-the-art topic segmentation algorithms that use transcribed speech (among other features) in their data representations [433].

In brief, this approach consists in defining vocalization events in terms of neighboring vocalizations and their speaker labels, as in equation equation (4.1), where $V_i$ is a nominal variable denoting the speaker role (or a pause type or group speech, in the cases of silences and vocalizations by more than one speaker, respectively) and $L_i$ is a continuous variable for the duration of the speech (or silence) interval.

$$s = (V_0, L_0, V_{-1}, L_{-1} \ldots, V_{-n}, L_{-n}, V_1, L_1 \ldots, V_n, L_n) \qquad (4.1)$$

Segmentation can then be implemented as a classification learning task to identify instances $s$ to be marked as boundary dialog states that start new topics. The approach successfully adopted in [626, 627] combined the conditional probabilities for the nominal variables into multinomial models and modeled the continuous variables' Gaussian kernels. In the full model, the probabilities to be estimated are simplified through Bayes' rule and the conditional independence assumption to:

$$
\begin{aligned}
P(b|S = s) \quad &\propto \quad P(V_0 = v_0, \ldots, L_n = l_n|b) \\
&= \quad \prod_{i=1}^{n} P(V_i = v_i|b)P(L_i = l_i|b) ,
\end{aligned}
\qquad (4.2)
$$

where $S$ denotes a random variable ranging over the vector representation of vocalization events, as defined in equation (4.1). In its simplest form, a classification decision employing a maximum a posteriori criterion will mark as a topic (patient case discussion) boundary all vocalizations $s$ such that $P(b|S = s) \geq P(\neg b|S = s)$.

Another application of these linked time-based data structures is in supporting automatic categorization of patient case discussions. Patient case discussions are higher level segments of medical team meetings and can be broadly categorized into two groups: *medical* and *surgical* case discussions. Once such case discussion segments are identified, they can be represented as *vocalization graphs* through the temporal links of vocalization event segments. These graphs encode patterns of speech duration and transitions between vocalization events which can be quantified and normalized for comparison. A typical vocalization graph is shown in Figure 4.4, where the node labels correspond to the participant's role (e.g. "sur" for surgeon, "rad" for radiologist, and so on) or to a general vocalization event (e.g. "Group" for group vocalization, "Floor" for silence, and so on).

In this representation, each patient case discussion is represented as a directed graph $G = (V, E)$ where $V$ is a set of vertices or nodes and $E$ a binary relation on $V$. Elements of $V$ are labeled by pairs $(s, p(s))$ representing the probability $p_s$ that the dialog is in state $s$ (e.g. a vocalization or a silence) at any given instant. Edges are labeled by conditional probabilities. A probability

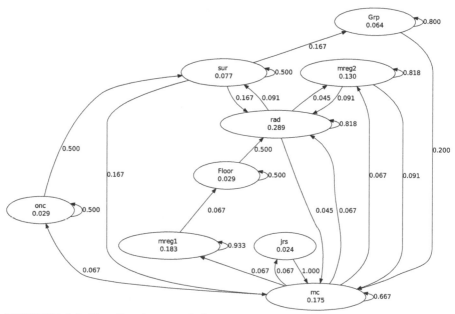

FIGURE 4.4: Vocalization graph for a patient case discussion extracted from a medical team meeting.

$p(t|s)$ labeling an edge corresponds to the likelihood that a dialog state $t$ (the terminal vertex) immediately follows dialog state $s$ (the initial vertex). Thus, in Figure 4.4, the numbers labeling nodes correspond to the steady state probabilities for those nodes.

Using this kind of graph structure as representations for case discussion segments in a "medical" vs. "surgical" categorization task we have been able to obtain high categorization accuracy (98.7%) using a k-NN (k nearest neighbor) classifier. This classifier operated in the usual way. In the "training" phase all training instances were converted into vocalization graphs and stored in the database. In the classification phase, given a new instance to classify, the algorithm selected from the database the $k = 5$ instances nearest to the unlabeled instance by graph similarity and took a vote among those instances. The majority class label was then assigned to the test instance.

In summary, the linked time-based model successfully characterizes speech interactions at medical team meetings for purposes of topic segmentation and classification of patient case discussions. In principle, the vocalization structures employed in these tasks could be augmented with gesture, image and text data and further acoustic features extracted from this type of meeting records, given appropriate data capture conditions. The proposed model can accommodate such extensions straightforwardly. Gesture and image events can be characterized as temporally linked segments in the same way as speech and text segments, in terms of Definition 1 and incorporated into an expanded temporal interaction graph. We are currently working on implementing and testing some of these extensions.

## 4.5 Conclusion

The need to combine complementary information coming from different media and sources in order to meet the challenges of multimedia information retrieval has been generally acknowledged in the literature [461]. Few applications illustrate this need better than those that aim to provide support for browsing and retrieving information from meeting records. The difficulties in these tasks stem not only from the divergent nature of the media and modalities involved, some of which result in parallel and permanent modes of access while others constrain the user to sequential and transient access to the content, but also from the lack of flexibility of existing systems to accommodate diverse sets of requirements.

This chapter discussed requirements for such systems against the background of complexity behind the apparent simplicity of real-time human communication. Lurking in this background one finds a variety of grounding mechanisms [233] which often manifest themselves in forms that lie beyond the analytic capabilities of current modality translation technologies, such as the use of silence, backchannels, facial expressions, etc., to negotiate shared meanings. One also finds subtle temporal and semantic relationships among segments within and between modalities and media as well as uncertainties as to what constitutes a segment in the first place. Building on this discussion, this chapter presented a simple model of linked events which seems general enough to accommodate the structures employed by current systems and shows promise with regard to the integration of new data sources and modalities. This can be seen as a step toward a characterization of multiparty interaction that can account for the semantic aspects of multimedia records of communication and collaboration.

So far the design of meeting browsing systems has been mainly driven by the capabilities of their modality translation and event analysis modules which are presented to the user through familiar time (in speech-centric systems) or space (in document-centric systems) interface metaphors. This chapter suggested that such metaphors may limit the system designer's as well as the user's ability to find richer contextual and semantic relations in the recorded media. This is particularly unsatisfactory given the new trends in data gathering technology. As ubiquitous sensors, processing devices, and methods for event detection, emotion recognition, improved speech processing, facial expression recognition and contextual information fusion begin to offer greater possibilities for creating truly rich records of human-human interaction, this design perspective may be ripe for a rethink.

## 4.6   Glossary

**Content salvaging:** the activity of sorting through the artefacts manipulated and produced during a meeting and the activities performed by its participants in order to reassemble these contents so as to make the meeting record more accessible.

**Grounding:** the process of establishing common ground, that is, of establishing a set of mutual knowledge, assumptions and beliefs that underpins communication.

**Meeting browsing:** the activity of visualizing multimedia meeting recordings and finding information of interest in such recordings.

**Modality translation:** the process of rendering one output modality into another. Examples include rendering speech into text by an automatic speech recognition system, text into speech by a speech synthesizer, video into key-frames, etc.

**Multimedia meeting record:** a digital recording of a meeting consisting minimally of a time-based data stream, typically speech, and a space-based data stream, typically text.

**Space-based media:** (or *static media*) a class of media for which space is the main structuring element. Data conveyed through space-based media are generally of a permanent and serial nature. Examples include: text and static graphics.

**Temporal link:** A relation between two media segments (e.g., two vocalizations, a vocalization and a text segment, etc.) that meet or co-occur in time.

**Temporal interaction graph:** A graph that encodes the temporal links of a multiparty interaction.

**Time-Based media:** (or *continuous media*) a class of media for which time is the main structuring element. Data conveyed through time-based media are generally of a transient and parallel nature. Examples include: audio and video.

**Vocalization:** a period of talk by a speaker (or group of speakers speaking together).

**Vocalization graph:** a graph encoding the structure of a dialog as transition probabilities from speaker to speaker, including group vocalizations and pauses.

# 5

# Semantic Enrichment for Automatic Image Retrieval

**Clement H. C. Leung**

*Hong Kong Baptist University,* `clement@comp.hkbu.edu.hk`

**Yuanxi Li**

*Hong Kong Baptist University,* `yxli@comp.hkbu.edu.hk`

## CONTENTS

A flourishing World Wide Web dramatically increases the amount of images uploaded, and exploring them is an interesting and challenging task. As there are definite relationships between the type of scenes and image acquisition parameters, image metadata and parametric dimensions, it is possible to determine from these the semantic content of images. By exploiting the judgment and expertise that has gone into the image capture process, and through the use of decision trees and rule induction, we can establish a set of rules which allows the semantic contents of images to be identified. When jointly applied with ontology-based and contextual feature-based expansion approaches, they are able to produce a new level of meaningful automatic image annotation, from which semantic image search may be performed.

## 5.1   Introduction

Diverse types of web images are increasingly prevalent, and their effective sharing is often inhibited by limitations in their search and discovery mechanisms,
which are particularly restrictive for images that do not lend themselves to
automatic processing or indexing. Web image search is often limited by inaccurate or inadequate tags, and many raw images are constantly uploaded with
little meaningful clue of their semantic contents, thus limiting their search and
discovery. While *content-based image retrieval*, which is based on the low-level
features extracted from images, has grown relatively mature, human users are
more interested in the semantic concepts behind or inside the images. Search
that is based solely on the low level features would not be able to satisfy users'
requirements. In order to increase the accuracy of web image retrieval, it is
necessary to enrich the concept index and semantic meaning of the images
as well as to overcome the semantic gap. With semantic search, meaning can
exist at different levels. In general, three levels of visual information retrieval
may be distinguished [305]:

1. **Level 1**: The lowest level is based on primitive features such as
   color, texture, shape, spatial location of image elements, or a combination of these.

2. **Level 2**: This level comprises retrieval by derived attributes or
   semantic content and corresponds to Panofsky's pre-iconographic
   level of picture description. Search requests on this level include the
   retrieval of objects of a given type or class as well as the retrieval
   of individual objects or persons.

3. **Level 3**: This level comprises retrieval by abstract attributes and
   includes search requests for named events or types of activity, corresponding with iconography, and those for pictures with emotional
   or symbolic significance, corresponding with iconology.

In this chapter, we shall explain the algorithms and mechanisms used for
semantic image indexing and retrieval on the Internet. The MPEG-7 Structured Annotation for images will be described. The measures of retrieval performance, as well as the limitations of current methods and approaches will
be assessed. The direction for future research in *semantic image retrieval* will
also be indicated.

## 5.2   Current Challenges in Image Retrieval

The number of web images is increasing at a rapid rate, and searching them semantically presents a significant challenge. Many raw images are constantly uploaded with little meaningful direct annotation of semantic content, limiting their capacity to be searched and discovered. Unlike in a traditional database, information in an image database is in visual form, which requires more space for storage, is highly unstructured and needs state-of-the-art algorithms to determine its semantic content.

As Web images tend to grow into unwieldy proportions, their retrieval systems must be able to handle multimedia annotation and retrieval on a web scale with high efficiency and accuracy. With the exception of systems that can identify or detect music, words, faces, irises, smiles, people, pedestrians, or cars, matching is not usually directed toward object semantics. Recent research studies show a large disparity between user needs and technological capabilities.

Vast numbers of web images are continuously added with few meaningful direct annotations of semantic content, limiting their search and discovery. While some sites encourage tags or keywords to be included manually, such is far from universal and applies to only a small proportion of images on the web. Research in image search has reflected the dichotomy inherent in the semantic gap [857, 396, 975], and is divided between two main categories: *concept-based image retrieval* and *content-based image retrieval.* The former focuses on retrieval by image objects and high-level concepts, while the latter focuses on the low-level visual features of the image. In order to determine image objects, the image often has to be segmented into parts. Common approaches to image segmentation include segmentation by region and segmentation by image objects. Segmentation by region aims to separate image parts into different regions sharing common properties. These methods compute a general similarity between images based on statistical image properties, [587, 85, 887] and common examples of such properties are texture and color where these methods are found to be robust and efficient. Some systems use color, texture, and shape [862, 225, 359, 1051, 243, 961, 610] as attributes and apply them for entire image characterization, and some studies include users in a search loop with a relevance feedback mechanism to adapt the search parameters based on user feedback, while various relevance feedback models and ranking methods for web search have been developed [222, 602, 48]. Segmentation by object, on the other hand, is widely regarded as a hard problem, which if successful, will be able to replicate and perform the object recognition function of the human vision system; although progress on this front has been slow, some advances in this direction have nevertheless been made [977, 787, 831]. In [589, 663, 599], semantic annotation of images combined with a region-based image decomposition is used, which aims to extract semantic properties of images based on the spatial distribution of color and texture properties. Such techniques have drawbacks, primarily due to their weak disambiguation and limited robustness in relation to object characterization. However, an advantage of using low-level features is that, unlike high-level concepts, they do not incur any indexing cost because they can be extracted by automatic algorithms. In contrast, direct extraction of high-level semantic content automatically is beyond the capability of current technology. Although there has been some effort in trying to relate low-level features and regions to higher level perception, these tend to be for isolated words, and they also require substantial training samples and statistical considerations [589, 304, 103, 102, 124, 476]. These methods, however, have limited success in determining semantic contents in broad image domains. There are some approaches which exploit surrounding and associated texts in order to correlate and mine these with the content of accompanying images [888, 184, 330, 511, 511, 981, 1003]. Text-based retrieval is often limited to the processing of tags, and no attempt is made to extract a thematic description of the picture. Some research focuses on implicit image annotations which involves an implicit, rather than an explicit, indexing scheme and, as a consequence, augments the original indexes with additional concepts that are related to the query [92, 696, 593], necessitating the use of some probabilistic weighting schemes.

With semantic search, meaning can exist at different levels. In [312], it has been suggested that image search requests may be categorized into unique and non-unique queries. Unique queries are those which can be satisfied by the retrieval of a unique person, object or event while non-unique ones cannot. Panofsky [725], based on picture identification in fine arts, categorizes images based on who, what, where and when.

Some of the most important notions in image retrieval are keywords, terms or concepts. Terms are used both from humans to describe their information need and from the system as a way to represent images. However, current image search systems, such as Yahoo! and Google, use some surrounding text description provided by humans in order to infer semantics. These techniques ignore the meaningful image features which can be extracted via image processing analysis. Further, as most of these images come without explicit semantic tags and those inferred from surrounding text are often unreliable, these models have limited effectiveness at present and they need further development and refinement. There lies the need to automatically infer semantics from raw images to facilitate semantics-based searches.

The effectiveness of image retrieval depends on meaningful indexing; the key problem of image retrieval is to organize them based on semantics. The word "semantic," which frequently appears in the content of this chapter, is the linguistic interpretation of multimedia objects, such as images and video clips, and is closely associated with the nature and meaning of the underlying objects.

From the viewpoint of image annotation and retrieval models, semantics can be textual descriptions attached to images [596]. Moreover, it can be high-level concepts describing the scene or relationships between images in a group that have a particular meaning for a user [938]. The process of assigning semantics to images is referred as image annotation, indexing or tagging in general for multimedia objects. The approach of manual annotation is followed by a number of commercial platforms and collaborative communities of multimedia, such as YouTube, Flickr and Pbase. However, the manual annotation of images is a laborious and error-prone task, as annotations can be biased towards the annotator's perspective, and it is difficult to define a strategy to ensure annotative consistency. This implicit connection is usually called the "semantic gap." The concept of the "semantic gap" is proposed and formalized by Smeulders et al [857]:

> "The semantic gap is the lack of coincidence between the information that one can extract from the visual data and the interpretation that the same data have for a user in a given situation."

They also conclude that:

> "A critical point in the advancement of content-based retrieval is the semantic gap, where the meaning of an image is rarely self-evident. The aim of content-based retrieval systems must be to provide maximum support in bridging the semantic gap between the simplicity of available visual features and the richness of the user semantics."

Despite continuous research efforts in developing and exploring new models, the gap between the expressive power of image features and semantic concepts is still a fundamental barrier. In simpler terms, the capability of current technology has no ability to fully understand the semantics of multimedia objects and direct extraction of semantic content automatically is not possible. Although there has been some effort in trying to relate low-level features and regions to higher-level perception, the success of such systems is limited. Although there has been some effort in trying to relate low-level features and regions to higher level perception, these are limited to isolated words. They also require substantial training samples and statistical considerations in order to partially or completely bridge these gaps.

## 5.3 Content-Based Image Retrieval

*Content-based image retrieval* (CBIR) [265, 197, 221, 697, 979, 667, 636, 707, 414] is a technique used for retrieving similar images from an image database. CBIR operates on retrieving stored images from a collection by comparing features automatically extracted from the images themselves. The most common current CBIR systems, whether commercial or experimental, operate at

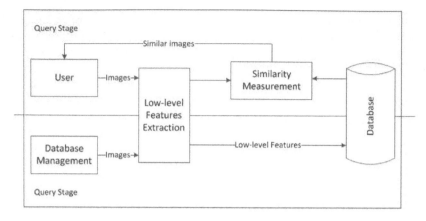

FIGURE 5.1: General framework of content-based image retrieval.

level 1. A typical system allows users to formulate queries by submitting an example of the type of image being sought, though some offer alternatives such as selection from a palette or sketch input. The system then identifies those stored images whose feature or signature values match those of the query most closely, and displays thumbnails of these images on the screen. The most challenging aspect of CBIR is to bridge the gap between low-level feature layout and high-level semantic concepts.

The general framework of *content-based image retrieval* is shown in Fig. 5.1.

Color, texture and shape features have been used for describing image content. Different CBIR systems have adopted different techniques. Few of the techniques have used global color and texture features [707, 414, 735, 884] whereas a few others have used local color and texture features [197, 221, 697, 590]. The latter approach segments the image into regions based on color and texture features. The regions are close to human perception and are used as the basic building blocks for feature computation and similarity measurement. These systems are called region based image retrieval (RBIR) systems and have proven to be more efficient in terms of retrieval performance. Few of the region based retrieval systems, e.g., [197], compare images based on individual region-to-region similarity. These systems provide users with rich options to extract regions of interest. But precise image segmentation has still been an open area of research. It is hard to find segmentation algorithms that conform to the human perception. For example, a horse may be segmented into a single region by an algorithm and the same algorithm might segment a horse in another image into three regions. These segmentation issues hinder the user from specifying regions of interest, especially in images without distinct objects. To ensure robustness against such inaccurate segmentation, the integrated region matching (IRM) algorithm [590] proposes an image-to-image similarity combining all the regions between the images. In this approach, every region is assigned the significance worth its size in the image. A region is allowed to participate more than once in the matching process till its significance is met with. The significance of a region plays an important role in the image matching process. In either type of system, segmentation close to human perception of objects is far from reality because the segmentation is based on color and texture. The problems of over-segmentation or under-segmentation will hamper the shape analysis process. The object shape has to be handled in an integral way in order to be close to human perception. Shape feature has been extensively used for retrieval systems [362, 458]

Image retrieval based on visually significant points [619, 675] is reported in literature. In [618], local color and texture features are computed on a window of regular geometrical shape surrounding the corner points. General purpose corner detectors [400] are also used for this purpose. In [99], fuzzy features are used to capture the shape information. Shape signatures are computed from blurred images and global invariant moments are computed as shape features. The retrieval performance is shown to be better than some of the RBIR systems such as those in [221, 590, 793].

The studies mentioned above clearly indicate that, in CBIR, local features play a significant role in determining the similarity of images along with the shape information of the objects. Precise segmentation is not only difficult to achieve but is also not so critical in object shape determination. A windowed search over location and scale is shown more effective in object-based image retrieval than methods based on inaccurate segmentation [420]. The objective of this chapter is to develop a technique which captures local color and texture descriptors in a coarse segmentation framework of grids, and has a shape descriptor in terms of invariant moments computed on the edge image. The image is partitioned into equal sized non-overlapping tiles. The features computed on these tiles serve as local descriptors of color and texture. In [413] it is shown that features drawn from conditional co-occurrence histograms using an image and its complement in RGB color space perform significantly better. These features serve as local descriptors of color and texture in the proposed method. The grid framework is extended across resolutions so as to capture different image details within the same sized tiles. An integrated matching procedure based on adjacency matrix of a bipartite graph between the image tiles is provided, similar to the one discussed in [590], yielding image similarity. A two level grid framework is used for color and texture analysis. Gradient Vector Flow (GVF) fields [1005] are used to compute the edge image, which will capture the object shape information. GVF fields give excellent results in determining the object boundaries irrespective of the concavities involved. Invariant moments are used to serve as shape features. The combination of these features forms a robust feature set in retrieving applications. The experimental results are compared with [221, 590, 99, 793] and are found to be encouraging.

### 5.3.1   Color Retrieval

Several methods for retrieving images on the basis of color similarity have been described in the literature [305], but most are variations on the same basic idea. Each image added to the collection is analyzed to compute a color histogram which shows the proportion of pixels of each color within the image. The color histogram for each image is then stored in the database. At search time, the user can either specify the desired proportion of each color (75% olive green and 25% red, for example), or submit an example image from which a color histogram is calculated. Either way, the matching process then retrieves those images whose color histograms match those of the query most closely. The matching technique most commonly used, histogram intersection, was first developed by Swain and Ballard [891]. Variants of this technique are now used in a high proportion of current CBIR systems. Methods of improving on Swain and Ballard's original technique include the use of cumulative color histograms [883], combining histogram intersection with some element of spatial matching [883], and the use of region-based color querying [195]. The results from some of these systems can look quite impressive.

### 5.3.2   Texture Retrieval

The ability to retrieve images on the basis of texture similarity may not seem very useful. But the ability to match on texture similarity can often be useful in distinguishing between areas of images with similar color (such as sky and sea, or leaves and grass). A variety of techniques has been used for measuring texture similarity; the best-established rely on comparing values of what are known as second-order statistics calculated from query and stored images. Essentially, these calculate the relative brightness of selected pairs of pixels from each image. From these it is possible to calculate measures of image texture such as the degree of contrast, coarseness, directionality and regularity [897], or periodicity, directionality and randomness [611]. Alternative methods of texture analysis for retrieval include the use of Gabor filters [644] and fractals [490]. Texture queries can be formulated in a similar manner to color queries, by selecting examples of desired textures from a palette, or by supplying an example query image. The system then retrieves images with texture measures most similar in value to the query. A recent extension of the technique is the texture thesaurus developed by [635], which retrieves textured regions in images on the basis of similarity to automatically-derived codewords representing important classes of texture within the collection.

### 5.3.3   Shape Retrieval

The ability to retrieve by shape is perhaps the most obvious requirement at the primitive level. Unlike texture, shape is a fairly well-defined concept — and there is considerable evidence that natural objects are primarily recognized by their shape [120]. A number of features characteristic of object shape (but independent of size or orientation) are computed for every object identified within each stored image. Queries are then answered by computing the same set of features for the query image, and retrieving those stored images whose features most closely match those of the query. Two main types of shape feature are commonly used — global features such as aspect ratio, circularity, and moment invariants [460], and local features such as sets of consecutive boundary segments [661]. Alternative methods proposed for shape matching have included elastic deformation of templates [735, 275]), comparison of directional histograms of edges extracted from the image [458, 72], and shocks, skeletal representations of object shape that can be compared using graph matching techniques [515, 914]. Queries to shape retrieval systems are formulated either by identifying an example image to act as the query, or as a user-drawn sketch [412, 1019].

Shape matching of three-dimensional objects is a more challenging task - particularly where only a single 2-D view of the object in question is available. While no general solution to this problem is possible, some useful inroads have been made into the problem of identifying at least some instances of a given object from different viewpoints. One approach has been to build up a set of plausible 3-D models from the available 2-D image, and match them with other models in the database [216]. Another is to generate a series of alternative 2-D views of each database object, each of which is matched with the query image [283]. Related research issues in this area include defining 3-D shape similarity measures [843], and providing a means for users to formulate 3-D shape queries [426].

### 5.3.4 Retrieval by Other Types of Primitive Feature

One of the oldest-established means of accessing pictorial data is retrieval by its position within an image. Accessing data by spatial location is an essential aspect of geographical information systems, and efficient methods to achieve this have been around for many years (e.g. [226, 791]). Similar techniques have been applied to image collections, allowing users to search for images containing objects in defined spatial relationships with each other [205, 204]. Improved algorithms for spatial retrieval are still being proposed [386]. Spatial indexing is seldom useful on its own, though it has proved effective in combination with other cues such as color [883, 860] and shape [429].

Several other types of image feature have been proposed as a basis for CBIR. Most of these rely on complex transformations of pixel intensities which have no obvious counterpart in any human description of an image. Most such techniques aim to extract features which reflect some aspect of image similarity which a human subject can perceive, even if he or she finds it difficult to describe. The most well-researched technique of this kind uses the wavelet transform to model an image at several different resolutions. Promising retrieval results have been reported by matching wavelet features computed

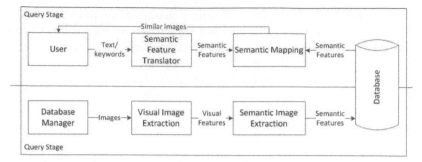

FIGURE 5.2: General framework of semantic-based image retrieval.

from query and stored images [453, 597]. Another method giving interesting results is retrieval by appearance. Two versions of this method have been developed, one for whole-image matching and one for matching selected parts of an image. The part-image technique involves filtering the image with Gaussian derivatives at multiple scales [775], and then computing differential invariants; the whole-image technique uses distributions of local curvature and phase [774]. The advantage of all these techniques is that they can describe an image at varying levels of detail (useful in natural scenes where the objects of interest may appear in a variety of guises), and avoid the need to segment the image into regions of interest before shape descriptors can be computed. Despite recent advances in techniques for image segmentation [187], this remains a troublesome problem.

## 5.4    Concept-Based Image Retrieval

Neither a single feature nor a combination of multiple visual features could fully capture a high-level concept of images [976]. Besides, because the performance of image retrieval based on low level features is not fully satisfactory, there is a need for retrieval based on semantic meaning by trying to extract the cognitive concept of a human to map the low level image features to high-level concept (semantic gap). In addition, representing image content with semantic terms allows users to access images through text query, which is more intuitive, easier and preferred by the users to express their mind compared with using images. For example, users' queries may be "Find an image of sunset" rather than "Find me an image that contains red and yellow colors".

The general framework of *concept-based image retrieval* is shown in Fig. 5.2.

Although CBIR and visual similarity techniques, discussed in the previous section, are important and widely applicable in the image retrieval and annotation domain, these studies indicate that the accuracy is subject to refinement. An advantage of using low-level features is that, unlike high-level concepts, they do not incur any indexing cost as they can be extracted by automatic algorithms. Some recent studies [215, 227, 1012] find that text descriptors, such as time, location, events, objects, formats, and topical terms are most helpful to users. In contrast, direct extraction of high-level semantic content automatically is beyond the capability of current technology. Although there has been some effort in trying to relate low-level features and regions to higher-level perception [304, 290], it is limited to isolated words, and also requires substantial training samples and statistical considerations [103, 151, 124, 476, 589, 724, 193].

The use of image-based analysis techniques is still not very accurate or robust and, after years of research, their retrieval performance is still far from users' expectations. Furthermore, the paramount challenge of *semantic image retrieval* and annotation, the semantic gaps, are not fully presentable in low-level features. It is inappropriate to fill and bridge the semantic gap by only the image pixel, but the effort should be made together with high-level semantic content. Thus *concept-based image retrieval* is still preferable in general commercial products.

The tagging process involves interpretation of the visual information given some context, either the context of the image or the context of the annotation or retrieval. For example, in the case that the image itself has no special meaning, we may still tag the place or the event of the image if it is available. The scenes of images, like indoor or outdoor scenes, are useful for the user-retrieval process as well.

Some studies [455, 410, 479, 252, 613, 428] unify concepts from the literature of diverse fields such as cognitive psychology, library sciences, art, and the more recent *concept-based image retrieval*. Then they present multiple-level structures for visual and non-visual information.

In [455], they begin by defining the distinction between visual content and nonvisual content:

1. The visual content of an image corresponds to what is directly perceived when the image is observed (i.e., descriptors stimulated directly by the visual content of the image or video in question, e.g. lines, shapes, colors, objects, etc).

2. The non-visual content corresponds to information that is closely related to the image, but that is not explicitly given by its appearance.

This is followed by the definition between percept and concept:

1. The percept refers to what our senses perceive — in the visual system these are light patterns. These patterns of light produce the perception of different elements such as texture and color. No

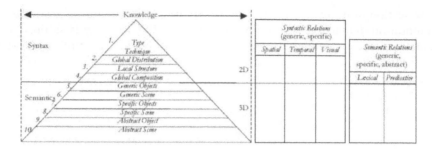

FIGURE 5.3: The indexing structure is represented by a pyramid.

interpretation process takes place when referring to the percept—no knowledge is required.

2. A concept, on the other hand, refers to an abstract or generic idea generalized from particular instances. As such, it implies the use of background or prior knowledge and an inherent interpretation of what is perceived. Concepts can be very abstract in the sense that they depend on an individual's knowledge and interpretation — this tends to be very subjective.

Their visual structure (Fig. 5.3) contains ten levels: the first four refer to syntax, and the remaining six refer to semantics. In addition, levels one to four are directly related to percept, and levels five through ten to visual concept. Some of these division may not be strict, but these are highly related to what the user is searching for and how they are finding images in a dataset.

Similarly, another study [431] defines a knowledge-based type of abstraction hierarchy with a three-layered image model to integrate the image representation. Meanwhile, some studies propose the thesaurus-based search model [214] to fulfill semantic image search.

Although keywords may be content-dependent, they are appropriate to express descriptive metadata [857]. In [103], two ways to link tags or concepts with images are suggested:

1. to predict annotations of entire images using all information present; we refer to this task as annotation.

2. to associate particular words with particular image substructures, that is, to infer correspondence.

Concept-based retrieval or semantic similarity is often limited to the processing of tags, and generally no attempt is made to extract a description of the picture and ignore the meaningful image features which can be extracted via image processing analysis. More important, it does not normally address the interaction of text and image processing in deriving semantic descriptions of a picture.

In current commercial image platforms, such as Flickr, concept-based retrieval approaches are always incorporated with a user-friendly user interface with suggestions of some similar keywords and terms to be entered by a user. Furthermore, such models may extract and recommend keywords and terms by extracting raw data from camera metadata.

Some research focuses on implicit image annotation which involves an implicit, rather than an explicit, indexing scheme and, as a consequence, augments the original indexes with additional concepts that are related to the query [92, 579, 991, 696], necessitating the use of some probabilistic weighing schemes.

## 5.5 Injecting Semantics into Image Retrieval

### 5.5.1 Automatic Semantic Image Annotation Scheme

In relation to image acquisition, many images can be broken down to few basic scenes [119, 573], such as nature and wildlife, portraits, landscapes and sports. The following are some examples.

1. A landscape scene comprises the visible features of an area of land, including physical elements such as landforms, living elements of flora and fauna, abstract elements such as lighting and weather conditions and human elements, for instance human activity or the built environment. Landscape photography is the normal approach to ensure that as many objects are in focus as possible, which commonly adopts a small aperture setting.

2. The goal of portrait photography is to capture the likeness of a person or a small group of people. Like other types of portraiture, the focus of acquisition is the person's face, although the entire body and the background may be included.

3. Nature scenes refer to a wide range of photography taken outdoors and devoted to displaying natural elements such as wildlife, plants, and closeups of natural scenes and textures. Nature photography tends to put a stronger emphasis on the aesthetic value of the photo than other photography genres, such as photo-journalism and documentary photography.

4. Sports photography corresponds to the genre of photography that covers all types of sports. The equipment used by a professional photographer usually includes a fast telephoto lens and a camera that has an extremely fast exposure time that can rapidly take pictures.

(a)          (b)          (c)          (d)          (e)

FIGURE 5.4: Each column shows the top matches to semantic queries of (a) "night scenes," (b) "outdoor portraits," (c) "day scenes," (d) "wildlife," and (e) "sports."

Definite relationships exist between the type of scenes and image-acquisition parameters. Some typical scene categories and scene images are given in Fig. 5.4, where scene images are a subset of the corresponding image categories.

The image file format standard, embedded in the images and established by the Japan Electronic Industry Development Association [34], makes use of the Exchangeable Image File Format (EXIF) and contains metadata specification for the image file format used in digital cameras. The Standard, defining image file system standards to enable image files to be exchanged among different recording media, was standardized in 1998 as a companion to the metadata standard. The most recent version, metadata standard version 2.2, was issued in 2004 with additional tag information and recording format options [34].

As Exif and the metadata standard are widely used by camera manufacturers, metadata tags are embedded into images by the majority of digital cameras automatically. The standard covers a broad spectrum including the following.

**Date and time information:** Digital cameras will record the current date and time and save this in the metadata.

**Camera acquisition parameters:** This includes static information including the camera model and make, and information that varies with each image such as orientation, aperture, shutter speed, focal length, metering mode, and ISO speed information.

**Location data:** Cameras having a built-in GPS receiver will automatically capture the location data and store it in the Exif header when the image is taken. In addition, GPS data can be retroactively added to any digital photograph, either by correlating the time stamps of the photographs with a GPS time-dependent record from a hand-held GPS receiver or manually using a map or mapping software.

**Makernote tag:** This contains interesting image information. It is normally in proprietary binary format.

**Image thumbnail:** A thumbnail for previewing the picture on the camera's LCD screen, in the file manager, or in photo manipulation software.

**File structure:** Layout rules for directory and file structures, so that a variety of files can be stored and managed efficiently on removable memory media.

**Descriptive information:** Descriptions and copyright information.

The metadata is shown to be useful for managing digital libraries [563] where only limited tags and comments should be entered manually. Some other common records of acquisition parameters include aperture ($f$), exposure time ($t$), subject distance ($d$), focal length ($L$) and flash activation ($h$). Location information can be included in the metadata, which could come from a GPS receiver connected to the image acquisition devices. With a given image

$$I_i = (d_{i1}, ..., d_{ik}) \qquad (5.1)$$

characterized by a number of dimensions $d_{ij}$ which correspond to the image acquisition parameters, each image may be represented by a point in multi-dimensional space. Fig. 5.5 shows the image points of a set of images in which particular types of image scenes tend to naturally cluster together. In [575], a rule-base induction method to separate images into elementary semantic categories has been developed. The induction rule is in the form of

$$R_{mn} : U_1 \wedge U_2 \wedge ...U_k \rightarrow c_{mn} \qquad (5.2)$$

where the antecedent consists of a range of dimensional values of the form

$$U_i = u_j | u_i \varepsilon D_i \qquad (5.3)$$

with $u_i$ representing a particular parametric dimension, and $D_i$ representing the associated restricted domain of values. This will produce a prior probability $P[c_{mn}]$ for the particular semantic content category. Such categorization will provide a large scale pruning of the search tree whereby highly

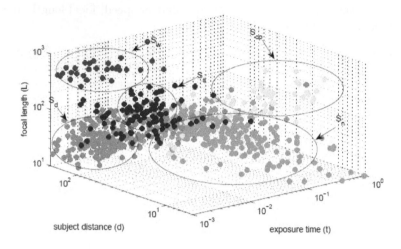

FIGURE 5.5: Image distribution in 3D space of some Flickr images.

selective procedures may be further applied differentially to different refined categories.

A sample rule set of Automatic Semantic Annotation is shown in Fig. 5.6.

## 5.5.2 Semantic Image Concept Expansion Algorithms

The presence of particular objects in an image often implies the presence of other objects. If term $U \to V$, and if only $U$ is indexed, then searching for $V$ will not return the image in the result, even though V is present in the image [576]. The application of such inferences will allow the index elements $T_i$ of an image to be automatically expanded according to some probability which will be related to the underlying ontology of the application. There are two main types of expansion.

(a)  Aggregation hierarchical expansion

This relates to the aggregation hierarchy of sub-objects that constitute an object. Associated with each branch is a tree traversal probability $t_{ij}$ (Fig. 5.7) which signifies the probability of occurrence of the branch index, given the existence of the parent index. In general, the traversal probabilities of different object classes exhibit different characteristics, with $t_{ij} > t_{mn}$ for $t_{ij}$ belonging to the concrete object class, and $t_{mn}$ belonging to the abstract object class.

(b)  Co-occurrence expansion

Rule 1)     $\forall i \in I, (t_i > 0.125) \wedge (d_i > 30) \wedge (EV_i \leq 8) \Rightarrow i \in S_n$

Rule 2)     $\forall i \in I, (d_i > 30) \wedge (EV_i > 8) \wedge (t_i \leq 0.125) \Rightarrow i \in S_d$

Rule 3)     $\forall i \in I, (f_i > 20) \wedge (d_i > 50) \wedge (EV_i > 11) \Rightarrow i \in S_{ss}$

Rule 4)     $\forall i \in I, [(f_i \leq 5.6) \wedge (5 < d_i \leq 8)] \wedge \{[(t_i \leq 0.00625)$
$\wedge (L_i \leq 30)] \vee [(30 < L_i \leq 182) \wedge (ISO_i \leq 250)]$
$\vee (L_i > 182) \vee (t_i \leq 0.003125)\} \Rightarrow i \in S_{op}$

Rule 5)     $\forall i \in I, (f_i > 5.6) \wedge (L_i \leq 25) \wedge (5 < d_i \leq 8)$
$\wedge (t_i > 0.003125) \Rightarrow i \in S_{oe}$

Rule 6)     $\forall i \in I, (f_i > 5.6) \wedge (0.003125 < t_i \leq 0.011111)$
$\wedge (5 < d_i \leq 8) \wedge (L_i > 25) \Rightarrow i \in S_{ip}$

Rule 7)     $\forall i \in I, (5 < d_i \leq 8) \wedge \{(f_i \leq 5.6) \wedge \{[(L_i \leq 30)$
$\wedge (t_i > 0.00625)] \vee [(ISO_i > 250)$
$\wedge (30 < L_i \leq 182)]\}\} \vee [(h_i = 1) \wedge (f_i > 5.6)$
$\wedge (L_i > 25) \wedge (t_i < 0.011111)] \Rightarrow i \in S_{ie}$

Rule 8)     $\forall i \in I, (d_i > 10) \wedge (150 < L_i \leq 400)$
$\wedge (t_i \leq 0.005) \Rightarrow i \in S_s$

Rule 9)     $\forall i \in I, (d_i \leq 5) \wedge (EV_i > 9) \Rightarrow i \in S_m$

Rule 10)    $\forall i \in I, (L_i > 450) \wedge (d_i > 20) \Rightarrow i \in S_w$

FIGURE 5.6: Sample rules of automatic semantic annotation.

This relates to the expectation that certain semantic objects tend to occur together. The relevant weighting is expressed as a conditional probability given the presence of other objects. Top-down traversal will lead to an expansion factor greater than 1, while bottom-up traversal will have an expansion factor smaller than 1 at each level of expansion [612].

Fig. 5.8 shows an example of an orchestra where the relevant objects are well defined in the real-life situation and common sense knowledge [577, 578].

In addition, such an expansion involves index scoring and ranking operations. The initial creation of an index term will result in a certain probabilistic score, which indicates its probability of being present in an image, and every index term $T_i$ in the index will have a score associated with it, which is used for the ranking and delivery of search results. Such a score is caused by the uncertainty of the index terms in earlier stages or by the sub-tree reliability factor responsible for generating it or both.

## 5.5.3 Semantic MPEG-7 Structured Image Annotation

Establishing a standard for visual content description is crucial, and it helps to efficiently discover and explore meaningful annotations of images. The Moving

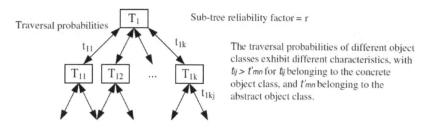

FIGURE 5.7: A tree traversal probability $t_{ij}$ which signifies the probability of occurrence of the branch index given the existence of the parent index.

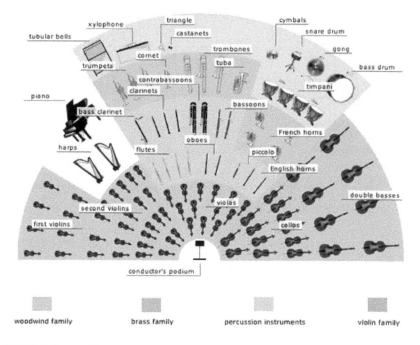

FIGURE 5.8: Example showing an orchestra with component objects.

Picture Experts Group (MPEG) is the name of a family of standards that aims to encode multimedia information in a digital compressed format [646].

This standard consists of different parts, and each part covers a certain aspect of the whole specification. For example, MPEG-1 defines the standard of compression for audio and video, while MPEG-21 describes the standard of multimedia framework [646, 650]. Also the MPEG Query Format (MPQF) has been developed for providing a standardized interface to multimedia document repositories by members of MPEG [763, 293, 707].

Among these standards, the MPEG-7 standard defines a comprehensive, standardized set for effective searching, identifying, filtering, and browsing in multimedia content including images, videos, audio files, and other digital or even analog materials [646, 532]. To support various types of descriptors, these are the four major MPEG-7 building blocks [650]:

**Descriptor:** a representation of a feature. A Descriptor defines the syntax and the semantics of the feature representation.

**Description Scheme:** the structure and semantics of the relationships between its components, which are both Descriptors and Description Schemes.

**Description Definition Language (DDL):** a language that allows the creation of new Description Schemes and, possibly, Descriptors. It also allows the extension and modification of existing Description Schemes.

**Systems Tools:** tools to support multiplexing of descriptions, synchronization of descriptions with content, delivery mechanisms, and coded representations (both textual and binary formats) for efficient storage and transmission and the management and protection of intellectual property in MPEG-7 Descriptions..

In terms of semantic richness for multimedia information, the MPEG-7 Structured Annotations Description Scheme is among the most comprehensive and powerful [895, 804], and it has been applied to image annotation [91, 896]. In practical situations involving image retrieval problems, as development of CBIR techniques have suffered from the lack of standardized means of describing image content, some researchers [534, 992, 994, 993] apply the visual content descriptors provided by MPEG-7 and compare the image annotation technique with a reference method based on vector quantization.

The Structured Annotation Data type is one that gives a textual description of events, people, animals, objects, places, actions, purposes, times, attributes and behavior. It provides a structured format that is a simple but expressive and powerful annotation tool. Syntactic relationships such as subject, object and verb modifiers between actions and objects are described. In addition, the modifiers are described for people, animals, objects, places, actions and times. The MPEG-7 Structured Annotations Data type plays a significant role in semantic multimedia database retrieval which allows a structured description of the image objects and related properties including such data fields as WhatObject, When, Where, Who, What Action, How and Why. Images will be annotated with predefined semantic concepts in conjunction with the methods and techniques of image processing, visual-feature extraction and semantic concept manipulation.

Using such an approach, some of these fields are automatically filled in a meaningful manner. In addition, semantic concepts have been enriched and expanded through semantic manipulation, which works in conjunction with specialization and generalization hierarchies [91]. Thus, the act of searching for an image with a particular object type can be specialized to a narrower type. For example, in searching for people, the portrait category is used, and in searching for animals, the wildlife category will fit the requirement. The more extensive and complete such hierarchies, the greater the scope for rich semantic manipulation.

## 5.6   Summary and Conclusion

Semantic image annotation and retrieval is an open-ended research topic in need of incremental improvement. Commercial Internet search engines have made significant progress with promising results in image searching using text-based search engines. While research on semantic image annotation and retrieval is developing at a rapid pace, it is still not widely deployed for public use due to the computational overhead associated with rapid index querying and updating on such an enormous scale. Research in this field still has a long way to go before it can be routinely utilized in our daily lives. Obviously, to fill the semantic gap, *content-based image retrieval* cannot solely be used without assistance from ontological mechanisms, and this is one of the directions research in this domain is taking.

One of the most important trends of *semantic image retrieval* research is making use of metadata, which includes a time stamp, location stamp and direction stamp. These are automatically associated with digital images and can supply very valuable information about the contents of the image. The Automatic Semantic Annotation approach, discussed in this chapter, takes this as an important step to lower computational cost and effective mechanism to semantic rich image retrieval. Undoubtedly, our approach with the CBIR techniques and ontology enables an advanced degree of semantic richness for semantic image annotation and retrieval. It has been shown that by the systematic analysis of embedded-image metadata and parametric dimensions, it is possible to determine the semantic content of images. Through the use of decision trees and rule induction, the semantic contents of images can be identified by a set of rules. From the joint application of ontology-based expansion, fully-automatic semantic annotations for specific images can be formulated, and images can be annotated purely by machine without any human involvement. When jointly applied with feature-extraction techniques, this produces a new level of meaningful image annotation. Using these semantic image annotation methods, we are able to provide semantic annotation for any unlabeled images in a fully automated manner, making *semantic image retrieval* a much more efficient process. The MPEG-7 Structured Annotation Datatype is often regarded as a powerful semantic information-bearing scheme for images and multimedia objects. Through the use of the Structured Annotation Datatype, semantic identification and search of images using search arguments that are meaningful to humans can be used. In the past, Structured Annotation descriptions were generally handcrafted by humans. This has become increasingly impractical due to the rapid rate at which images are captured, created, and uploaded. Through the use of the Automatic Semantic Annotation approach, such manual and laborious involvement will become increasingly unnecessary.

## 5.7 Key Terms and Definitions

**Image retrieval:** An image retrieval system is a computer system for browsing, searching and retrieving images from a large database of digital images. Most traditional and common methods of image retrieval utilize some method of adding metadata such as captioning, keywords, or descriptions to the images so that retrieval can be performed over the annotation words. Manual image annotation is time-consuming, laborious and expensive; to address this, there has been a large amount of research done on automatic image annotation.

**Content-Based image retrieval:** *Content-Based image retrieval* (CBIR), also known as query by image content (QBIC) and content-based visual information retrieval (CBVIR) is the application of computer vision techniques to the image retrieval problem.

**Concept-Based image indexing:** also variously named "description-based" or "text-based" image indexing/retrieval, refers to retrieval from text-based indexing of images that may employ keywords, subject headings, captions, or natural language text.

**Automatic image annotation:** (also known as automatic image tagging or linguistic indexing) is the process by which a computer system automatically assigns data in the form of captioning or keywords to a digital image. This technique is used in image retrieval systems to organize and locate images of interest from a database.

**Semantic similarity:** Semantic similarity or semantic relatedness is a concept whereby a set of documents or terms within term lists are assigned a metric based on the likeness of their meaning /semantic content.

**Semantic gap:** The semantic gap characterizes the difference between two descriptions of an object by different linguistic representations, for instance languages or symbols. In computer science, the concept is relevant whenever ordinary human activities, observations, and tasks are transferred into a computational representation.

**Query expansion:** Query expansion (QE) is the process of reformulating a seed query to improve retrieval performance in information retrieval operations. In the context of web search engines, query expansion involves evaluating a user's input (what words were typed into the search query area, and sometimes other types of data) and expanding the search query to match additional documents or multimedia data objects.

**MPEG-7:** MPEG-7 is a multimedia content description standard. It was standardized in ISO/IEC 15938 (Multimedia content description interface). This description will be associated with the content itself, to allow fast and efficient searching for material that is of interest to the user. MPEG-7 is formally called Multimedia Content Description Interface. Thus, it is not a standard which deals with the actual encoding of moving pictures and audio, like MPEG-1, MPEG-2 and MPEG-4. It uses XML to store metadata, and can be attached to timecode in order to tag particular events, or synchronize lyrics to a song.

# Part II

# Semantic Knowledge Exploitation and Applications

# 6

# Engineering Fuzzy Ontologies for Semantic Processing of Vague Knowledge

**Panos Alexopoulos**

*iSOCO S.A., palexopoulos@isoco.com*

## CONTENTS

Fuzzy Ontologies comprise a relatively new knowledge representation paradigm that is being increasingly applied in application scenarios where the treatment and utilization of vague knowledge is important. Nevertheless, while such ontologies have been developed and utilized in a number of different systems and domains, their development has been performed in rather ad hoc and incomplete manners. As a consequence, knowledge engineers that need to develop a fuzzy ontology do not have available a central point of reference that will effectively guide them throughout the development process and help them produce a high quality result. In this context, this chapter provides a comprehensive guide for developing a fuzzy ontology, covering all required stages from specification to validation. The skeleton of the guide is a fuzzy ontology engineering methodology, called IKARUS-Onto, which we complement with techniques, tools and best practices for each of the development process's steps.

## 6.1 Introduction

Ontologies are formal conceptualizations of domains, describing the meaning of domain aspects in a common, machine-processable form by means of concepts and their interrelations [200]. As such they have been realized as the key technology for modeling and utilizing domain knowledge for purposes like semantic annotation [809], document clustering [338], decision support [146] and knowledge management [486]. Recently, however, classical ontology representation formalisms, such as the Ontology Web Language (OWL)[110], have been deemed as inadequate for the semantic expression of imprecise and vague knowledge which is inherent to several real-world domains [1028] [912].

Indeed, the phenomenon of vagueness, manifested by terms and concepts like *Tall, Expert, Modern, Expensive* etc., is quite common in human knowledge and language and it is related to our inability to precisely determine the extensions of such concepts in certain domains and contexts. That is because vague concepts have typically fuzzy, or blurred, boundaries which do not allow for a sharp distinction between the entities that fall within the extension of these concepts and those which do not. This is not usually a problem in individual human reasoning, but it may become one, i) when multiple people need to agree on the exact meaning of such terms and ii) when machines need to reason with them. For example, a system could never use the statement *"This process requires many people to execute"* in order to determine the number of people actually needed for the process.

Consequently, classical representation formalisms, in which all predicates have well-defined extensions (i.e they cannot have fuzzy boundaries), cannot accommodate the representation of vagueness. Yet, such a representation in ontology-based systems is important, not only because vagueness is present in many domains but also because, in many application scenarios, the consideration and exploitation of vagueness can significantly enhance the systems' effectiveness [57] [185].

Traditionally, vague terms and concepts in expert and knowledge-based systems have been dealt with through the utilization of Fuzzy Set Theory [524]. The latter is a mathematical theory that enables the capture of vague knowledge through the notions of **Fuzzy Set** and **Fuzzy Relation**. Informally, a fuzzy set is a set to which an object may belong to a certain degree (a real number from the interval [0,1]). In a similar fashion, a fuzzy relation associates two objects to a certain degree. These degrees practically allow the definition of facts and statements that are "less strict" and which are considered true to some given degree (e.g., *"someone who is 29 years old is considered young to a degree of 0.4"*).

A more recent knowledge representation paradigm that has been proposed for the same problem is **Fuzzy Ontologies** [58] [128], extensions of classical ontologies that, using the same principles as Fuzzy Set Theory, allow the assignment of truth degrees to vague ontological elements, in an effort to quantify their vagueness. Thus, for example, whereas in a traditional ontology one would claim that *"Tiffany's is a moderately priced shop"* or that *"A book is a typical birthday gift,"* in a fuzzy ontology one would claim that *"Tiffany's is a moderately priced shop to a degree of 0.6"* and that *"A book is a typical birthday gift to a degree of 0.8."*

Fuzzy ontologies have been developed and utilized in a number of different systems and application areas including information retrieval [185] [733], thematic categorization [974], semantic matchmaking [49], decision support [59] and data mining [313] [213]. In all these cases there are two common issues that fuzzy ontology developers have to deal with:

1. How to perform the ontology development in the most effective way.

2. How to ensure that the final product is of good quality.

In traditional ontology development these two issues have been typically tackled by means of corresponding methodologies which provide structured processes, guidelines and best practices that knowledge engineers may use to develop better ontologies [372] [939] [825].

The development of fuzzy ontologies, on the other hand, as reported in the literature, has so far been performed in rather implicit and ad hoc manners. As a consequence, knowledge engineers that need to develop a fuzzy ontology do not have available a central point of reference that will effectively guide them throughout the development process and help them produce a high quality result. In particular, several conceptual formalisms for representing (and reasoning with) fuzzy ontologies have been proposed in the literature [1034] [911] [128] [877] [639] [564] and various software tools have been developed for defining fuzzy ontologies through such formalisms and for performing corresponding reasoning tasks [847] [128]. However, these formalisms and tools do not address the following issues:

1. That the ontology engineers and the domain experts can easily and correctly identify vague knowledge in a domain that needs to be modeled as fuzzy.

2. That the ontology engineers can determine the appropriate fuzzy ontology elements for representing this knowledge.

3. That the domain experts can intuitively decide what the values of

the degrees of the various fuzzy ontology elements should approximately be.

4. That the fuzzy degrees of the ontology approximate, as accurately as possible, the vagueness of the domain.

5. That the developed fuzzy ontology is reusable and shareable by having the meaning of its fuzzy elements and their degrees explicitly defined and commonly accepted.

With that in mind, in this chapter we provide a comprehensive guide for developing fuzzy ontologies, covering all the required stages from specification to validation. The skeleton of the guide is a fuzzy ontology engineering methodology, called **IKARUS-Onto** [58], which considers existing crisp (i.e. non-fuzzy) ontologies and guides the knowledge engineer in transforming them into fuzzy ones. The methodology provides concrete steps and guidelines for i) correctly identifying vague knowledge within a domain (e.g., by not mixing vagueness with other notions such as uncertainty or ambiguity) and ii) modeling this knowledge by means of fuzzy ontology elements in an explicit and as much as possible accurate way. In this chapter, we provide a detailed description of IKARUS-Onto, augmented with suggested techniques, tools and best practices for each of the development process's steps. Of particular importance is our suggested generic framework for automating the generation and maintenance of the ontology's fuzzy degrees.

Given the above, the structure of this chapter is as follows: In the next section we provide some basic definitions and clarifications regarding the phenomenon of vagueness, we discuss the way it is typically manifested in ontologies and we describe the typical elements a fuzzy ontology uses in order to represent it. In Section 6.3 we describe in detail the steps of the IKARUS-Onto methodology and the tasks involved within them along with concrete guidelines on how to perform them. In Section 6.4 we practically demonstrate the way to apply the methodology through a concrete application scenario regarding the development of a fuzzy enterprise ontology. In Section 6.5 we address the difficult and resource-consuming task of generating fuzzy degrees for the ontology's elements and we outline a generic framework for automating it. Finally, in Section 6.6 we summarize several key points and discuss potential future directions.

## 6.2   Preliminaries

### 6.2.1   Vagueness and Ontologies

Vagueness as a semantic phenomenon is typically manifested through predicates that admit borderline cases [441] [837], namely cases where it is unclear whether or not the predicate applies. For example, some people are borderline tall: not clearly tall and not clearly not tall. In a more formal way an object $a$ is a borderline case of a predicate $P$ if $P(a)$ is "unsettled," namely if it is not determinately true that $P(a)$ nor is it determinately true that $\neg P(a)$ [837]. The characterization "determinately" for the truth of a predicate means that the thoughts and practices in using the language have established truth conditions for it [657].

Obviously, having borderline cases is related to having fuzzy boundaries. For example, on a scale of heights there appears to be no sharp boundary between the tall people and the rest. Therefore two equivalent ways of drawing the distinction between vague and non-vague (or precise) predicates are to say that i) vague predicates can possibly have borderline cases while precise predicates do not, or that ii) vague predicates lack sharp boundaries.

In the relevant literature two basic kinds of vagueness are identified: *degree-vagueness* and *combinatory vagueness* [441]. A predicate $P$ has degree-vagueness if the existence of borderline cases stems from the lack (or at least the apparent lack) of precise boundaries between application and non-application of the predicate along some dimension. For example, *Bald* fails to draw any sharp boundaries along the dimension of hair quantity and *Tall* along the dimension of height. Of course it might be that a predicate has degree-vagueness in more than one dimensions (e.g., *Red* can be vague along the dimensions of brightness and saturation).

On the other hand, a predicate $P$ has combinatory vagueness if there is a variety of conditions all of which have something to do with the application of the predicate, yet it is not possible to make any sharp discrimination between those combinations which are sufficient and/or necessary for application and those which are not. A classical example of this type is *Religion* as there are certain features that all religions share (e.g., beliefs in supernatural beings, ritual acts etc.), yet it is not clear which of these features are able to classify something as a religion.

It is important that vagueness is not confused with the notions of inexactness, uncertainty and ambiguity. For example, stating that someone is between 170 and 180 cm is an inexact statement but it is not vague, as its limits of application are precise. Similarly, the truth of an uncertain statement, such as *"Today it might rain,"* cannot be determined due to lack of adequate information about it and not because the phenomenon of rain lacks sharp boundaries. Finally, the truth of a statement might not be determinable due to the ambiguity of some term (e.g., in the statement *"Yesterday we went to the bank"* the term *bank* is ambiguous), yet again this doesn't make the statement vague.

Finally, vagueness is context-dependent as the extension of a vague term may vary depending on the context in which it is being applied. For example, a person can be tall with respect to the average population height and not tall with respect to professional basketball players. Similarly, a person can be wealthy with respect to the local community but poor with respect to the boss. This does not mean that a term may be vague in one context and non-

**TABLE 6.1**
Crisp Ontology vs. Fuzzy Ontology.

| Crisp Ontology Statement | Fuzzy Ontology Statement |
|---|---|
| Jane is an expert at Artificial Intelligence | Jane is an expert at Artificial Intelligence to a degree of 0.8 |
| The film "Notting Hill" is a comedy | The film "Notting Hill" is a comedy to a degree of 0.6 |
| John is 20 years old | John is young |

vague in another but rather that the interpretation of its vagueness may be different.

## 6.2.2 Fuzzy Ontologies

Classical ontologies are typically represented by means of concepts, instances, attributes and relations. A concept represents a set or class of objects within a domain, while the objects that belong to a particular concept are called instances of this concept. An attribute in turn represents some characteristic of a concept (for which instances of this concept have values) and a relation describes the relationship that can be established between concepts (and consequently between their instances). Fuzzy ontologies, on the other hand, are represented by similar to the above elements which, in addition, allow for representation of fuzzy degrees that express the vagueness of the knowledge. Table 6.1 illustrates the difference between classical ontological statements and fuzzy ones.

In particular, given the various existing formalisms for fuzzy ontologies [128] [877] [1034], the basic elements a fuzzy ontology consists of can be summarized as follows:

- **Fuzzy Concepts:** A fuzzy ontology concept is a concept whose instances may belong to it at certain degrees. Such a degree practically denotes the extent to which a given entity should be considered as being an instance of the concept. As an example consider the concept *TallPerson* whose instances are meant to be people whose height classifies them as tall. Since *tall* is a vague predicate, the concept is also vague and therefore can be represented as a fuzzy one by allowing the expression of statements such as *Person X is an instance of TallPerson at a degree of 0.8*. In languages based on fuzzy Description Logics such statements are called *fuzzy concept assertions* [128] [877].

- **Fuzzy Relations:** Similar to fuzzy concepts, a fuzzy ontology relation links concept instances at certain degrees. Such a degree practically denotes the extent to which the relation between the two instances should be considered as true. For example, the relation *isExpertAt*, which contains the vague predicate *expert*, can be represented as a fuzzy one by allowing the expression of statements like *John is expert at Knowledge Management at a degree of 0.5.* In languages based on fuzzy Description Logics such statements are called *fuzzy role assertions* [128] [877].

- **Fuzzy Attributes:** A fuzzy attribute assigns literal values to concept instances at certain degrees. Such a degree denotes the extent to which the value is applicable to the instance for the given attribute. For example the attribute *category* of concept *Film* which takes as values string literals denoting the category a film may belong to (e.g., "science fiction," "horror," "comedy" etc.), can be represented as a fuzzy one by allowing the expression of statements like *The category of film "High Fidelity" is "comedy" at a degree of 0.7.* In practice, fuzzy attributes are like fuzzy relations with the difference that the second relatum is a literal value rather than an instance.

- **Fuzzy Datatypes:** A fuzzy datatype consists of a set of vague terms which may be used within the ontology as attribute values. For example, the attribute *performance*, which normally takes as values integer numbers, may in a given domain, context or application scenario be required to take as values terms like *very poor, poor, mediocre, good* and *excellent.* What a fuzzy datatype does is to map each term to a fuzzy set that defines its meaning by assigning to each the of datatype's potential exact values a fuzzy degree. This degree practically indicates the extent to which the exact value and the vague term express the same. Fuzzy datatypes are also known as fuzzy concrete domains [128] or fuzzy linguistic variables [1034].

The above description of fuzzy ontology elements is not exhaustive and it may vary among different fuzzy ontology formalisms and languages. Nevertheless, it is a convenient terminology for use within this chapter as it makes it easier for both the engineers and the domain experts to identify and express fuzzy information as they are natural extensions of the elements of classical ontologies.

## 6.3 Developing a Fuzzy Ontology with IKARUS-Onto

As suggested in Section 12.1, IKARUS-Onto methodology defines a set of concrete steps and guidelines for representing vague knowledge by means of fuzzy ontology elements. The focus of the methodology is not so much on the structure of the fuzzy ontology but rather on the process followed for its development and the content it ultimately has. The process aims to make it easier for the ontology engineer and the domain experts to identify and model the vagueness of the domain while the content reflects this vagueness as accurately

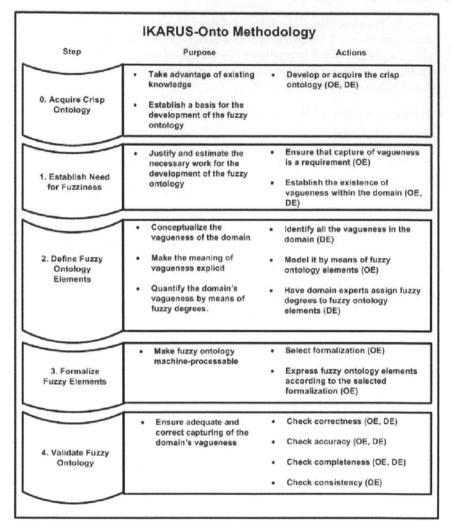

FIGURE 6.1: The IKARUS-Onto Methodology.

as possible. IKARUS-Onto assumes prior knowledge of classical ontology development from those who use it because it is practically a methodology for the transformation of existing conventional ontologies into fuzzy ones.

The life cycle of IKARUS-Onto and the steps involved in it are depicted in figure 6.1. For each step we list the purposes it serves, the actions that need to be done in order to fulfil them and who typically performs each action, i.e., ontology engineers (OE) or domain experts (DE).

The starting point of the methodology is some domain which is already modeled by means of some crisp ontology. If no such ontology exists then it needs to be developed (by using some conventional methodology) as it will be the basis upon which the development of the fuzzy ontology will be performed. The main reasons we follow this approach (i.e. not mixing the development of the crisp part of the ontology with the fuzzy one) are because:

- It is easier for the ontology engineers and domain experts to identify vagueness in already conceptualized and structured knowledge.

- In many domains the crisp part of the fuzzy ontology is already available as a separate ontology.

In the following paragraphs we present each step of the methodology in a more detailed manner.

### 6.3.1 Step 1: Establishment of the Need for Fuzziness

Establishing the need for fuzziness practically means determining whether and to what extent vagueness is present in the domain at hand, as well as whether the intended uses of the ontology require the capturing of this vagueness. This is necessary, as it justifies and estimates the work that will be required for the development of the fuzzy ontology.

In particular, the execution of this step is performed by first having the ontology engineer ensure that the capturing and modeling of the domain's vagueness is required by the ontology's application scenario. This is usually the case when there is (or there needs to be) some knowledge-based system which handles and utilizes vague knowledge, by means of fuzzy ontologies, for purposes such as information retrieval, thematic categorization, knowledge management, etc. In rarer cases there might not be such a system but merely the requirement to represent the domain's vagueness in an explicit way.

Given the requirement for vagueness, the engineer then establishes whether the latter is actually present within the domain. This is done by identifying elements of the (acquired or developed) crisp domain ontology whose meaning may be interpreted as vague in the given domain and/or application scenario. In particular, given the definitions about vagueness in Section 6.2.1 and the elements of Section 6.2.2, this identification is performed as follows:

- **Identification of vague concepts:** A concept is vague if, in the given domain, context or application scenario, it admits borderline cases, namely if there are (or could be) individuals for which it is indeterminate whether they instantiate the concept. Primary candidates for being vague are concepts that denote some phase or state (e.g Adult, Child) as well as attributions, namely concepts that reflect qualitative states of entities (e.g., Red, Big, Broken, etc.).

- **Identification of vague relations and attributes:** A relation is vague if, in the given domain, context or application scenario, it admits borderline cases, namely if there are (or could be) pairs of individuals for which it is indeterminate whether they stand in the relation. The same applies for attributes and pairs of individuals and literal values.

- **Identification of vague attribute value terms:** Such terms are identified by considering the ontology's attributes and assessing whether their potential values can be expressed through vague terms. Primary candidates for generating such terms are gradable attributes such as size or height which give rise to terms such as *large, tall, short*, etc.

Obviously, during this step, the identification of vagueness need not be exhaustive but merely sufficient for establishing the need for the development of the fuzzy ontology. What is important, however, is making sure that the identified-as-vague elements do actually convey a vague meaning and not something else (e.g., uncertain or inexact meaning). To that end, the above guidelines of our methodology are important.

## 6.3.2  Step 2: Definition of Fuzzy Ontology Elements

The second step of IKARUS-Onto involves the comprehensive identification of the vague knowledge of the domain and its explicit description by means of the fuzzy ontology elements of Section 6.2.2. The goal of this description is to ensure that the defined fuzzy elements i) have a clear and specific vague meaning which makes them shareable and reusable and ii) approximate, through their fuzzy degrees, this vagueness as accurately as possible. For that, IKARUS-Onto defines a specific procedure and description template for describing each type of fuzzy ontology element.

In particular, step 2 starts by identifying and describing the ontology's fuzzy relations and fuzzy attributes. As suggested in the previous sections, these two elements are quite similar, differing only in that fuzzy relations link instances to each other while fuzzy attributes link instances to literal values. Therefore, the procedure for defining these two types of elements is the same and comprises the following tasks:

1. The identification in the crisp ontology of <u>all</u> the relations/attributes that convey vague meaning using the guidelines described in step 1.

2. The determination for each relation/attribute of the type of its vagueness (combinatory or degree-vagueness). If the element has degree-vagueness then the dimensions along which the it is vague need to be identified.

3. The usage of the above information for defining for each element the exact meaning of its vagueness. In case the element has degree-vagueness along multiple dimensions then the distinction between the dimensions might or might not be important. In case it is, then it is necessary to define a distinct fuzzy element for each dimension.

4. The definition of the expected interpretation of each element's fuzzy degrees. If fuzziness is due to degree-vagueness, then the fuzzy degree of a related pair of instances (or instances and literal values) practically approximates the extent to which the pair's value for the given dimension places it within the elements's application boundaries. If, on the other hand, fuzziness is due to combinatory vagueness, then the fuzzy degree practically approximates the extent to which the pair's set of satisfied application conditions of the relation/attribute is deemed sufficient for the relation/attribute to apply.

5. The assignment of specific fuzzy degrees to pairs of instances (or instances and literal values) that instantiate each element. These degrees should approximate as accurately as possible the already defined element's degree interpretation for the given pair. In general, this assignment by domain experts can be a quite time and resource consuming task; that's why in Section 6.5 we suggest a generic way to do it automatically.

In the end of the above procedure, a set of element descriptions similar to those of Table 6.2 shall be produced. Through this template, the nature of the fuzzy element's vagueness and the expected interpretation of its fuzzy degrees are made explicit. This ensures not only the common understanding of the elements' vague meaning by all the fuzzy ontology users but it also helps the domain experts to assign fuzzy degrees to the instances of these elements in a more intuitive and accurate way.

After the specification of all the ontology's fuzzy relations and attributes, the specification of fuzzy datatypes takes place. The procedure for this involves:

1. The identification of the ontology attributes whose values may be expressed by means of vague terms.

2. The identification of these terms and their grouping into fuzzy datatypes (usually one datatype for each attribute). It should be noticed that the same term may belong to more than one datatype.

3. The definition for each vague term within a fuzzy datatype of a fuzzy set that defines its meaning. Again it is important to note that the same term may be associated with different fuzzy sets in different datatypes. In all cases, the definition of these fuzzy sets can be automated (see Section 6.5).

Figure 6.2 shows a sample fuzzy datatype for the attribute *Performance* which comprises the vague terms *very poor, poor, mediocre, good* and *excellent*. As shown in the figure each term corresponds to a fuzzy set that defines its meaning. Thus for example, a performance of 45 is considered *mediocre* to a degree of 1.0, while a performance of 38.75 is considered *poor* at a degree of 0.5 and *mediocre* to the same degree.

**TABLE 6.2**

Sample Description Template for Fuzzy Ontology Relations and
Attributes.

| Relation or Attribute | Vagueness Nature | Degree Interpretation |
|---|---|---|
| *isNearTo* | Degree-vagueness along the dimension of distance. | The extent to which the distance between the related instances classifies them as being near to each other. |
| *isFunctionalPartOf* | Degree-vagueness along the dimension of the part's contribution to the functionality of the whole. | The extent to which the part's contribution to the functionality of the whole classifies the part as functional. |
| *isCompetitorOf* | Degree-vagueness along the dimension of the competitor's business areas and the dimension of the competitor's target markets. | The extent to which the relation subject's business areas and/or target markets classifies it as a competitor of the object. |
| *belongsToCategory* | Combinatory vagueness due to the lack of sharp discrimination between those conditions that are necessary for something to belong to a given category. | The extent to which the subject's set of satisfied category's conditions classifies it as belonging to this category. |
| *isExpertAt* | Degree-vagueness along the dimension of the level of knowledge on a subject | The extent to which the level of someone's knowledge on a subject classifies him as expert on it. |

The final type of fuzzy ontology element to be defined within this step is that of fuzzy concept. The process followed for this definition is similar to the one for fuzzy relations and attributes with, nevertheless, an important difference. In many cases, vague concepts "owe" their vagueness to some vague relation, attribute or term which has been already defined by means of a corresponding fuzzy ontology element. If that is the case then the definition

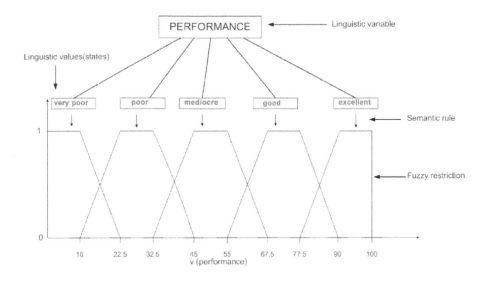

FIGURE 6.2: Fuzzy datatype example.

of the concept's vagueness can be directly derived from the one of its causal element. Thus, the definition process for fuzzy concepts includes:

1. The identification of all the ontology concepts that convey vague meaning using the guidelines described in step 1.

2. The determination for each vague concept of whether its vagueness is caused by one or more already defined fuzzy elements (fuzzy relation, fuzzy attribute, fuzzy datatype). If this is the case then the meaning of the concept's vagueness as well as the interpretation of the fuzzy degrees of its instances are directly derived from the ones of the causal element.

3. The definition, otherwise, of each concept in the same way as in the case of fuzzy relations and fuzzy attributes (including the assignment of fuzzy degrees to pairs of instances and concepts).

Table 6.3 shows a sample output of the above process. Again, the benefit of such a description is related to the explicitness of the concepts' vague meaning and the ease and accuracy of the assignment of fuzzy degrees to their instances.

### 6.3.3 Step 3: Fuzzy Ontology Formalization

The formalization step involves the transformation of the defined fuzzy ontology elements into a formal machine-interpretable form through some corresponding fuzzy ontology language. As with conventional ontology languages, fuzzy ontology languages typically vary in terms of the representation and

**TABLE 6.3**

Sample Description Template for Fuzzy Ontology Concepts.

| Concept | Vagueness Nature | Degree Interpretation |
|---|---|---|
| *BaldPerson* | Degree-vagueness along the dimension of hair quantity. | The extent to which the person's hair quantity classifies it as bald. |
| *MediocrePerformer* | Degree-vagueness derived from the *Performance* fuzzy datatype. | The same as in the term *mediocre* of the *Performance* datatype. |
| *Religion* | Combinatory vagueness due to the lack of sharp discrimination between those conditions that are necessary for something to be a religion. | The extent to which the instance's set of satisfied religion-related conditions classifies it as a religion. |
| *ScienceFictionFilm* | Combinatory vagueness derived from the *belongsToCategory* fuzzy ontology relation | The same as in the *belongsToCategory* relation. |

reasoning capabilities they provide. Therefore, the ontology engineer needs to consider the particular characteristics of each language and the capabilities it provides for representing (and reasoning with) vague knowledge. In particular, important parameters of a fuzzy ontology language include:

- **The range of fuzzy ontology elements it supports:** Not all languages support the whole range of Section's 6.2.2 fuzzy ontology elements and in the same way. For example, the fuzzy description logic f-SHIN [878] does not support fuzzy datatypes and the language in [128] supports fuzzy datatypes by means of fuzzy concrete domains while the one in [1034] by means of fuzzy linguistic variables.

- **The range of fuzzy reasoning capabilities it has support for:** Certain reasoning services provided by crisp ontology reasoners require adaptation when the ontology is fuzzy. An example is entailment which, in the fuzzy case, requires from the reasoner to be able to determine whether an individual belongs to a concept in a specific degree. Some, but not all, fuzzy ontology languages are accompanied by corresponding reasoners [128] [847] which typically vary in the range of reasoning capabilities they provide.

The above practically means that, depending on the application scenario within which the fuzzy ontology is utilized, a certain language might be more appropriate than another. A complete coverage of all languages and formalisms that may be used to formalize a fuzzy ontology falls outside the scope of this work. Nevertheless, in order to take a glimpse of how formalization may be

```
<owl:Axiom>
    <fuzzyLabel>
        <fuzzyOwl2 fuzzyType="axiom">;
            <Degree value="0.6"/>
        </fuzzyOwl2>
    </fuzzyLabel>
    <owl:annotatedSource rdf:resource="CompanyX"/>
    <owl:annotatedTarget rdf:resource="Competitor"/>
    <owl:annotatedProperty rdf:resource="&rdf;type"/>
</owl:Axiom>

<rdfs:Datatype rdf:about="HighProjectBudget">
    <fuzzyLabel>
        <fuzzyOwl2 fuzzyType="datatype">
            <Datatype type="rightshoulder" a="350000.0" b="500000.0"/>
        </fuzzyOwl2>
    </fuzzyLabel>
</rdfs:Datatype>
```

FIGURE 6.3: Sample fuzzy OWL 2 definitions.

performed, one may use the Fuzzy OWL 2 framework [130] which enables the formalization of fuzzy ontologies through the OWL 2 ontology [246] [968] language and provides querying and reasoning services over them through a corresponding reasoner [128].

In practice, this framework uses the ontological elements that OWL 2 supports and extends them with annotation properties to represent the features of the fuzzy ontology that OWL 2 cannot directly encode. The basic annotation property used is **fuzzyLabel** which contains all the fuzzy definitions related to a given element. Each fuzzy element definition is delimited by a start tag **<fuzzyOwl2>** and an end tag **</fuzzyOwl2>** while a **fuzzyType** attribute is used to denote the element's type. Figure 6.3 depicts two sample definitions in Fuzzy OWL 2, the first representing the fuzzy statement *"Company X is a Competitor to a degree of 0.6"* and the second the fuzzy term *"high"* for the datatype projectBudget.

For the easier creation and management of these definitions, Fuzzy OWL 2 is accompanied by a corresponding graphical editor, implemented as a plugin for the well known ontology editor *Protégé* 4.1[1]. Figure 6.4 depicts the main menu of this plugin with the whole range of supported fuzzy ontology elements.

## 6.3.4 Step 4: Fuzzy Ontology Validation

After the fuzzy ontology has been built, a validation process needs to take place in order to ensure that the developed artifact captures and represents the vagueness of the domain in an adequate and correct way. This is practically translated into evaluating the following four properties of the fuzzy ontology:

- **Correctness:** A fuzzy ontology is correct when all its fuzzy elements convey a meaning which is indeed vague in the given domain, context or application

---

[1]http://protege.stanford.edu/

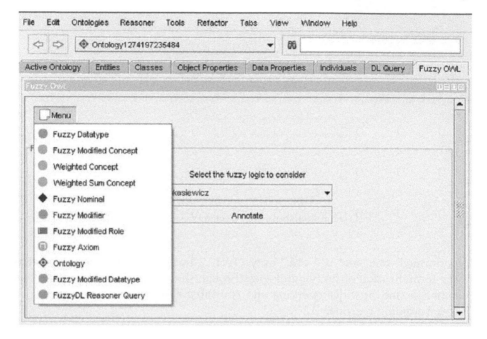

FIGURE 6.4: Fuzzy OWL 2 elements.

scenario. This means that the fuzziness of each element is actually caused by the potential existence of borderline cases and not because of some other reason (e.g. uncertainty or ambiguity).

- **Accuracy:** A fuzzy ontology is accurate when the degrees of its fuzzy elements approximate the latter's vagueness in an intuitively accurate way for the given domain, context or application scenario. This doesn't mean that the fuzzy degrees should have specific values but merely that these values should be perceived as natural by those who use the ontology. For example, the fuzzy statement *Obama is a BlackPerson at a degree of 0.2* is highly unintuitive (and therefore inaccurate) while the statement *Sergey Brin is a RichPerson at a degree of 0.8* makes much more sense.

- **Completeness:** A fuzzy ontology is complete when all the vagueness of the domain has been represented within the ontology. This primarily means that the ontology should not contain crisp elements which in the given domain, context or application scenario convey vague meaning. It also means that for each vague element all the (required) dimensions of its vagueness should have been identified and modeled.

- **Consistency:** A fuzzy ontology is consistent when it does not contain controversial information about the domain's vagueness as this is expressed by fuzzy degrees. For example, the fuzzy statements *Obama is a BlackPerson at a degree of 0.9* and *Obama hasSkinColor black at a degree of 0.4* are controversial.

The first three properties are typically evaluated by humans (ontology users and domain experts) while the fourth may be checked by means of some fuzzy ontology reasoner. Also, it should be noted that the evaluation of completeness is performed with respect to the domain as this has been captured by the initial crisp ontology. This means that vague knowledge that was not originally contained as crisp within the initial domain ontology will not be contained in the final fuzzy ontology either and, therefore, should not be included in the evaluation process.

Finally, the validation process does not include the task of checking the applicability of the developed ontology because i) the applicability of the utilized crisp ontology has already been checked in step 0 of the methodology as part of the task of deciding what ontologies to reuse and ii) fuzziness does not really affect the applicability of the original ontology, it merely makes the vagueness of the domain explicit. Such a check would be important if someone considered reusing an existing fuzzy ontology but this task is out of the scope of this chapter.

## 6.4 Practical Example: Developing a Fuzzy Enterprise Ontology

To illustrate the way IKARUS-Onto can be practically applied to the construction of fuzzy ontologies, we present in this section an indicative portion of a real life application scenario that involves the development of a fuzzy enterprise ontology for a consulting firm. According to the scenario, the ontology needs to model knowledge about the firm's operation which is to be utilized by a knowledge-based decision support system the consulting firm has. The system is described in more detail in [59] and it is capable of handling and exploiting vague knowledge in the form of fuzzy ontologies for helping the firm to decide whether it should write a proposal for a tender call.

In the following paragraphs we describe how each step of IKARUS-Onto is executed in order to produce the aforementioned ontology in such a way so as to effectively capture and express the vagueness of its domain. We omit from the example step 3 (formalization) as this would require the analytical description of some specific fuzzy ontology language, which falls outside the scope of this work.

As suggested in Section 6.3, the starting point of the fuzzy ontology development is the acquisition of a crisp ontology that models the target domain. For the needs of this demonstration and given the publication's length and clarity considerations, we limit the ontology to the following elements:

## Concept Taxonomy

- Company
    - Competitor
- Employee
    - HighPotentialEmployee
- ConsultingArea
    - CompanyCoreConsultingArea: Refers to those consulting areas that constitute the firm's core competencies.
- Project
    - IT_Project
    - HighBudgetProject

## Relations

- isExpertAtConsultingArea: Relates employees to the consulting areas they are expert at.
- regardsConsultingArea: Relates projects to the consulting areas they fall into.

## Attributes

- projectBudget
- employeeExperience

## Concept Instances

- Companies: Accenture, McKinsey, Bain, Boston Consulting Group
- Competitors: Accenture, McKinsey
- Employees: Jane, John, Karen, Ian
- High Potential Employees: Jane, Ian
- Consulting Areas: Information Technology, Human Resources, Strategy, Marketing
- Company Core Consulting Areas: Information Technology, Strategy
- Projects: P1, P2, P3

- IT Projects: P1, P2

**Relation Instance Pairs**

- *isExpertAtConsultingArea*(Jane, Information Technology)

- *isExpertAtConsultingArea*(Jane, Strategy)

- *isExpertAtConsultingArea*(Ian, Strategy)

- *isExpertAtConsultingArea*(Ian, Marketing)

- *isExpertAtConsultingArea*(John, Marketing)

- *isExpertAtConsultingArea*(Karen, Strategy)

- *regardsConsultingArea*(P1, Information Technology)

- *regardsConsultingArea*(P1, Human Resources)

- *regardsConsultingArea*(P2, Information Technology)

- *regardsConsultingArea*(P3, Strategy)

### 6.4.1   Applying Step 1: Establishing the Need for Fuzziness

The first premise for fuzziness, namely the existence of some knowledge-based system which handles and utilizes fuzzy ontologies, is satisfied within the scenario through the firm's intelligent decision support system. For the second premise, namely the existence of vagueness within the domain, the ontology engineer tries, in collaboration with the firm's consultants who play the role of the domain expert, to detect borderline cases (for relations, attributes and concepts) and potential vague terms (for attributes) in the above ontology. The result of this process are the following vague elements:

- **Vague concepts:** Competitor (because some companies are borderline competitors), HighPotentialEmployee (because some employees are borderline high potential), CompanyCoreCompetenceArea (because some areas are borderline core), IT_Project (because some projects are borderline IT) and HighBudgetProject (because some projects are borderline high valued).

- **Vague relations:** *isExpertAtConsultingArea* (because some employees are borderline expert at some areas) and *regardsConsultingArea* (because some projects are borderline relevant to some areas).

- **Vague attribute value terms:** {low, average, high} for the attribute projectBudget and {junior, senior, veteran} for the attribute employeeExperience.

**TABLE 6.4**

Fuzzy Relations for the Consulting Ontology Example.

| Relation | Vagueness Nature | Degree Interpretation |
|---|---|---|
| *isExpertAtConsultingArea* | Degree-vagueness along the dimension of the level of knowledge on a consulting area. | The extent to which someone's knowledge on a consulting area classifies him as expert on it. |
| *regardsConsultingArea* | Combinatory vagueness due to the lack of sharp discrimination between those conditions that are necessary for a project to belong to a given consulting area. | The extent to which the project's set of satisfied consulting area's conditions classifies it as belonging to this area. |

### 6.4.2 Applying Step 2: Defining the Fuzzy Ontology Elements

Given the vague ontology elements of step 1, step 2 proceeds with the definition of each of these elements as fuzzy ones by following the corresponding procedures and templates of paragraph 6.3.2. This means that the ontology engineer, in collaboration with the domain experts, produces the following:

- The descriptions of the two fuzzy ontology relations (Table 6.4) and their corresponding fuzzily related instance pairs (Table 6.5). The fuzzy degrees

**TABLE 6.5**

Fuzzy Relations Statements for the Consulting Ontology Example.

| Relation Statement | Fuzzy Degree |
|---|---|
| *isExpertAtConsultingArea*(Jane, Information Technology) | 0.8 |
| *isExpertAtConsultingArea*(Jane, Strategy) | 0.9 |
| *isExpertAtConsultingArea*(Ian, Strategy) | 0.6 |
| *isExpertAtConsultingArea*(Ian, Marketing) | 0.4 |
| *isExpertAtConsultingArea*(John, Marketing) | 0.75 |
| *isExpertAtConsultingArea*(Karen, Strategy) | 0.8 |
| *regardsConsultingArea*(P1, Information Technology) | 0.9 |
| *regardsConsultingArea*(P2, Human Resources) | 0.5 |
| *regardsConsultingArea*(P2, Information Technology) | 0.5 |
| *regardsConsultingArea*(P3, Strategy) | 0.8 |

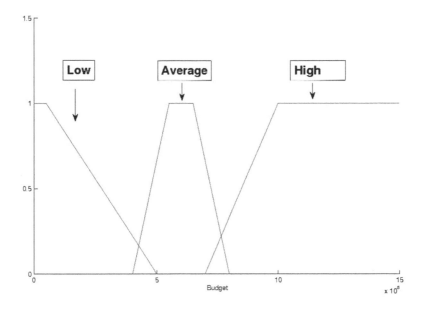

FIGURE 6.5: Fuzzy datatype for project budget attribute.

are assigned by the domain experts based on the degree interpretations that they have themselves assigned to the fuzzy ontology relations (second column of Table 6.4).

- Two fuzzy datatypes (Figures 6.5 and 6.6), each defining the meaning of the vague value terms of the attributes projectBudget and employeeExperience. The definition of the terms' fuzzy sets for each fuzzy datatype is performed by the domain experts based on their own understanding about when, for example, a budget is considered high or a consultant veteran.

- The descriptions of the fuzzy ontology concepts (Table 6.3) and their corresponding fuzzily assigned instances (Table 6.6). For those fuzzy concepts that are defined through some fuzzy relation or fuzzy datatype (IT_Project and HighBudgetProject respectively), the fuzzy degrees of their instances are left to be determined by the system's reasoner. The others are determined by the domain experts, again by considering the degree interpretations that they have themselves assigned to the fuzzy concepts (second column of Table 6.3) .

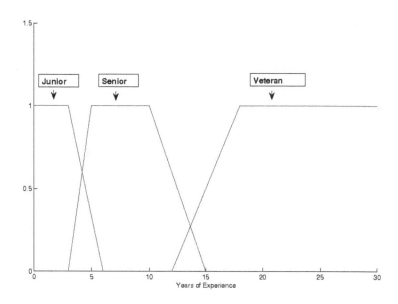

FIGURE 6.6: Fuzzy datatype for employee experience attribute.

### 6.4.3    Applying Step 3: Formalizing the Fuzzy Ontology

As suggested in paragraph 6.3.3, the formalization of the fuzzy ontology requires the transformation of the defined fuzzy ontology elements into a formal machine-interpretable form through some corresponding fuzzy ontology language. In this case, we used the Fuzzy OWL 2 language to do that, coming up with definitions like the ones of Figure 6.3.

**TABLE 6.6**

Fuzzy Concept Instances for the Consulting Ontology Example.

| Concept Instance | Fuzzy Degree |
|---|---|
| Competitor(Accenture) | 0.8 |
| Competitor(McKinsey) | 0.5 |
| CompanyCoreConsultingArea(Information Technology) | 1.0 |
| CompanyCoreConsultingArea(Strategy) | 0.6 |
| HighPotentialEmployee(Jane) | 0.9 |
| HighPotentialEmployee(Ian) | 0.5 |

**TABLE 6.7**

Fuzzy Concepts for the Consulting Ontology Example.

| Concept | Vagueness Nature | Degree Interpretation |
|---|---|---|
| Competitor | Degree-vagueness along the dimension of the competitor's business areas and the dimension of the competitor's target markets | The extent to which the instance's business areas and target markets classify it as a competitor. |
| CompanyCoreConsultingArea | Degree vagueness along the dimensions of the company's expertise and experience in the given area | The extent to which the company's expertise and experience in the given area classify it as core. |
| HighPotentialEmployee | Combinatory vagueness due to the lack of sharp discrimination between the necessary characteristics for an employee to be considered as high potential. | The extent to which the employees's characteristics classify him/her as high potential. |
| IT_Project | Combinatory vagueness derived from the *regardsConsultingArea* vague relation | The same as in the *regardsConsultingArea* relation. |
| HighBudgetProject | Degree vagueness derived from the *projectBudget* fuzzy datatype | The same as in the term *high* of the *projectBudget* datatype. |

## 6.4.4 Applying Step 4: Validating the Fuzzy Ontology

Validating the above fuzzy ontology in the specific scenario means ensuring that:

- The fuzzy ontology is correct, namely all its fuzzy elements convey a meaning which is indeed vague in the given domain. This is done by having another team of domain experts verify that the definitions of Tables 6.7 and 6.4 are correct.

- The fuzzy ontology is accurate, namely the degrees of its fuzzy elements approximate the latter's vagueness in an intuitively accurate way. This is done by having another team of domain experts assess that the degrees of Tables 6.6 and 6.5 are intuitively correct.

- The fuzzy ontology is complete, namely all the vagueness of the domain has been represented within the ontology. This is done by having another team of domain experts check that the non-fuzzy parts of the ontology do not convey vague meaning.

- The fuzzy ontology is consistent, namely it does not contain controversial information about the domain's vagueness. This is done by using some fuzzy ontology reasoner that is compatible to the language the ontology has been formalized with.

## 6.5 Automating the Vague Knowledge Acquisition Process

An important bottleneck in the process of developing and maintaining a fuzzy ontology is that of vague knowledge acquisition, namely the problem of determining the optimal fuzzy degrees the fuzzy ontological elements should have in order to accurately reflect the vagueness of the domain. The causes of this bottleneck are three main phenomena related to the interpretation of vague terms:

1. **Subjectiveness:** The same given vague term might be differently interpreted by different users. For example, for John, 100 euros for a dinner in a restaurant might be too expensive, while for Jane it might be ok.

2. **Context Dependence:** The same given vague term might be differently interpreted in different contexts even if the user is the same. For example, celebrating an anniversary is different than celebrating birthday when it comes to judging a restaurant as expensive or not.

3. **Changing Interpretations:** The same user in the same context may change his/her interpretation of a particular vague notion due to changing circumstances. For example, if the user gets a better paying job, his/her notion of expensive restaurant might change.

All these three characteristics of vagueness make the creation and maintenance of a fuzzy ontology a quite time- and resource-consuming task. With that in mind, we outline in this section a generic approach for automating the generation of fuzzy degrees and membership functions.

## 6.5.1 Problem Definition

The problem to be tackled can be informally defined through the following question: *Given a fuzzy ontology, what are the optimal fuzzy degrees and membership functions that should be assigned to its elements (concepts, relations and datatypes) in order to represent their vagueness as accurately as possible?* Thus, the goals are:

- Given a fuzzy concept (e.g., CompanyCompetitor) and a set of its instances (e.g., a set of companies), to learn the degree to which each of these instances belongs to this concept (e.g., to what degree each company is considered a competitor).

- Given a fuzzy relation (e.g., isExpertAt) and a set of related through it pairs of instances (e.g., persons related to business areas), to learn the degree to which the relation between these pairs actually stands.

- Given a fuzzy datatype (e.g., ProjectBudget) and the terms it consists of (e.g., *low, average, high*), to learn the membership functions of the fuzzy sets that best reflect the meaning of each of these terms.

## 6.5.2 Proposed Framework

Since vague pieces of knowledge are characterized by the existence of blurry boundaries and by a high degree of subjectivity they are expected to provoke disagreements and debates among people or even among people and systems. For example, it might be that two product managers disagree on what the most important features of a given product are or that two salesmen cannot decide what amount of sales is considered to be low. Similarly, it might be that the user of a recommendation system does not agree with the system's classification of "expensive" and "non-expensive" restaurants. Given that, our approach for automating the generation of fuzzy degrees is based on the capture and statistical analysis of such disagreements by means of appropriate mechanisms.

More specifically, the generic process for performing vague knowledge acquisition consists of the following steps:

1. Development of the fuzzy ontology with IKARUS-Onto without, nevertheless, having domain experts define the fuzzy degrees and fuzzy membership functions of the ontology's elements.

2. Deployment of an appropriate mechanism that facilitates the generation and gathering of vague knowledge assertions, namely statements related to the elements already defined in the fuzzy ontology. For example, the assertion *"A budget of 100,000 euros is low"* is related to the fuzzy datatype "ProjectBudget," while the assertion

**TABLE 6.8**

Example Vague Concept Assertions.

| Concept | Instance |
|---|---|
| Competitor | Autonomy |
| Competitor | IBM |
| Competitor | Indra |
| Competitor | Atos |
| Strategic Client | Coca Cola |
| Strategic Client | Endesa |

> *"John is expert in ontologies"* is related to the fuzzy relation "is-ExpertAt."

3. Use of the vague assertions to generate fuzzy degrees and membership functions for the respective elements.

Vague assertion gathering mechanisms can take many forms, depending on the application context of the fuzzy ontology. A concrete implementation is described in [56], where in order to learn the fuzzy degrees of a fuzzy enterprise ontology, we used an enterprise microblogging framework in which the members of the enterprise were participating and performing discussions and information exchange on aspects regarding the enterprise and its environment. The framework, through appropriate semantic information extraction techniques [191], was able to detect within the users' microposts statements related to the fuzzy ontology and extract them in the form shown in Tables 6.8 and 6.9.

Then, through a set of corresponding methods, (described in [56]) we measured the strength of the extracted assertions and generated the degrees of the corresponding fuzzy elements. Figure 6.7 shows the outcome of this process for the fuzzy datatype "ResearchProjectBudget" while Table 6.10 shows the calculated fuzzy degrees for some of the ontology's statements.

**TABLE 6.9**

Example Vague Datatype Assertions.

| Datatype | Vague Term | Crisp Value |
|---|---|---|
| ProjectBudget | High | 3 million |
| ProjectBudget | Fairly high | 1.2 million |
| ProjectBudget | Fairly high | 1 million |
| ProjectBudget | Fairly high | 0.8 million |
| ProjectBudget | Average | 0.85 million |
| ProjectBudget | Low | 0.6 million |
| ProjectBudget | Average | 0.6 million |
| ProjectBudget | Low | 0.2 million |

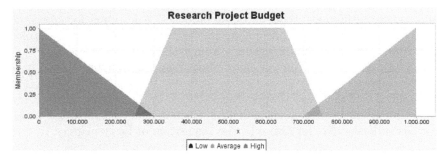

FIGURE 6.7: Example-generated fuzzy data type.

**TABLE 6.10**
Examples of Generated Fuzzy Statements.

| Concept | Instance |
|---|---|
| Competitor(Autonomy) | 0.3 |
| Competitor(IBM) | 0.2 |
| Competitor(Indra) | 0.8 |
| Competitor(Atos) | 0.6 |
| StrategicClient(Coca Cola) | 0.9 |
| StrategicClient(Endesa) | 0.8 |

Generalizing this example, there are two main ways by which vague assertions may be gathered, namely mining them from existing structured or unstructured data or eliciting them, in the form of explicit or implicit feedback, from the users of applications that are based on the fuzzy ontology. In all cases, the prior use of a methodology like IKARUS-Onto for the identification and conceptualization of vagueness, makes this process much more accurate and effective. As for the transformation of the vague assertions into fuzzy membership functions, this is a well studied problem in the relevant literature [423] [220] [1011] [601] and many existing methods may be used.

## 6.6 Conclusions

Although fuzzy ontologies have been developed and utilized in a number of different systems and domains as a way to enable the semantic processing of vague knowledge, this development has been typically performed in ad hoc and incomplete manners. With that in mind, we provided in this chapter a step by step guide for developing fuzzy ontologies, covering all required stages from specification to validation and discussing, for each stage's execution, appropriate techniques, tools and best practices. Additional focus was given to the resource-consuming task of generating fuzzy degrees for the ontology's elements where we outlined a generic framework for automating it.

In future work based on this chapter, one may seek the direction of making the development of fuzzy ontologies an increasingly easier task by i) developing a graphical tool that may support the engineers in performing all the relevant tasks described in this guide and ii) by automating, as much as possible, various subtasks like, for example, the acquisition of fuzzy degrees.

# 7

# Spatiotemporal Event Visualization on Top of Google Earth

**Chrisa Tsinaraki**

*Technical University of Crete, chrisa@ced.tuc.gr*

**Fotis Kazasis**

*Technical University of Crete, fotis@ced.tuc.gr*

**Nikolaos Tarantilis**

*Technical University of Crete, nicktaras@gmail.com*

**Nektarios Gioldasis**

*Technical University of Crete, nektarios@ced.tuc.gr*

**Stavros Christodoulakis**

*Technical University of Crete, stavros@ced.tuc.gr*

## CONTENTS

Several types of information objects (like, for example, events, sites of interest, etc.) can be better visualized on top of spatial representations, since the location of these objects is an important object feature. Several types of spatial representations (i.e., canvases, maps, diagrams, spatial plans, etc.) have been used throughout human history for the visualization of such information.

The use of interactive maps on the Web has become very popular nowadays, especially after the development of robust interactive web map infrastructures like the OpenLayers, the Google Earth and Google Maps.

In addition, the exploitation of several types of mobile devices (i.e., phones, tablets, etc.) has allowed the development of location based services. A requirement for those applications is the strong support for geographic context including map distances, paths, time, location of objects and presentation of the objects on the map, awareness (location of services or friends nearby), etc. The *geovisualization* (essentially the visualization on top of maps) is of special interest, since it conveys geolocation information.

In this chapter we will review the research that is relevant to the geovisualization of events and will present our recent research for the geovisualization of events on top of the Google Earth 3D interactive maps, with respect to their spatial and temporal features.

## 7.1   Introduction

The spatial representations (i.e., canvases, maps, spatial plans, etc.) that have been used throughout human history allow for the better visualization of several types of information objects (like, for example, events, sites of interest, etc.), since they also depict the (absolute or relative) location of these objects.

An important type of visualization on top of spatial representations is the *geovisualization*, short for Geographic Visualization, which refers to a set of tools and techniques supporting geospatial data analysis through the use of interactive visualization [989]. The widespread use of computers in everyday life in the last decades has allowed using such spatial representations in electronic format. The development of the web and the need for interaction with spatial representations have led to the increasing use of interactive maps, especially after the development of robust interactive web map infrastructures like the OpenLayers[1], Google Earth[2] and Google Maps[3]. Information geovisualization allows to exploit the interactivity of these infrastructures and is facilitated from them.

In addition, the exploitation of several types of mobile devices (i.e phones, tablets, etc.) has allowed the development of location based services. A requirement for those applications is the strong support for geographic context including map distances, paths, time, location of objects and presentation of the objects on the map, awareness (location of services or friends nearby), etc.

---

[1]OpenLayers, http://www.openlayers.org/
[2]Google Earth, http://earth.google.com/
[3]Google Maps, http://maps.google.com/

Events and sites of interest may be also associated with multimedia content that is easily annotated with contextual information, since such information is automatically captured by modern GPS-enabled devices (like cameras, mobiles, tablets, etc.).

In this chapter, we focus on event geovisualization, since many important practical applications in several domains (like, for example, education, culture, tourism, real estate, etc.) are associated with events or they can be described by events. In addition, event processing has recently attracted a lot of attention in both the business and academia [315]. We also present our recent research for event geovisualization on top of the Google Earth 3D interactive maps, with respect to their spatial and temporal features.

The rest of the chapter is structured as follows: The state of the art in event modeling is discussed in Section 7.2, information geovisualization is surveyed in Section 7.3, the *EVISUGE (Event VISUalization on Google Earth)* system that we have developed for the geovisualization of events on top of the Google Earth 3D interactive maps is described in Section 7.4 and the chapter concludes in Section 7.5.

## 7.2 Event Modeling

In this section we examine the state of the art in event modeling. Formally, an *event* is an occurrence within a particular system or domain; the term event is used to describe both the real-world occurrence and its computerized representation [315]. Examples of real-world events include accidents, marriages, baptisms, invited talks, debates, lectures, disasters like fires and floods, battles, experiments, product presentations, etc.

Typically, a real-world event is recorded using only media such as video or pictures and audio; however, the events are usually associated with some spatial information that can be described on top of spatial representations and also have some associated metadata (i.e., actors, time, etc.) which are used for further processing or filtering these events. Today's multimedia capturing devices are often equiped with a GPS receiver and can automatically capture a lot of contextual information (like, for example, GPS location, compass, azimuth), which can be manipulated to automatically register the 3D natural environment of the picture or video captured [229], [228]. This eliminates the need for extensive manual editing for the systematic capturing of multimedia events and their context.

Systematic real-world event capturing should associate and exploit all the contextual data with the captured audiovisual content in order to offer better cognitive clues to the viewer and allow exploitation of the metadata for workflow processing. For example, in a battle event the multiple media may capture the scene of the battle (armies involved, weapons, etc.); spatial information such as maps may record the location and movement of the armies involved and the metadata may be used to record date and time as well as names of people involved in the battle.

Complex events are important in modern applications of business processes like business intelligence. In business intelligence, events or event patterns may be tracked to determine certain conditions that are important to the intelligence of the organization. A very promising research direction in business applications and business intelligence is the *semantic modeling of events* [622] [805] [489] [427] [926] [236] [437]. Event Processing Architectures are becoming very important in business processing applications [622]. Although Event Processing Architectures may seem to contrast to Service Oriented Architectures (SoA), researchers mostly agree for their complementarity, with the Event Processing Architectures covering better the asynchronous requirements of business processing. An Event Driven SoA is a hybrid form combining the intelligence and proactiveness of the Event Driven Architectures with the SoA organizational capabilities [669], [622].

Semantic modeling of events is also an active research area in multimedia. The *MPEG-7 semantic model* allows for the modeling of events and includes, among others, modeling of actors that are participating in the event as well as the event time and place. The MPEG-7 semantic model is in a high level, but the extensibility mechanisms of MPEG-7 can be used to give powerful domain specific semantic descriptions of events in specific domains while still remaining completely within MPEG-7 [926]. The semantic MPEG-7 descriptions can be also transformed in logic based languages such as OWL for further processing and inference [927]. A related recently expanding area of research is the automatic extraction of semantic events from video sequences [118].

Modeling such aspects as camera parameters, camera location, camera movement, subject movement, light and sound locations, etc. is necessary for the representation of the multimedia capturing of events. The use of such parameters for capturing quality multimedia is studied in the cinematography area, where a rich bibliography exists [77] [826], [159], [653], [788], [500]. These principles can be taken into account when producing guidelines for taking specific shot types. Some research related to cinematography tries to capture the rules for good cinematography into language or expert system constructs, often with the objective to automate the presentation of virtual reality scenes in games and elsewhere [296], [369].

While rich literature exists in the area of semantic event modeling and event-based multimedia modeling, as well as in cinematography principles, very limited research exists in generic tools and methodologies for systematically capturing multimedia events and their context, including the spatial context and the context of capturing. Such tools should be based on models that associate composite events with the locations where the events took place and the time of the event evolution. In addition, they should facilitate the modeling of the capturing processes in space and time and its associations with the event evolution. This will facilitate the visualization of the events, the shot capturing processes and the actual multimedia captured. Without such models the capturing process is very expensive and slow, and the visualization possibilities are limited, overtaxing the cognitive capabilities of the users.

Should the above discussion be taken into account, the representation of a real-world event on a map should include:

1. The spatial information of the event, which is represented by its position on the map.

2. The event participants (both actors and objects).

3. The temporal features of the event.

4. The representation of any relevant events (including sub- events and preceding/following/parallel events).

These real-world event representation requirements are satisfied by the *MOME (MObile Multimedia Event Capturing and Visualization)* event representation model [726]. Interrelated real-world events may be combined in MOME to form scenarios; the latter are visualized through the visualization of their component events, with respect to their spatiotemporal order. An important feature of the MOME model is that it supports the real-time multimedia event capturing that exploits smart devices (like, for example, cameras with automatic GPS coordinate capturing capabilities). It can also accommodate domain specific descriptions for the events, which include domain-specific knowledge systematically captured as described in [926]. In addition, the MOME event model can be mapped to the MPEG-7 [805] and MPEG-21 [737] standards (used, respectively, for multimedia content description and for the interaction with multimedia content).

## 7.3 Information Geovisualization

The spatial representations are useful for information visualization, since several types of information objects have a spatial component. In this section we will review the research literature regarding information geovisualization. The research challenges on which we will focus are information visualization on top of GIS (in Section 7.3.1) and event geovisualization (in Section 7.3.2).

### 7.3.1 Visualization on Top of GIS

A *Geographic information system (GIS)* is a system designed to capture, store, manipulate, analyze, manage, and present all types of geographical data. A GIS uses spatiotemporal (space-time) location as the key index variable for all other information [988]. An important application of the GIS is the presentation of information on top of it that also allows the exploratory spatio-temporal visualization [71]. An example of such an application is the city exploration application of [731] that helps users identify interesting spots in a city using spatio-temporal analysis and clustering of user contributed photos.

Another important research direction is the *Time Series Data visualization* on top of maps. Such an application is the TimeMapperWMS, which allows users all over the world to visualize the dynamics inherent in the Antarctic Iceberg time series [111]. The TimeMapperWMS has been built on top of the *Web Map Service (WMS)* [716] and uses the *Scalable Vector Graphics (SVG)* [967] in order to produce *animated maps* (i.e., maps on which animation has been applied in order to add a temporal component displaying change in some dimension [987]). The authors of [906] present a 3D visualization approach to address some of the challenges in effective visual exploration of geospatial time-varying data; their system also provides a holistic display of the spatio-temporal distributions of the data on a geographic map and employs standard visual-analytical tools (like, for example, interactive data mapping and filtering techniques) to support exploratory analysis of multiple time series.

The modern cameras provide integrated GPS technology, thus allowing the automatic capture of the geographical context of the images taken; in addition, they allow wireless connection to computers. This way, the *geotagged images* (i.e., images annotated with their geographical context) may be easily integrated in GIS and associated with spatial objects. A *spatial object* may be an individual which relates to a semantic concept in an ontology [229], [228]. Through this relationship, the user can see semantic information about the visited space (for example, when visiting a room, the user may see who resides in this room or the room functions).

The geotagged images have allowed the development of spatial information processing and retrieval frameworks [229], [228], [519]. The *SPIM (SPatial Image Management)* framework [229], [228] provides a picture database that allows users to store and view their photos. Along with the photos, the users can view personalized semantic maps, annotated with semantic objects described using ontologies. These maps are supplied from a remote server. Photos with position only information as well as photos with both position and direction information can be visualized on top of the maps and be associated with semantic objects. Several algorithms that allow interactive exploration of the picture contents have been implemented. In addition, the framework employs the use of image processing and other algorithms to enable automatic annotation of the photos. In [519] a conceptual framework and a methodology have been developed that allow analyzing events and places using geotagged photo collections shared by people from several different countries. These data are often semantically annotated with titles and tags that are useful for learning facts about the geographical places and for detecting events that occurr in these places. The knowledge obtained through the analysis performed on them may also be utilized in sociological and anthropological studies as well as for building user centric applications like tour recommender systems.

Among the latest popular developments of the GIS technology were *Google Map Mashups*, i.e., the Google Earth and the Google Maps interactive web map infrastructures. Since Google has provided Application Programmers Interfaces (APIs) for them (the Google Earth API[4] and the Google Maps API[5], respectively), Google Map Mashups have been extensively used for the presentation of multimedia content related to sites and events of interest. Some examples of this type of functionality are listed below:

- The Natural History museum of Crete allows presenting information about several types of birds on top of Google Maps[6].

- The Virtual Bulgaria project allows presenting, on top of Google Maps, panoramic views of several points of interest (cultural, touristic, etc.) in Bulgaria[7].

- The DC Crime Visualization[8], which is an event visualization mashup of violent crimes in the District of Columbia that has been developed using the Google Maps API.

- The data-mining system presented in [237], which deals with very large spatio-temporal datasets has exploited Google Earth in order to display the data mining outcomes combined with a map and other geographical layers.

An important type of GIS that has practical applications in several domains (i.e., environmental sciences, military, computer assisted cartography, etc.) is the *Historical Geographic Information System* (also written as Historical GIS or HGIS). An HGIS is a geographic information system that may display, store and analyze data of past geographies and track changes in time [990]. The Google Earth interactive web maps are the GIS infrastructure utilized for the China Historical GIS [116].

Another recently popular web map infrastructure is OpenLayers, an open source JavaScript library for displaying maps on any web page. It provides a JavaScript API for building rich web-based geographic applications, similar to the Google Maps' APIs. OpenLayers is highly extensible and it serves as the foundation of all the web mapping interfaces. OpenLayers accesses data through industry standards and it may overlay multiple standards-compliant map layers into a single application. Some examples of OpenLayers-based applications are listed below:

- The *Community Almanac*[9], which is a collaborative community building and story telling application developed for the Orton Foundation.

---

[4]Google Earth API, https://developers.google.com/earth/
[5]Google Maps API, https://developers.google.com/maps/
[6]The Natural History Museum of Crete, http://www.nhmc.uoc.gr/
[7]The Virtual Bulgaria Project, http://www.bulgaria-vr.com/
[8]DC Crime Visualization, http://www.geovista.psu.edu/DCcrimeViz/app/
[9]The Community Almanac, http://www.communityalmanac.org/

- The *Vespucci*[10] collaborative mapping tool, which takes advantage of the versioning capability of GeoServer[11] (a software server written in Java that allows users to share and edit geospatial data). Vespucci has been used by Landgate, the government authority responsible for land and property information in Australia.

- The *Styler*[12] interactive styling application for geospatial data.

- The *RAE Geospatial Map*[13], developed by the *Greek Regulatory Authority for Energy (RAE)*, which provides user-friendly tools for navigating, querying, searching, measuring and selecting areas of interest. Using the map, RAEs customers may be informed about the status of their energy facilities and can search, visualize and retrieve information from RAEs resources.

### 7.3.2 Event Geovisualization

The events have a spatial component; thus, a challenging research direction is that of real-world event visualization on top of spatial representations.

Some event visualization applications developed on top of spatial representations are presented below:

- The application presented in [356], which allows the spatiotemporal visualization of battles on top of a canvas.

- The Trulia Hindsight[14], which animates, on top of the Microsoft Virtual Earth platform, housing construction by year, providing prospective buyers a quick overview of the historical development of a neighborhood. The events are represented by circles of different colors, which are not interactive.

- The AsthMap application[15][478], which is a mashup for mapping and analyzing asthma exacerbations in space and time, developed on top of the Microsoft Virtual Earth[16] platform. The events are also represented by circles, but the circles are now interactive and allow the retrieval of event attribute data.

- The DC Crime Visualization that allows the interactive presentation of the violent crimes in the District of Columbia on top of Google Maps. The DC Crime Visualization was the basis for the development of an event animation code library that extends the Google Maps API [789] [790].

---

[10]Vespucci, http://demo.opengeo.org/vespucci/
[11]Geoserver, http://opengeo.org/technology/geoserver/
[12]Styler, http://suite.opengeo.org/geoserver/www/styler/index.html
[13]Greek Regulatory Authority for Energy, http://opengeo.org/publications/rae/
[14]Trulia Hindsight, http://hindsight.trulia.com/
[15]The AsthMap Application, http://indiemaps.com/asthMap/
[16]Microsoft Virtual Earth, http://www.viawindowslive.com/VirtualEarth.aspx

The above applications can be distinguished in the ones that need advanced event support for complex events, like [356], and the more "lightweight" ones that are visualized on top of maps like the HidInsight, the AsthMap, and the DC Crime Visualization.

The importance of the event geovisualization applications has led to the development of generalized frameworks that allow event geovisualization for different application domains.

The SpatialKey[17] [477] is a collection of templates, built on top of the MapQuest[18] platform, which provide a suite of geovisualization techniques for spatiotemporal information, including event animation. In addition, it allows the users to upload their own data. The most important limitation of the SpatialKey is that it cannot easily accomodate applications like that of [356], which need advanced event support.

The MOME mobile multimedia event capturing and visualization framework [726] implements the MOME event model and allows the spatiotemporal visualization of complex events on top of spatial representations. Based on the MOME model, we have implemented the EVISUGE system [902], which allows complex event visualization on top of Google Earth (see Section 7.4 for details).

## 7.4 The EVISUGE (Event VISUalization on Google Earth) System

In this section we present *EVISUGE (Event VISUalization on Google Earth)* [902], a system that allows the visualization and management of real-world scenarios on Google Earth. An EVISUGE scenario is composed of events, which are represented according to the MOME event representation model that we have developed. These events are visualized on top of Google Earth 3D interactive maps, with respect to their spatial and temporal features. We demonstrate the EVISUGE system through real-world scenarios: The specification, navigation and visualization of a naturalistic route, the spatiotemporal representation and visualization of battles and the visualization of weather conditions.

Compared with existing applications developed over interactive web maps (see Section 7.3 for details), the major advantage of the EVISUGE system is that it allows complex event geovisualization in addition to the presentation of multimedia information and/or simplified event visualization. Compared with earlier event visualization applications like [356], it allows event visualization

---

[17]The SpatialKey, http://www.spatialkey.com/
[18]MapQuest, http://www.mapquest.com/

and integration on top of interactive web maps and not only canvases or other proprietary diagrams.

In the following sections we discuss in detail the conceptual model of the EVISUGE system (Section 7.4.1) and the EVISUGE system architecture and functionality (Section 7.4.2) and we present the real-world application scenarios that we have already implemented on top of the EVISUGE system (Section 7.4.3).

## 7.4.1 The EVISUGE Conceptual Model

In this section we detail the *EVISUGE conceptual model*, which allows the scenario and event representation in the EVISUGE system.

The *scenario* is the central entity in the EVISUGE conceptual model. An EVISUGE scenario has a name and a textual description, is composed of events and is associated with spatial objects that represent the scenario participants. An EVISUGE event is associated with the event time, the actor roles involved in it and its sequencing information.

The event capturing requirements that were taken into account in the design of the conceptual model of the EVISUGE system include:

1. Capturing of the event's *spatial aspects* such as integrating maps and diagrams with additional relevant details.

2. Capturing and representation of the event's *actors*, their spatial representations, locations and movements during the events.

3. Capturing of the *real scenes* that relate to the events, using audio-visual material (photos, video, and sound).

4. Capturing the event *time* considering absolute and relative times with respect to other events, as well as demarcating the beginning and the end of shots that relate to a specific event for easy event browsing and extraction.

5. Representation of the *multimedia capturing process* with respect to space and time (camera location, direction, angle, movement, etc.) on top of the maps in order to facilitate the visualization of the spatial context.

6. Representation of the *event workflows* and mechanisms to facilitate the scheduling and preparation of the event capturing and the complete coverage of all the component events.

Since the MOME event representation model satisfies the above-referred requirements [726], we have decided to adopt it for the representation of EVISUGE events. The MOME model includes concepts that belong to the following four major categories:

1. Concepts that describe the *events* and the *multimedia shots* taken to capture those events. An event may be associated with structured

*Spatial Objects*, the event *Time*, and *Event Shots* that represent the audiovisual capturing of the event. An event shot is a multimedia item (i.e., image, video or audio segment) capturing the event. The time associated with an event refers to a specific time interval and it takes the relative sense of time.

2. Concepts used for the *logical spatial event representation*. The events are associated with *physical spaces* that may be described by *spatial objects*. These objects may be represented in the form of diagrams on a spatial representation of the event site (e.g., map, canvas, etc.) with a semantic meaning provided by the spatial object itself or without any semantic meaning indication at all. The events are also associated with *Roles* that semantically describe the event participants.

3. Concepts used for the *logical spatial shot representation*. A shot spatial representation describes the original and final locations and the camera parameters, the sound and light equipment on top of the spatial event site representation, in the same way that the event concepts were represented on top of it. The representation shape may change during the capturing process.

4. Concepts used for the *physical representation of the events and shots* described on top of maps or other spatial representations like diagrams or 3D views. The concepts of this part of the model are responsible for mapping the logical event representation to (one or more) other representations as required. To do this, rotation, translation and scaling of the logical representations may be required. The logical representations mapped on the site map are called physical representations. A physical representation has a location on the site map as well as shape and possible movement.

In the MOME model, the events belong to complex event types that may be composed of events belonging to simpler event types. The metadata captured for each event depend on its type. Events are captured by multimedia shots that belong to shot types. Events and multimedia shots may be associated with maps and/or diagrams. Shots are described by the position (captured by a GPS device or otherwise), direction (compass) and camera movement of the shot placed on a diagram of a scene that describes the event, as well as with other parameters. The shot types are associated with visual descriptions; these can be used for learning and preparation for multimedia capturing of events of a particular type.

The concepts used for the representation of scenarios and events in the EVISUGE system are depicted in the UML class diagram of Figure 7.1. This diagram essentially contains the MOME concepts together with the concepts that are necessary for the representation of the EVISUGE scenarios and shows the relationships between them.

According to the class diagram of Figure 7.1, an EVISUGE *Scenario* is

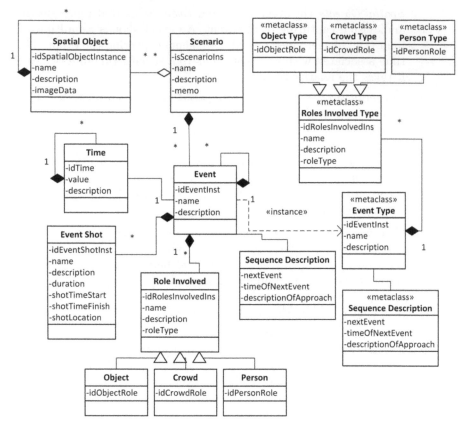

FIGURE 7.1: The EVISUGE conceptual model. The instantiation of a scenario results in the instantiation of all the events associated with it.

composed of *Events* and is associated with *Spatial Objects*. The *Events* may be complex, since an EVISUGE *Event* may be composed of simpler ones. The *Events* are also associated with the *Time* that they occur, the *Event Shots* that capture them, their *Sequence Description* that specifies their temporal order and the *Roles* (of type *Object*, *Person* or *Crowd*) that are involved in them. The *Spatial Objects* may also be complex, since they may comprise other *Spatial Objects*.

Notice also the MOME meta-modeling facilities (i.e., metaclasses) on the upper right part of Figure 7.1; the metaclasses include *Event Type*, *Sequence Description* as well as the *Roles Involved Type* and its sublcasses *Object Type*, *Crowd Type* and *Person Type*. The EVISUGE meta-modeling facilities allow the instantiation of a scenario to result in the instantiation of all the events associated with it.

The instantiation of the EVISUGE conceptual model that was outlined in

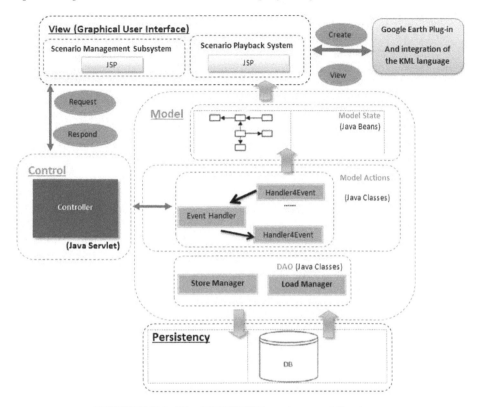

FIGURE 7.2: The EVISUGE system architecture.

this section is demonstrated in the real-world scenarios presented in Section 7.4.3.

## 7.4.2 The EVISUGE System

In this section we present the EVISUGE system in terms of system architecture and functionality.

*System Architecture.* The EVISUGE system architecture (depicted in Figure 7.2) is based on the *MVC (Model-View-Controller)* [609] design pattern.

The *View* layer is the Graphical User Interface (GUI) of the EVISUGE system and interacts with the end users for scenario management and playback. Both scenario management and playback are performed on top of the interactive 3D Google Earth maps. In order to achieve this, we have utilized the Google Earth plug-in[19] and the KML (Keyhole Markup Language) [715].

The *View* layer sends the requests that are generated from the user actions to the *Control* layer, which encodes the EVISUGE system logic. As a

---
[19]The Google Earth Plug-in, http://code.google.com/intl/el-GR/apis/earth/

consequence, the complex event processing that takes place in EVISUGE is performed in the *Control* layer, which is implemented as a Java servlet.

The *Control* layer also interacts with the *Model* layer, which allows for the transparent handling of the information stored in the persistency layer (i.e., the system database).

As is shown in Figure 7.2, the EVISUGE system comprises two subsystems:

1. The *Scenario Management Subsystem*, which allows scenario and event management (creation/editing/deletion), event capturing, as well as the specification of scenarios that are composed of existing events.

2. The *Scenario Playback Subsystem*, which allows:

   (a) The playback of both scenarios and events.

   (b) The navigation of the routes specified in the scenarios and the visualization of the associated events. The users are also allowed to interact with the scenario visualizations by clicking on the (2D or 3D) representations of the objects that are visualized on the maps and view the information associated with them.

*Functionality.* Table 7.1 presents the most important use cases that are related to the core functionality of the EVISUGE system. We distinguish two main actors that may be interacting with the EVISUGE system:

- The *scenario observers*, who are accessing the system in order to navigate through the existing scenarios. The scenario observers may also annotate the existing scenarios.

- The *scenario creators*, who specify the scenarios, register the scenario events and appropriately represent these events on Google Earth.

A snapshot of the EVISUGE user interface during scenario creation is shown in Figure 7.3 (the EVISUGE user interface currently supports Greek and English speaking users). The functionality offered by the system is accessible through the navigation bar (on the upper part of Figure 7.3).

The user interface of the EVISUGE system has been designed according to the following principles:

1. *Ease of navigation*, which is achieved through the navigation bar.

2. *Functionality grouping*, in the navigation bar, of the categories and subcategories of the functionality offered in the system.

3. *User notification*, in the navigation bar, of the category/subcategory of the system functionality that the user is currently using.

4. *User guidance*, through the links offered to the user in the navigation bar.

**TABLE 7.1**
The Most Important Use Cases of the EVISUGE System.

| Primary Actor | Goal | Description |
|---|---|---|
| Observer | Scenario navigation/visualisation | Interactively navigate through a scenario that contains one or more events. |
| Observer | Scenario search | Perform a keyword-based search to retrieve the most related scenarios. |
| Observer | Scenario annotation | Manage (create/delete) textual annotations/comments for a scenario during its visualisation. |
| Observer | Scenario metadata personalization | Select the scenario metadata information that will be projected during its visualization/navigation. This metadata refer to the events and the objects that are related to the specific scenario. |
| Observer | Scenario browsing | Browse through the existing (already stored) scenarios. |
| Creator | Scenario management | Manage (create/edit/delete) a scenario. A scenario comprises one or more events. |
| Creator | Event management | Manage (create/edit/delete) an event. The manipulation of an event is related to its spatial-temporal characteristics, its descriptive information (along with any multimedia information) and its representation (schema, fill/outline color, etc.) in Google Earth. |
| Creator | Create parallel events | Create events with temporal overlapping. |
| Creator | Manage event temporal information | Manage (create/edit/delete) the temporal information (i.e., begin/end time) in order to compute the event duration. |
| Creator | Manage event image information | Manage (attach/delete) an image to an event. |
| Creator | Manage event 2D/3D representation | Manage (create/edit/delete) the 2D/3D representation schema of the event. The schema is selected from a draw suite and may have a 3D representation. During the editing, the background and the outline colors are also specified. |

5. *Consistency*, since the same user interface layout has been followed in all the system screens.

6. *Appropriate color selection*, so that the user interface is not tiring for the end-users and has the look and feel of a robust application.

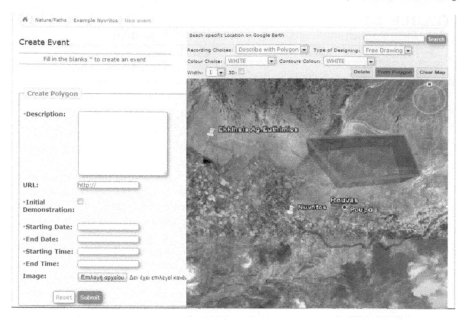

FIGURE 7.3: The EVISUGE Scenario Creator Interface.

## 7.4.3   Real-World Application Scenarios

In this section we describe the specification and visualization of real-world application scenarios that we have developed, as case studies, on top of EVISUGE. These scenarios belong to the three types of scenarios that we have thoroughly examined and are fully supported in EVISUGE : The *naturalistic route scenario* type, the *battle scenario* type and the *weather condition scenario* type.

### 7.4.3.1   Naturalistic Routes

We describe here the specification, navigation and geovisualization of a naturalistic route scenario in EVISUGE. In particular, we present the scenario of the Nyvritos-Gergeri route navigation (Nyvritos and Gergeri are two villages located in southeastern Crete, that attract a lot of tourists due to their natural beauty).

The structure of the Nyvritos-Gergeri route navigation scenario is depicted in the object diagram of Figure 7.4. As is shown in Figure 7.4, the scenario is represented by the Scenario class instance "Navigation of the Route Nyvritos-Gergeri." Since the route from Nyvritos to Gergeri passes through the Sellia village, this Scenario instance comprises the Event instances "Movement to Sellia village" and "View of Nyvritos village." The "Movement to Sellia village" event is a complex event that contains the event "Movement to Gergeri village."

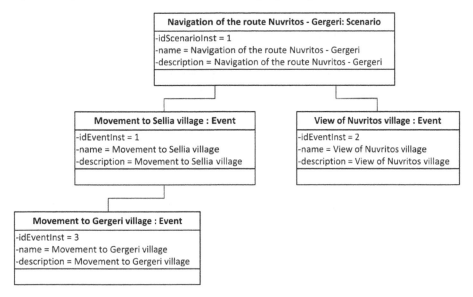

FIGURE 7.4: Object diagram that shows the structure of the Nyvritos–Gergeri route navigation scenario.


For the representation of this scenario in the EVISUGE system, several MOME concepts have been instantiated; for example, the road that is followed is represented in EVISUGE as an instance of the "Participating Spatial Object Shot" MOME class. Table 7.2 presents the MOME concepts and their associated instances that are used for the representation of the "Movement to the Sellia village" event contained in the scenario.

In the object diagram of Figure 7.5 we provide a more detailed view in the corresponding part of the structure of the scenario of Figure 7.4. In particular, we focus on the Event instance "Movement to Sellia village" that is part of the Scenario instance "Navigation of the Route Nyvritos-Gergeri." As is shown in Figure 7.5, the Event instance "Movement to Sellia village" is associated with the road that should be followed (represented by the "Road" instance of the SpatialObject class), the car used (represented by the "Car" instance of the Object class) and the time that this event takes place (represented by the "Day of the movement to Sellia village" instance of the Time class).

The spatial objects that participate in an event may also be structured; for example, in the object diagram of Figure 7.6 we present the "Village" instance of the SpatialObject class, which represents a village (i.e., Gergeri, Sellia or Nyvritos). The "Village" instance comprises other SpatialObject instances:

- The "Buildings" instance, which represents the village buildings. The "Buildings" instance is also structured, and comprises the "Church" in-

**TABLE 7.2**
Association of the MOME Event Representation Model Concepts with Their Instances in EVISUGE That Represent the Movement to the Sellia Village in the Nyvritos–Gergeri Route Navigation Scenario.

| MOME | EVISUGE /Route Navigation |
|---|---|
| Event Shot | Movement to the Sellia village |
| Participating Spatial Object Shot | Road |
| Logical Spatial Shot Representation | Representation of the car on the road |
| GPS Point | GPS points for the representation of the car and the route |
| GPS Representation | Set of GPS points for the representation of the car and the route |
| Physical Representation | Representation of the movement to the Sellia village using a line and the car representation |
| Shot Site Map | Google Earth Map |
| Geographic Map | Google Earth Map |
| Digital Elevation Model | Google Earth Map |
| Shape | The route line and the car representation shape |
| Point | GPS representation points for the Google Earth shape representation |
| Movement | Utilization of GPS points on the route representation, from the route beginning to its end |
| Color | Line color |

stance, which represents the church of the village and the "House" instance, which represents the village houses.

- The "Roads" instance, which represents the village roads.

- The "Square" instance, which represents the square of the village.

The representation discussed above can be specified by a scenario creator using the EVISUGE scenario management subsystem. Once the scenario is specified and stored in the EVISUGE database, it can be visualized by a scenario observer, who may also navigate the route Nyvritos-Gergeri, using the EVISUGE scenario playback subsystem.

The visualization of the Nyvritos-Gergeri route navigation scenario in the EVISUGE system is presented in Figure 7.7. As is shown in Figure 7.7, some textual information about the route is placed on the top of the web page presented to the user, while the route is visualized in Google Earth. The route navigation scenario events are visualized according to their temporal order

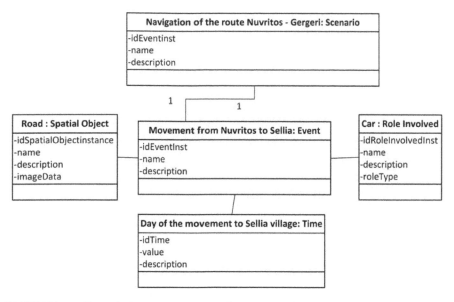

FIGURE 7.5: Detailed object diagram for a part of the Nyvritos–Gergeri route navigation scenario.

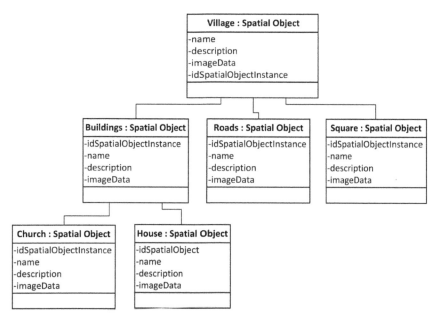

FIGURE 7.6: Object diagram for the detailed description of the structure of a village.

FIGURE 7.7: Visualization of the Nyvritos–Gergeri route navigation scenario.

and are represented by polygons that allow interacting with them in order to view relevant information as well as the event shots. The event shots provide multimedia information about the events that is available in specific points of the route. The current point in the route navigation is denoted by the position of the 3D model of a car that follows the route. The observer may browse the scenario using the scenario playback bar and the buttons associated with it, which are placed at the bottom of the Google Earth window.

### 7.4.3.2   Battles

As a case study of the battle scenarios, we have worked out the scenario of the battle of Marathon. This scenario is associated with the Marathon valley and the time of the battle and has as participants the Greek army and the Persian army. The initial positioning during the playback of the battle of Marathon is presented in Figure 7.8

The two participant armies are essentially crowds, which are represented on the map using polygons that change shape according to the movements that makes every army (or a part of it) in specific time points (see Figure 7.9). Notice that, since several parts of the armies may move simultaneously, the parallel event visualization support offered by EVISUGE is required for

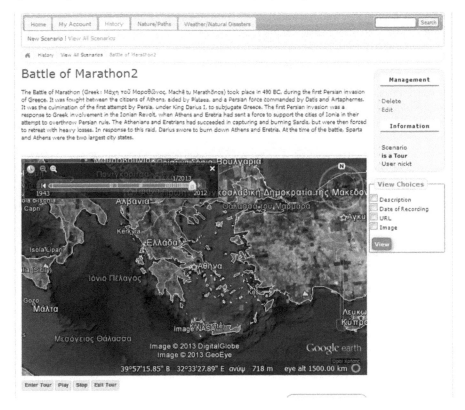

FIGURE 7.8: Initial positioning of the Battle of Marathon.

this scenario type. The event shots include photos of the event locations as well as photos of cultural heritage objects (e.g., amphora drawings, wall paintings, etc.) that depict the events.

### 7.4.3.3 Weather Conditions

As a case study of the weather condition scenarios, we have worked on the spatial representation of the temperature in an area in specific time points. In the current version of the EVISUGE system we do not support "live" events (like, for example, the real-time visualization of the temperature changes) because Google Earth does not yet provide support for the real-time update of objects visualized on top of Google Earth maps. As a consequence, we cannot "feed" the scenario visualization with real-time data and have an up-to-date map of the temperatures of the area that is presented.

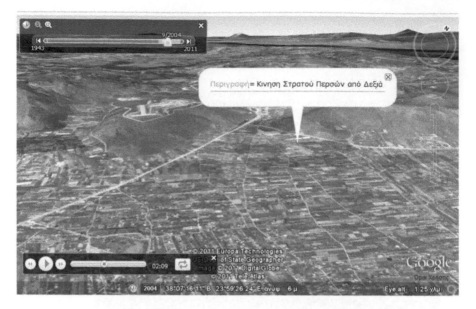

FIGURE 7.9: Visualization of the scenario of the Battle of Marathon.

## 7.5    Conclusions

The use of interactive maps on the Web has become very popular nowadays, especially after the development of robust interactive web map infrastructures like the OpenLayers, the Google Earth and Google Maps.

Since many important practical applications in several domains are associated with events or they can be described by events, we focus in this chapter on event geovisualization.

In particular, we have reviewed the relevant research and we have presented our recent research in the EVISUGE system for the visualization of events on top of the Google Earth 3D interactive maps, with respect to their spatial and temporal features.

# 8

# Enrich the Expressiveness of Multimedia Document Adaptation Processes

**Adel Alti**

*Ferhat ABBAS University of Setif,* `altiadel2002@yahoo.fr`

**Sébastien Laborie**

*University of Pau - LIUPPA - T2i,* `sebastien.laborie@iutbayonne.univ-pau.fr`

**Philippe Roose**

*University of Pau - LIUPPA - T2i,* `Philippe.Roose@iutbayonne.univ-pau.fr`

## CONTENTS

ABSTRACT One of the key aspects of any mobile multimedia application is
the management of multimedia documents. Currently, these documents are accessible on a wide variety of devices, such as laptops, tablets and smartphones.
The (hardware and software) heterogeneity of such devices and the diversity
of user preferences require adaptation of multimedia contents. Since the last
decade, a fair amount of research has been conducted in order to develop
adaptation frameworks. These frameworks generally transform documents in
order to comply with all target device constraints. For that purpose, they
generally exploit device and user profiles in order to detect conflicts between
the target execution context and the document characteristics that are going
to be played. Furthermore, when conflicts have been detected, they usually
search and compose several relevant adaptation components that will perform
the document transformations in order to solve these conflicts. However, we
have noticed that current adaptation frameworks do not fully exploit the semantic benefits for describing the profiles, the adaptation components and
the quality of services. In this chapter, we propose a model for specifying rich
profiles containing explicit constraints expressions with qualitative and quantitative information. Moreover, we propose a generic ontology that handles
semantic rules allowing the automatic generation of a dynamic and a quality
composition of heterogeneous adaptation components. Our proposal has the
great advantage offering to users a global flexible adaptation infrastructure

exploiting semantic information at multiple levels, i.e., from profiles to the computation of adapted multimedia documents.

## 8.1 Introduction

Nowadays, mobile technology has been widely accepted by users and is still evolving very fast. Actually, mobile phones are no longer simple text or voice communication devices, PDAs are no longer planning and organization gadgets, and many handheld devices are no longer isolated from the Internet. The development and the fusion of multiple functionalities in Smartphones yield to information systems called "pervasive," i.e., to make information accessible at anytime, anyhow and anywhere. Typical information that users want to play on their devices is multimedia documents that are composed of several contents, such as videos, audios, images and texts. However, devices have heterogeneous capabilities and characteristics in terms of hardware (e.g., screen size, battery level, memory available) and software (e.g., players, codecs). Moreover, user's preferences or handicaps may prevent playing specific multimedia contents. For instance, a user may avoid reading texts written in French and/or avoid playing audio contents while he is participating at a meeting. Hence, a user context may imply some constraints that have to be specified in a profile.

In a profile, various categories of information have to be managed: (1) device characteristics (hardware and software), (2) context information related to the user's environment and/or interactions between the user and the device, such as the preferred languages, the bandwidth, the user location or the background noise, and (3) document structure like the types of contents that could be played or the preferred presentation organization (e.g., layout, duration of document, multimedia contents synchronization). Consequently, if a multimedia document does not comply with some constraints that are specified inside a target profile, the document might not be correctly executed on the target device. Thus, in order to display multimedia documents on any devices, these have to be adapted, i.e., transformed in order to comply with the target profiles.

Since the last decade, a fair amount of research has been conducted on multimedia document adaptation, e.g., [47, 549, 571, 82, 464, 50]. Considering some target profiles, these approaches combined multiple operators: transcoding (e.g., AVI to MPEG), transmoding (e.g., text to speech) and transformation (e.g., text summarization). Of course, each profile expressiveness is exploited by these approaches in order to determine a combination of these operators, such as in [570], or to optimize their deployments (e.g., for saving battery energy), such as in [555] and [367].

However, each proposal exploits a specific profile format, which usually contains a list of multiple descriptive information values, such as the screen size, the user languages and the battery power. Consequently, an adaptation process has to interpret such profile values and to deduce implicitly some constraints. For instance, if a battery power is lower than 10%, one adaptation process may avoid playing any videos, while another one may summarize the videos. Obviously, each adaptation mechanism may deduce different constraints that in many situations might be wrong, thus providing to users incorrect adapted documents. Furthermore, current context modeling languages, e.g., [133, 339], do not consider expressing rich explicit constraints (such as if my battery power is low do not play hi-quality videos), while they might be very useful to guide the adaptation process.

Besides, current adaptation frameworks are using simple descriptions of adaptation operators (i.e., transcoding, transmoding and transformation operators). Generally, they are analyzing the inputs and the outputs of these operators in order to determine some dependencies between them. It is important to point out that the input/output mapping between operators are mainly syntactically-based [280]. Then, from all dependencies, they are trying to determine a combination of these operators (i.e., a chain of adaptation transformations) in order to adapt the original document. To select the "best" combination of adaptation operators, several approaches have defined different quality of service (QoS) properties. For instance, some of them minimize the computation cost for producing an adapted document, while others may maximize the proximity between the original and the adapted content. Usually, the quality of service of an adaptation framework is not customizable, in other words, the quality of the adaptation process is usually predefined on a fixed set of properties.

In this chapter, we propose a global flexible adaptation infrastructure which exploits semantic information at multiple levels, i.e., from profiles to the computation of adapted documents:

1.  We define a generic model for specifying rich semantic profiles containing explicit constraint expressions. In such a profile, information is organized into facets (e.g., device characteristics, context information and document structure). Facets can be composed of services that either provide data or require modifications, and explicit constraints can be defined by specifying actions under rich sets of conditions.

2.  We develop a generic ontology that handles semantic rules allowing the automatic generation of a dynamic and a quality composition of heterogeneous adaptation components. This ontology provides a semantic description of adaptation components (e.g., name, location, category, role), allows a dynamic selection of components based on context metadata parameters (user, environment, device, service provider context) and describes semantic relationships between adaptation components, such as the dependency, the substi-

tution (i.e., a service has the same role as another one) and the equivalence.

3. We have specified an adaptation architecture that exploits our profile structures and our ontology in order to compute adapted multimedia documents. This architecture uses a semantic inference engine that is responsible for planning and optimizing the selection, the deployment and the execution of adaptation processes based on a customizable quality of service.

The rest of this chapter is organized as follows: Section 8.2 presents an overview of current multimedia document adaptation approaches. In this section, we highlight the lack of expressiveness of current profile modeling approaches and current adaptation frameworks. Section 8.3 details how semantic information can be added for describing the profiles, the adaptation components and the quality of service. Section 8.4 shows how a global adaptation infrastructure may exploit the semantic descriptions for adapting multimedia contents. Section 8.5 illustrates some experimental results on some use-cases. Finally, we conclude and give some future perspectives in Section 8.6.

## 8.2 Overview of Multimedia Document Adaptation Approaches

In this section, we give an overview of current multimedia document adaptation approaches focusing on two aspects: (1) profile structures exploited by adaptation approaches (Section 8.2.1) and (2) mechanisms that select adaptation components, i.e., transcoding, transmoding and transformation operators (Section 8.2.2). In Section 8.2.3, we summarize the main limitations of these works and promote a global flexible adaptation framework which exploits semantic information at multiple levels, i.e., from the profiles to the computation of adapted multimedia documents.

### 8.2.1 Profiles Exploited by Multimedia Document Adaptation Approaches

Since the last decade, a lot of research has been proposed in order to model devices' characteristics and users' contexts that are further exploited by multimedia document adaptation processes. We have noticed that some of these approaches provide exclusively a descriptive view of context information (e.g., §8.2.1.1, §8.2.1.2), while others propose enhancements with some constraints expressions (e.g., §8.2.1.3, §8.2.1.4). In the following, we present an overview of these approaches.

### 8.2.1.1 Composite Capability/Preference Profiles (CC/PP)

CC/PP (Composite Capability/Preference Profiles) [525] is a World Wide Web Consortium (W3C) recommendation for specifying device capabilities and user preferences. This profile language is based on RDF (Resource Description Framework) [112] and was maintained by the W3C Ubiquitous Web Applications Working Group (UWAWG). The profile structure is very descriptive since it lists sets of values which correspond to the screen size, the browser version, the memory capacity, etc. However, the CC/PP structure lacks functionality, for instance it limits complex structure description by forcing a strict hierarchy with two levels. Furthermore, it does not consider the description of relationships and constraints between some context information.

### 8.2.1.2 User Agent Profile (UAProf)

UAProf (User Agent Profile) [341] is based on RDF and is a specialization of CC/PP for mobile phones. More precisely, its vocabulary elements use the same basic format as the one used in CC/PP for describing capabilities and preferences for wireless devices. Thus, it describes specific items, such as the screen size, the supported media formats, the input and output capabilities, etc. UAProf is a standard adopted by a wide variety of mobile phones and it provides detailed lists of information about the terminal characteristics. However, this standard is limited to the description of wireless telephony equipment. Hence, it does not allow a user to express his/her requirements, such as avoiding playing videos while the mobile phone battery level is lower than 15%.

### 8.2.1.3 Comprehensive Structured Context Profiles (CSCP)

CSCP (Comprehensive Structured Context Profiles) [171] uses RDF and is also based on CC/PP. In contrast to CC/PP, CSCP has a multilevel structure and models alternative values according to predefined situations. Even if CSCP provides a description of the context, which is not limited to two hierarchical levels, this proposal does not allow the specification of complex user constraints mixing qualitative and quantitative information (e.g., avoiding playing hi-quality videos while the mobile phone battery level is low). Moreover, [445] stated that this model proposal was developed as a proprietary model for specific domains.

### 8.2.1.4 Context-ADDICT

Context-Aware Data Integration Customization and Tailoring proposes the Context Dimension Tree [134]. In other words, the context can be represented with a hierarchical structure composed of a single root and some level nodes. The authors propose constraints and relationships among values. In Context-ADDICT, the data sources are generally dynamic, transient and heterogeneous in both their data models (e.g., relational, XML, RDF) and schemas.

The Context-ADDICT approach lacks some relevant features for the data-tailoring problem, such as context history, context quality monitoring, context reasoning, and ambiguity and incompleteness management. Moreover, this model strongly depends on the application used and does not permit portability between platforms.

### 8.2.1.5 Wireless Universal Resource File (WURFL)

WURFL (Wireless Universal Resource File)[1] is an XML description of mobile device resources. WURFL contains information about the capabilities and the functionalities of mobile devices with more than 500 "capabilities" for each device (divided into 30 groups). This project is intended to adapt web pages on mobile devices. Unfortunately, users cannot specify rich explicit constraints, such as decreased screen luminosity if the battery power level is below 10%.

### 8.2.1.6 Generic Profiles for Information Access Personalization

Chevalier et al. [224] propose a generic UML profile for describing the structure and semantics of any type of user profile information. This contribution is used to describe the semantic links between elements and incorporate weighting on elements. The semantic graph is described thanks to a logic-oriented approach with RDF, RDFS and OWL. However, this model does not express explicitly adaptation actions that an adaptation process may exploit to transform the multimedia document, such as increase audio volume and avoid playing hi-quality videos.

## 8.2.2 A Brief Survey of Adaptation Mechanisms

In the last decade, several works have been done for multimedia document adaptation and the composition of adaptation services. Some approaches are presented in this section.

### 8.2.2.1 Negotiation Adaptation Core (NAC)

In [570], we can find all techniques that consider structural and syntactically based adaptation, such as adaptation based on XML documents, like SMIL. Those adaptations depend largely on the particularities of the multimedia document model. Several tools are used to perform these transformations, such as XSLT. This work lacks the richness of user profiles and the semantic descriptions of services is also severely limited.

### 8.2.2.2 Knowledge-Based Multimedia Adaptation

Jannach et al. [464] uses the OWL-S language for the automatic detection of suitable services and their compositions into service chains. The proposed ap-

---

[1]http://wurfl.sourceforge.net

proach takes into account only semantic information necessary for web service composition, and does not consider QoS parameters and their semantics for multimedia services composition and selection. Moreover, this work lacks some relevant features for solving the explosion of service-space planning problems. In contrast, our approach reduces the search space by suggesting mechanisms to provide a semantic and run-time composition of heterogeneous services based on context metadata parameters (user, environment and device) and to include both quality of service and qualitative context features.

### 8.2.2.3 Ontologies and Web Services for Content Adaptation

Forte et al. [340] introduce an ontology framework and web services for automatic and generic multimedia content adaptation. This framework described an adaptation policy specified through ontologies in OWL-S from a set of data expressed via profiles upon a set of adaptation rules. It offers a flexible solution for the content adaptation domain but does not succeed in the suggesting mechanisms to provide a semantic and dynamic composition of heterogeneous services based on context metadata parameters (user, environment, and device) and the consideration of richness and subjectivity of semantics in user profiles and does not consider how to enrich services with explicit context qualitative semantics. This work did not address any issue related to adaptation plan quality and optimization.

### 8.2.2.4 MultiMedia Software Architecture (MMSA)

Derdour et al. [280] proposed the UML profile for application-based multimedia components. This profile is based on multimedia software architectures. It proposes a generic solution to solve the problem of incompatibility of services, according to media types and media formats. This work treats syntactic interoperability between heterogeneous services and does not take into account the semantics of services and the location, category, QoS, and explicit user constraints. It therefore lacks supporting automatic discovery and retrieval of services to address the necessary adaptation.

### 8.2.2.5 Semantic Adaptation of Multimedia Documents

Laborie et al. [549] addressed the organization of various multimedia objects of documents that executed in multi-device platforms within temporal and spatial constraints, but it does not take into consideration multimedia document types and formats nor semantic information related services (category, role, location) and semantic service qualities (adaptation time, transfer time, etc.). This work did not address any issue related to service composition in limited devices.

### 8.2.2.6 DynamicCoS

Silva et al. [250] developed a framework for dynamic service composition; the framework provides an automatic composition process, targets the use of ontologies and semantic models to runtime service delivery through automated service composition. This work provides mechanisms to allow end-users to play a central role in the creation process of their services. This work does not consider the use of non-functional properties (context and QoS) and explicit user context and constraints in the composition and composition selection process.

### 8.2.2.7 Automatic Discovery and Composition of Multimedia Adaptation Services

Moiss [682] proposed an ontology framework for automatic discovery and composition of multimedia adaptation services. It introduced a new service categorization ontology for describing of top-level categories that he used as a basis for the definition of services' classes. The proposed approach takes into account only semantic information and adaptation directives necessary for service composition, and does not consider QoS parameters and their semantics for multimedia services composition and selection. It does not profit from the potential of semantic representation techniques to express high level explicit constraints, while they may be useful to guide the selection and adaptation process.

## 8.2.3 Importance of Expressiveness in Multimedia Adaptation Processes

As we have shown in Section 8.2.1, many existing profiles list quantitative values related to device characteristics and user preferences. This type of profile has many drawbacks:

- None of them are portable, i.e., it is not possible to migrate a profile between different platforms without modifying multiple quantitative profile values. For instance, if a profile value corresponds to a device screen size, this information has to be modified from platform to platform. Consequently, updating such profiles is time consuming in order to maintain correct information.

- None of them express rich constraints. For instance, a user cannot specify the following constraint in his profile: "If I'm participating at a meeting, do not execute audio contents." In other words, current profiles do not support some specific desired actions under several conditions.

As shown in Section 8.2.2, many existing adaptation frameworks do not describe adaptation operators with semantic information. For instance, they did not take into account the location, the category, the QoS, the provider and the execution constraints related to an adaptation component. Consequently, existing frameworks lack supporting the automatic discovery and the dynamic composition of relevant adaptation components needed to address the necessary adaptation.

In order to better guide the adaptation process and to better discover and compose adaptation components, we propose to enhance the expressiveness of profiles and the semantic descriptions of adaptation components. In the next section, we will detail these enhancements.

---

## 8.3  Enhancing Expressiveness in Profiles and Adaptation Processes

In this section, we present the Semantic Generic Profile (SGP), which allows rich constraint specifications and permits profile portability and genericity (Section 8.3.1). Then, we present ASQ (Adaptation Service Quality) ontology for describing the adaptation components, the adaptation process and the quality of service useful for selecting relevant adaptation transformations (Section 8.3.2, Section 8.3.3, and Section 8.3.4).

### 8.3.1  Modeling Context with Rich Explicit Constraints

A profile should provide information about some device capabilities, the user context, and the document characteristics that the target device is able to support. Currently, profile descriptions mainly contain quantitative and/or expressive values, such as the device screen size, the user preferred languages, the device model, etc. However, if this type of profile migrates on different platforms, many profile characteristics have to be reconfigured. An easy evolving characteristic is for instance a screen resolution or a set of available codecs. Therefore, to ensure profile portability, our profile structure is based on service references.

In Figure 8.1, we have illustrated an example of the proposed profile. Each profile refers to a set of services, like the user location service, the codec service, the battery power service... Of course, these services can be grouped into sets of services in order to better retrieve them. For instance, our profiles could handle services about the user, the weather, the audio or video contents, and the hardware and software capabilities. Several hierarchies of services providing information may be constructed and consequently sets of services can be categorized into profile facets. In the figure above, we have illustrated three profile facets: the context facet, the document facet and the device facet.

We propose to express in such a profile rich constraints, like the following one "if the user location is located inside an area defined by some GPS coordinates and if the battery power level is less than 10%, turn off the sound of audio contents." As you can see in Figure 8.1, this profile model is generic; it may handle several kind of explicit constraints with complex sets of disjunctive

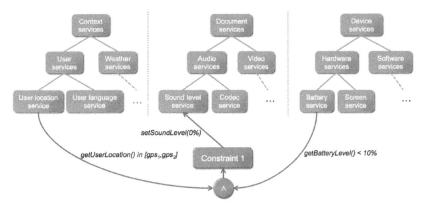

FIGURE 8.1: A profile containing rich explicit constraints.

and conjunctive conditions. Consequently, these constraints will better guide the adaptation processes in order to provide relevant adapted documents. Another advantage of this profile model is that we do not need to update the profile at any time; only specific services may be called for providing the only needed information. Finally, since the above profile refers to services, we can migrate the profile from platform to platform by preserving the constraints.

We have encoded such profiles using RDF, which is a standard model for data interchange on the web. Several reasons motivate us to use this formalism: First, RDF allows us to perform aggregations of descriptions, which can be useful if several applications describe a profile. Second, RDF can handle semantic concepts described in ontologies, thus enhancing the SGP semantics. For instance, semantics allow us to state that "Sensor" service is equivalent to "Captor" service. For example, a semantic query on "Captor" will also refer to "Sensor." Third, RDF does not force us to express hierarchies of data as defined in other languages (it has a graph-based structure thanks to triples), it facilitates us to describe constraints between different profile information. Finally, other languages and proposals whose objectives are to describe profiles are based on this formalism (e.g., CC/PP, UAProf and CSCP).

As you may notice in the previous example, the explicit constraints expressed in SGP profiles currently refer to pure quantitative values, such as 10% and some GPS coordinates. This is also the case for several standard profiles, such as CC/PP and UAProf. In fact, one may want to specify inside a profile that if the battery level is "Low" and if the user location is "Close" to his office, then do not play audio contents. In order to enhance our profile expressiveness, we have defined in the following a framework that bridges the gap between quantitative and qualitative information specified in profiles.

Adding qualitative terms inside profiles is not straightforward. Indeed, a qualitative term, like "Low," may be applied on several profile aspects, such as the device battery level, the sound output level, or the bandwidth. Hence, the meaning of a term can be completely different depending on the context used. Moreover, even in a specific context, for several domains, a term may have different interpretations. For instance, the term "Low" in the context of the battery level corresponds to different quantitative values if we consider a smartphone or a laptop.

In order to bridge the gap between qualitative and quantitative information contained in profiles, we have defined the following multilevel approach (Figure 8.2):

- Level 1: This is the higher semantic level and it refers to the qualitative term that one may want to use in a profile. In the figure, we want to define the qualitative term "Low."

- Level 2: This level corresponds to the potential services which are associated to the qualitative terms. In the figure, the term "Low" is attached to the battery level service.

- Level 3: For a particular qualitative term and for a specific service, multiple application domains may be specified in this level. In the figure, a smartphone and a tablet may have different interpretations of a low battery level.

- Level $i$ ($i < n$): Subdomains may be defined in order to associate a qualitative term to more specific application domains. Moreover, depending on subdomains, the qualitative terms may refer to different quantitative values. For instance, one may define different low level batteries for particular device types or brands.

- Level $n$: This is the lower level and it refers to the corresponding quantitative values which are associated to the qualitative term for a specific context and a particular application domain. For instance, in the example, we have defined that a low battery level on a tablet corresponds to a value which is less than 15%.

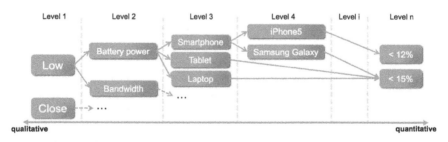

FIGURE 8.2: Bridging the gap between quantitative and qualitative values in profiles.

This multilevel approach for bridging the gap between qualitative and quantitative information described in profiles is generic. Actually, this proposal may be used for a wide variety of qualitative terms. For instance, one may define qualitative terms for spatial information, such as "close," "far," "intersect," etc. These qualitative terms may be attached to particular services, like location or surrounding devices. It may be associated with particular domains, e.g., GPS, street names, hotspot identifiers, etc. Furthermore, one may also define qualitative terms for temporal information, such as "before," "after," "during," etc.

We have proposed to encode the information contained in Figure 8.2 in an RDF/XML description. Consequently, thanks to URIs (Unique Resource Identifiers), it is now possible to exploit qualitative terms inside profiles presented in Figure 8.1. More precisely, it is possible to use these qualitative terms in order to express rich semantic constraints inside profiles, such as if my battery level is "Low" do not play videos.

## 8.3.2   Modeling the Multimedia Adaptation Services

Adaptation services usually require specific types of media and they also generally transform the multimedia content characteristics. In this context, it may lead to several input and output incompatibilities, such as media resolution problem, data types and formats, etc. Moreover, an adaptation mechanism should be able to retrieve a service through its semantic description, independent of the service technology used, to identify which service context is evolving, and who the service provider is. The resource properties and constraints' nature makes the service qualities and sharing even more difficult. The ontology presented herein aims at providing a common model to the heterogeneous technologies usually available in pervasive environments, to share their execution devices properties and constraints, to provide semantic service determination facilities and to provide a better solution for quality assembly of heterogeneous multimedia adaptation services fitting as much as possible the device constraints and user preferences.

Unfortunately, there are currently few tools dedicated to the description of adaptation services. Some services description languages like WSDL (Web Service Description Language) can be used if semantically enriched. Indeed if the syntax is important for the service invocation and composition, semantic information of service parameters and its functionalities (semantic of input/output parameters, cost of service, context constraint, execution time, quality of service, etc.) is essential for its practical uses. One of our contributions is to improve the semantic service description to ease service selection and to automatically generate more accurately an adaptation process.

Several techniques for enhancing the semantic description of services have been proposed: OWL-S [274], WSMO[2], DAML-S [165], ConteXtML [552], CWSC4EC [154]. Those languages provide the means for describing services at a high level of abstraction, techniques for verifying composition of two services and the semantic assembly of services. However, neither automatic

---

[2]http://www.wsmo.org

adaptation of heterogeneous services are discussed from the technical point of view (type and format of multimedia data) and the semantic aspects (service location, service role and service category). Our approach addresses the limitations of the existing approaches by suggesting mechanisms to provide a semantic and dynamic composition of heterogeneous services based on context metadata parameters (user, environment, device) and the management of the heterogeneity of service needs, of mobile devices capacities as well as the media variety (sound, video, text and image).

We identify five reasons to investigate the importance of ontology for the adaptation of multimedia documents:

- Homogenization: describing different contextual information and different qualities of service using the same ontology confers to the ontology the role of homogenization for at least the service quality and the context information that it describes [934].

- Interoperability: ontologies provide a good way to treat semantic interoperability between the heterogeneous representations of web services and thus QoS parameters and context parameters of services by sharing common representation and semantics [681].

- Expressiveness: most ontology languages, such as RDF and OWL-S, with high expressiveness are efficient and perform much better for expressing semantics of various portable device properties (Tablet, smart phone, iPod, etc.) and various QoS parameters.

- Easy heterogeneity problems detection: ontologies make the heterogeneity problems of web business quality services easier, faster and automatic. The main mechanism for service heterogeneity problems is service registry, and semantics can be used for an automatic generation of quality assembly of services.

- Optimization: In practice, the multimedia document adaptation algorithms generate a very large number of services at each step of the adaptation process; we must check a large amount services. To reduce the checking time, we define the semantic adaptation rule (e.g., Transmoding $\rightarrow$ Transcoding) from the relation between a service and other ones having the same adaptation role. We use these semantic relations to easily find the next adaptation action. We can also restrict the scope of the search into the range of finite sets of relevant adaptation services that correspond semantically to adaptation directives resulting from explicit user constraints. Adaptation services with higher potential benefit are selected earlier and find a limited number of composition solutions to reduce the computation complexity.

Figure 8.3 presents the relevant classes of our ASQ (Adaptation Semantic Quality) ontology. It allows the service providers to publish metadata about their services within their context constraints in the UDDI-like registry.

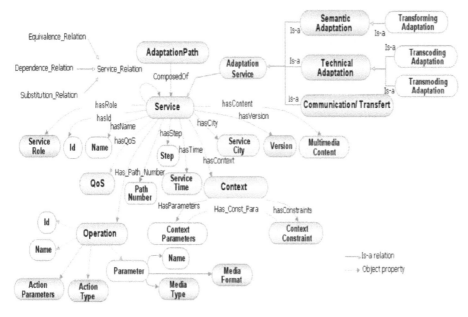

FIGURE 8.3: Service ontology.

Providers would instantiate classes from the ontology and publish the resulting individual results as OWL files on their websites. Our ontology defines common concepts as well as relationships among those concepts to represent multimedia data, service parameters (compression ratio, resolution, color number, etc.), semantic service information (service_role, media_type, action_type, version, resources needs, QoS parameters, semantics of input/output, location and time) and context constraints.

One of our contributions is to improve the semantic service description to give more precision on a resource adaptation and to facilitate the search of services. An adaptation service is represented by a set of information describing the necessary contextual constraints. It also describes the parameters for customizing the quality of service. The semantic information service enables comparison with other services of the same type. Semantic information on the input/output service represents a criterion for service selection and allows its assembly with other services for generating an adaptation process. We distinguish the following concepts:

- Service has a name, a version and unique ID (e.g., URI).

- Service_Role = {Transmoding, Transcoding, resize, compression, etc.}.

- Service_Location: the location where the service facilities are located.

- Service_Time: time when the service is executed.

- multimedia_content = {text, images, sound, video}.

- Step: Start, Current, Passed, Finish.

- Service_Semantic_Relation: A semantic relation is an ordered relationship between two adaptation services. The semantic relations are used to guide the automatic assembly in heterogeneous applications. When the configuration of the service is not possible, the adaptation can be done by replacing one or more components composing the service or switching to another available service (or set of services playing this role) offering the same functionalities within the same location as the mobile user with the best quality of service. A semantic relation can be one of the following types:

  - Service dependency relation: is a semantic relation between two adaptation services which means that the successor is linked to its predecessor.

  - Service equivalence relation: is a semantic relation between two services that play the same role within the same location and the same category.

  - Service substitution relation: is a semantic relation permitting for example dynamic changes affecting QoS properties while a service is running. The choice of substituted service is based on the semantics and parametric QoS properties (e.g., customize the QoS parameters in order to fulfill substitution needs).

- QoS_parameters: the list of the QoS parameters. QoS parameters concern not only the services but also the device resources where the execution is taking place.

- Adaptation Service: allows the transformation of a multimedia object into another multimedia object satisfying a given profile. Two types of adaptations can be distinguished :

  - Technical adaptation: This adaptation is related to the capacity of mobile devices (memory, display, etc.). It has two parts:
    * Transcoding: conversion of format, e.g., JPEG to PNG.
    * Transmoding: conversion of types, e.g., text to sound.

  - The semantic adaptation: This adaptation is related to the constraints of the data types handled by mobile devices. It allows the content change without changing the media type and format, e.g., text summarization, language translation, etc.

  - Communication service: allows the transformation of communication protocol/gateway, e.g., WIFI to GSM.

- Context_constraint: consists of simple or composite context expression. For example, a context constraint can be bandwidth=10000 or memory_size= 'less' and User_Location ='Near House'.

- Context-parameters: each context category has specific context parameters. For example, the device related context is a collection of parameters (memory size, CPU power, bandwidth, battery life cycle, etc.). Some context parameters may use semantically related terms, e.g., CPU power, CPU speed.

- Context-expression: denotes an expression which consists of context parameter, logic operator and logic value.

### 8.3.3   Modeling the Semantic Adaptation Process

The semantic adaptation process is composed of semantic adaptation services. We can model the semantic adaptation process as a (possibly infinite) set of finite adaptation paths, where each path defines a sequence of adaptation services in which the first element in the path is the service of the current context, and the final element is a desired target service. Links between successive nodes in a path are associated with transitions that are selected from a set of multimedia adaptation services (transcoding, transmoding, transforming).

A suitable evaluation of adaptation quality can be obtained by customizing and analyzing two main factors (benefit/cost) making a better selection of adaptation paths. The primary use of such evaluation is allowing us to explore the consequences of different decisions about the path. For example, the user may select a few adaptation services to use for reducing the time to reach the target adaptation. Alternatively, if time is not a constraint and output quality is an important factor under constrained resource, the user may decide to stretch the adaptation out over a larger number of adaptation services.

As you may see in the figure below, the semantic adaptation process is thus defined by adaptation services and adaptation path subclasses (Figure 8.4):

- Adaptation services: is a dynamic set of adaptation services matching user preferences and its context usage. Classification is useful for reasoning a quality and intelligent decision making to optimize adaptive content delivery over the across users access. The purpose of classifying adaptation service is to improve the efficiency of adaptation delivery by either prioritizing them according to quality output level, media type, adaptation role (compression, resize, audio removal, text summary, format conversion, keyframe extraction, etc.), media transcoding, transfer time level, adaptation time level, data transcoding, media prioritization, media domain. If the semantic adaptation process has been divided correctly to suitable adaptation components, the semantic adaptation process can be analyzed more easily. At the semantic level, this factor can be measured with criteria, named cost and benefit.

- Adaptation paths: describe various adaptation service compositions, while the output service matches another service input (e.g., semantic output parameters of a given service share the semantic input parameters of another service). The first service (Step = Started) of the path must be able to process the original media. The path must contain an adaptation service that solves the conflict. The last service (Step = finished) of the path does not have to introduce new conflicts. The quality function is used for selecting the best adaptation path that has a higher ratio Benefit/Cost. In this case the values of a quality formula are used for classifying the relevant adaptation paths that have potential benefit. The evaluation of the adaptation paths having the same mark of benefice that maximize users' qualities (expressed as preferences). That will only modify the mark of the adaptive criterion (response time, adaptation effort, etc.). So, reasoning is specified by analyzing finite sets of adaptation paths having the same mark of benefice metric and differs only by their adaptability cost to the context.

$$Quality(User, Env, Event) = \frac{W_{Benefits} * Benefits}{W_{Cost} * Cost}$$

Where $W_{Benefits}$ and $W_{Cost}$ are weights associated respectively to the mark of benefit and cost. These weights cannot be set directly by the user. A user may specify in advance that the execution time is not a constraint and that he prefers a hi-quality video, and these weights are then computed automatically during the profile creation. All these weights are set between [0, 1]. The higher the weights, the more important the values are. The quality function is calculated for each adaptation service in the adaptation path and for each adaptation path. The comparison of adaptation path scores allows us to select the best path. The quality function is customizable and parametrizable according to different user needs and context parameters. Cost is parametrizable according to context parameters (like CPU load, energy saving, low bandwidth, etc.) and the benefit is parametrizable by specific media parameters (like compression ratio, frame rate, resolution, etc.).

### 8.3.4  Modeling the Quality of Service of the Adaptation Process

Service capability describes the operational features of an adaptation service. The service confidence level is very interesting when having two services providing the same functionalities. They may have varying values for their QoS. The service confidence level can contain several QoS according to the user objectives (Figure 8.5).

- Cost: we distinguish two criteria (1) average time needed to deliver service adaptation and (2) average time to transfer the adaptation document.

- Benefit: degree of output quality compared to input quality.

There is two types of QoS management: static and dynamic [279]. The QoS management is achieved by a process of selecting between several services,

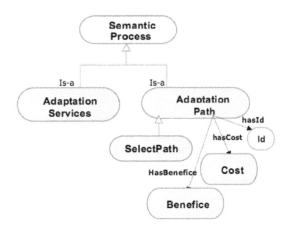

FIGURE 8.4: Adaptation process ontology.

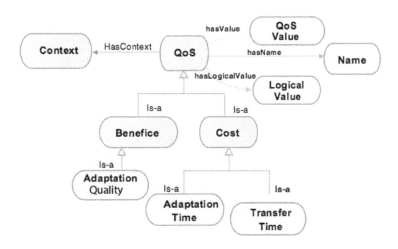

FIGURE 8.5: Adaptation quality ontology.

while the dynamic QoS management is provided by the dynamic adaptation process control that handles the adaptation as parameters change (location, time, battery level, memory size, compression ratio, image resolution, image size, etc.) to provide adequate quality that satisfy context execution. The quality model can contain several performance measures matching with service objectives. Different media and their features, as well as their variety of uses suggest that there is a need to define a generic ontology of services with common quality and common media types annotated with concepts that are semantically close to each other. The evaluation of QoS is defined through an evaluation of the adaptation service (Cost) and the evaluation of output quality (Benefit). The first one depends on the service features, such as execution time, available resources, portability, reliability, flexibility, etc. The second one depends on the media features, such as image resolution, image size, video speed and video size, etc. The QoS depends on user preferences and its context usage. The quality of outputs is related to the type of media (text, image, audio and video). For images adaptation, for example, the parameters that we measure in this case are:

- TA (Adaptation Time): average time to run service.

- TT (Transfer Time): average time to transfer the adaptation image.

- QA (Adaptation Quality): degree of output image quality compared to input image quality.

The adaptation plan can be selected according to some end-users' profiles (e.g., quality preferences). The end-user can define a quality oriented constraint for each quality dimension, hence there are constraints regarding both the static and the dynamic quality of composition. In our work, we focus on the use of an ontology to facilitate the semantic selection of adaptation service that best meets users' needs and a quality evaluation of adaptation paths. Our ontology allows the automatic computation of static and dynamic QoS on the basis of the adaptation model and the quality model.

The consideration of customization makes the evaluation more suitable from different users' viewpoints and becomes appropriate for evaluation of particular media type (e.g., video with high quality). The quality evaluation of adaptation paths is customized by assigning weights to the quality criteria by experts. The user is able to customize the quality parameters in order to fulfill his/her needs. For example, one user may want a high quality image, while another one may want an energy saving adaptation with an average quality of media.

To select an adaptation service (or adaptation path), we must take into account the quality of the process (execution time, energy, CPU speed, etc.) and quality of outputs (resolution, size, compression ratio, etc.). The elaboration of the quality criterion done for each service and for each output quality, is a given parameter in the context of associated need, for a given criteria with associated weight. Currently, values of a weight are used for ordering adaptation paths and selecting the best adaptation path. The adaptation path is evaluated according to the greatest values of the quality criterion. If more than one adaptation path have the same values, one adaptation path is randomly selected for the adaptation of the multimedia document.

## 8.4 The Dynamically Semantic-Based Assembling of Adaptation Services

The proposed approach so far is a generic, dynamic and expressive system regarding the management of adaptation process qualities. The framework facilitates automatic and semantic dynamically assembling of multimedia adaptation services. It offers two operations: the first one is the semantic user's constraint analysis and translation. This takes as input the user profile, and then translates qualitative constraints to quantitative ones. The second one is to allow a semantic service-based adaptation of multimedia documents based on the ontology as illustrated in Figure 8.6, which makes adaptation easier with the constraints of the context exploitation. There are some issues that should be mentioned before designing the architecture:

1. The approach must have an engine to annotate the semantics of the deployed services. Without this feature, our approach will not be useful for large scale and real world multimedia services.

2. It must be independent of any particular technology for the discovery and the composition of adaptation services.

3. It must be portable, generic and expressive.

In the following sections, we will discuss the architecture of our proposed approach.

### 8.4.1 Proposed Architecture for Multimedia Adaptation

- The **Semantic Profile Analyzer** component must be independent of any particular technology and language. The semantic profile analyzer component provides an interface needed to analyze and translate the semantic profile constraints. More precisely, it converts the semantic profile constraints expressed with qualitative terms to quantitative values. For example, suppose the following explicit constraint: "If the battery is low, exclude videos and include images instead." This component translates this qualitative constraint into a quantitative constraint, e.g., "if my battery is less than 15%, exclude videos and include images instead." Of course, the semantic profile analyzer may control the validity and the consistency of a profile. For in-

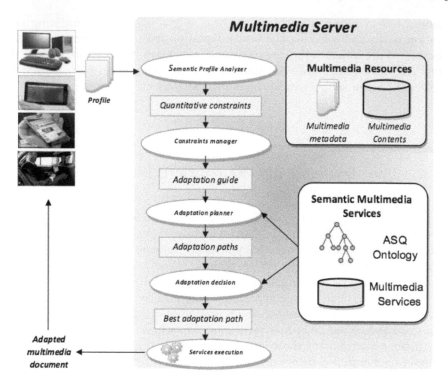

FIGURE 8.6: Proposed architecture for multimedia adaptation.

stance, a descriptive word may appear in a profile but it is not well-defined (i.e., the qualitative term is not associated to a valid context).

- The **Constraints Manager** component compares the profile quantitative constraints (implicit and explicit) with the multimedia content metadata properties. It produces an adaptation guide that contains some adaptation directives for solving the detected constraints. For instance, from a profile and a current situation, if we confirm that we must exclude videos and include images (because the current battery level is less than 15%), the adaptation process should start by executing a transmoding operator under a valid device codec. Of course, the constraint manager evaluates dynamically the profile (the user preferences and its usage context) and consequently may continuously update the adaptation guide.

- The **Adaptation Planner** component provides the necessary interface to implement an inference engine for determining a combination of adaptation services based on the semantic information described in the services (e.g., service_role, media_type_in, media_type_out, format_in, format_out, memory size, bandwidth level, service_location, service_time, QoS_service, etc.)

and the constraints manager information (e.g., start a video-to-image trans-moding operator, available output format image=JPEG and battery<15%). There are two subcomponents in this component: the discovery component and the automatic adaptation plan generation component.

- *Discovery component*: The adaptation guide processed by the constraint manager is used by the discovery component. This component discovers some services that will match the semantic requirements of the adaptation guide: inputs/outputs, action types (transcoding, transmoding, transforming: image resizing, language translation, video compression/decompression, video summarization, etc.), media type (image, video, sound, text), a specific service context information (user location, user language, user age, screen resolution, battery level, memory size, etc.). After finding the matched services, it sends them to the automatic adaptation plan generation component. To aid in the determination of an appropriate set of services, the ontology may contain rules specified in SWRL. This enables the discovery component to be independent of variation caused by dynamic context changes and inclusion of new services and rules.

- *Automatic adaptation plan generation component*: From the set of services discovered by the discovery component and the adaptation guide, it generates possible chains of services. In order to establish the sequence of services, our ASQ ontology is used. This ontology provides the semantic relationships (dependency, substitution, equivalence) between adaptation services and roles. It also specifies the correspondence between the roles and the service semantic information.

• The **Adaptation Decision** component is responsible of selecting the best path after calculating the benefit and cost of each adaptation path. The benefice and cost is customizable and parametrizable according to different user needs. For instance, it may select an adaptation path that produces rapidly an adapted content with a good quality (execution time of services is fast and compression ratio is low).

• The **Services Execution** component runs a sequence of transformation actions on the original media, and also uses the service repository, offering service qualities and context descriptions. It also monitors the execution of selected services of the semantic adaptation process for a better performance according to adaptation process under constraints changes.

• The **Semantic Multimedia Services** component consists of a service repository annotated with context and service quality ontologies. All needed services description will be accessed through this repository. For each service in this repository, there are three descriptions: functional service description, quality and context ontologies descriptions. Our approach allows providers to publish metadata about their services dynamically and to define their own specializations of the default classes based on the ASQ ontology.

- The **Multimedia Resources** component provides the description of the multimedia documents (source, nature, etc.).

### 8.4.2   An Algorithm for Determining Combinations of Adaptation Services

The discovery and adaptation process is a 4-step process:

**Inputs**: User Profile, Multimedia document, Service repository.

**Output**: Adapted Document.

**Step 1 – User Profile Analyzing**

The semantic profile analyzer component interprets the profile, which expresses users' preferences (e.g., explicit constraint), context information about the device (memory_size, battery_level, CPU_speed, screen resolution, location, etc.), supported documents (media_format, media_type, content_size, etc.), network characteristics (bandwidth, protocol type, etc.). Moreover, it converts some semantic users' constraints specified in qualitative terms to quantitative values. Furthermore, it evaluates the explicit constraints with the current situation, in order to confirm some constraints or not. If no constraints have been confirmed, no adaptation is needed. Otherwise, if some constraints have been confirmed, it produces an adaptation guide containing adaptation directives:

- If an action corresponds to a media inclusion or exclusion, then the adaptation directive contains at least one transmoding operator.

- If an action corresponds to a media update, then the adaptation directive contains at least one transcoding and/or transforming operator.

- Add other adaptation directives related to implicit profile constraints, such as the supported codecs, the supported media types, the battery level, etc.

**Step 2 – Finding adaptation services from the adaptation guide**

The discovery component is responsible to communicate with the constraint manager in order to initiate the relevant service identification. This is made thanks to a semantic-based matching process, which exploits the categorization of the ontology hierarchy to find suitable matches. The adaptation guide enables the discovery component to find appropriate services. For that purpose, it matches its functional and non-functional properties (service_role, service_location, multimedia_content, QoS_parameters, service_time, media_type, media_format, etc.). Each requirement of the adaptation guide can be expressed as an AND-OR clause of a simple semantic expression, like the following one: (serviceRole ='Resizing' OR serviceRole ='Conversion_BW') AND serviceRole ='Transcoding' AND serviceRole ='Transmoding' AND mediaType='Image' AND mediaFormat ='PNG'.

**Step 3 – Assembling dynamic adaptation services**

Depending on the resources availability of the device, the environment context and the service provider context, the function of the adaptation planner is to generate adaptation paths and to adapt a service from the primary list identified in the previous step within the equivalent functionalities. When the adaptation plan returned does not contain just one adaptation service, there is an atomic service that can satisfy the requested constraint.

**Step 3.1 – Find dynamically relevant adaptation services**

The adaptation planner provides dynamically a combination of adaptation services based on the semantic equivalent service relationships defined above. In case of changes according to the client context (e.g., less available memory, less processor power, user location) or the environment context (e.g., less bandwidth) or the user preferences, the Discovery component will be notified by a set of contexts' metadata in order to dynamically:

- Filter adaptation services which match the context usage from the list of candidate services. The adaptation service is matched from the new changes in the profile constraints (e.g., the client context).

- Order the result list of candidate services by QoS level and bound it by the user preferences with maximum potential benefit. Relevant adaptation services with higher potential benefit are selected earlier and find a limited number of composition solutions to reduce the computation complexity.

**Step 3.2. – Semantic relationships between Adaptation Services and Inference Rules**

This step consists of constructing all possible combinations between adaptation services based on the list of adaptation services in Step 3.1. To determine adaptation paths, the ontology may contain rules specified in SWRL. This enables the component responsible for the automatic adaptation plan generation to be independent of the variation caused by changes and inclusions of new rules, service and new semantic relationships among services. The adaptation planner uses SWRL rules to automatically infer all possible combination possibilities from the adaptation services list in Step 3.1 considered to be relevant for an adaptation guide. This is based on semantic relationships between adaptation services; those are used to implement combining rules based on their semantics. For example, assume that the following problem is JPEG $\rightarrow$ PNG; the adaptation planner has to deduce two rules due the service's dependencies: JPEG $\rightarrow$ BMP and BMP $\rightarrow$ PNG. In order to find a good reasoning result, the reasoning mechanism should find the quality relation between the candidate relations. For that purpose, the adaptation planner uses an ordered relationship which exists between semantic service relations: dependency $\leq$ equivalent $\leq$ substitution, when the rightmost reflects a better quality relationship between services than the one on its left. This step returns a list of adaptation paths.

**Step 4 – Selection of quality adaptation path**

The goal of this step is to select a better optimal path from the primary adaptation paths identified in the previous step. At first, we calculate the cost and benefit of each path. Once all possible compositions have been computed, we select the optimal path in terms of cost, with the best quality of service. The selection of the best adaptation path from the chain of adaptation services is in the case where there are several services within the equivalent functionalities. The user is able to customize the quality parameters in order to fulfill his/her needs. For example, one user may want a fast adaptation with quality outputs, while another one may want an energy saving adaptation with an average quality of media.

## 8.5    Use-Cases and Evaluation Scenarios

SGP profiles have been specified using RDF, especially since we have created and managed RDF/XML descriptions from context-aware application data, independent of their programming language. Moreover, as shown in Section 8.3, these descriptions have been enhanced with more semantic information. We implemented all the major functionalities of the dynamic service assembly with the IDE NetBeans and the Protégé-OWL API (to manipulate the ontology at run time). Standard DL reasoning inference, such as substitution, is achieved by means of the reasoner JENA and SWRL language (Protégé Ontology Editor 4.7, 2011) in order to generate the adaptation graph defined within the ontology, whereas quality oriented constraints (required for establishing the adaptation guides) are performed by SGP prototype [295].

### 8.5.1    Examples of Scenarios

The proposed platform is able to automatically combine dynamic quality and semantic adaptation services. Once our SGP profile is specified, all its explicit constraints will be available, and will be contributed by the actions and qualitative-values. The analyzer then checks to remove any constraint violation. We carried out test runs with dynamic context changes. For each new context changes, the context model was tested and validated with the semantic constraints defined by the profile at runtime, displayed in the actionType (Include, Exclude, Increase, Decrease), serviceRole, ParameterName, mediaType, mediaFormat properties. The adaptation planner then inferred the appropriate adaptation path.

Let's suppose that the user A has realized an oral conference presentation video, illustrated in Figure 8.7. He made this presentation for some computer desktops. User A would like to send this document to another user named B, who has a Smartphone. Video should be delivered in quality and within a period no longer than one minute from the request. For evaluating our approach using the repository above, several scenarios have been experimented.

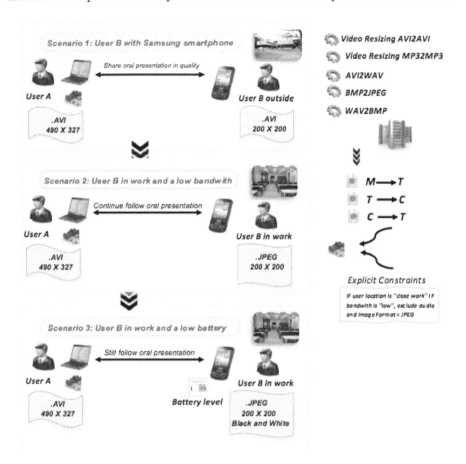

FIGURE 8.7: Scenario examples: On-the-fly adaptation.

A first simple example is made of a user B, who wants to share oral conference presentation using a Samsung Smartphone. There is an atomic adaptation service that can semantically fit the device constraint (e.g., a low screen resolution of the Samsung Smartphone = 200 x 200, inferior to the document pictures resolution 400x400). The adaptation planner is used as a basis for finding adaptation services that semantically match the Samsung Smartphone context (i.e., the smaller screen size). The Discovery component returned two resizing services: a black and white picture and a colored picture. To limit the search results when browsing the oral presentation on the laptop, only the first one, that semantically matches the laptop context (i.e., the smaller screen size), and the quality oral conference presentation (i.e., colored picture) is returned.

A second scenario is when the adapted document is ready to be executed. User B receives a phone call to join his colleagues at work. The constraint manager has noticed a problem of bandwidth, and the profile is dynamically maintained. Of course, user B wants to exclude audio during his work, and include pictures by using a specific codec (e.g., JPEG). This constraint "sound excluded while close to work" is an explicit constraint between a context information (user close to work) and a document (audio and video sounds not allowed). In this situation, the semantic analyzer checks the constraints of the user and interprets a close level of user location as longitude = 14, latitude = 25 and altitude = 40. Then, the adaptation planner looks for adaptation services that satisfy this constraint in order to built the adaptation chains. Several adaptation chains are generated and can be used to adapt the oral presentation. The adaptation plan can be selected according to some quality preferences. Each path is associated with an aggregate score. On the one hand, the end-user can define a quality oriented constraint for each quality dimension. In case the matching of a video quality is required to be higher than a given user threshold [0; 1], such constraint can be defined by the compression ratio > 0.5. Similar cost matching can be applied on low adaptation time and low transfer time (e.g., less than 1500 ms). These are the main properties that have to be taken into account for selecting an adaptation plan that meets user quality constraints. These constraints are enforced during service selection. Those adaptation plans which violate them are filtered from the list of candidate services, reducing the number and the size of the constraint satisfaction.

The user may control weights for each quality criteria (e.g., resolution, compression ratio) according to their needs in a graphical form, and our tool transforms it to a formal quality metric used by the inference engine. By using SWRL rules for calculating the QoS for each service and the QoS for each adaptation path, we can select the best adaptation path in a faster and more useable way. Figure 8.8 illustrates the search result of the best path preserving the shortest time and the quality image resolution. Consider the case shown in Fig. 8.8 again, adaptation path 1: AVI→MP3→WAV→TXT→JPEG is a solution that solves the format constraint AVI and JPEG. This path is selected as the best path based on user quality preferences, while the adaptation path Resize→AVI is a solution that solves resolution constraint.

A third scenario, for instance, is when a current service delivers a high density color video. A notification is sent to the Inference Engine Component telling it that there is not enough available battery to continue displaying such a video. The semantic analyzer checks the constraints of the user and interprets that there is a low battery level on smartphones between 0% and 15%. To alleviate too many changes (i.e., minimum adaptation cost) in the current adaptation plan, the adaptation planner can switch to the ideal adaptation plan if the video stream of data can be supported for long enough time (depending on the size of the buffer). The ideal management on battery degradation is to follow a minor change by the replacement of an adaptation service (picture resizing with high quality) by another adaptation service (picture resizing with lower quality: black and white) using a semantic substitution.

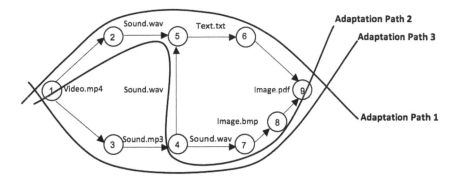

FIGURE 8.8: Possible adaptation paths for specific picture format constraint.

As shown in this real life example, adaptation processes should take into account two categories of constraints: qualitative and quality explicit constraint (user preferences) and implicit constraints (device and document properties). This situation demonstrates the capacity of our SGP profile to deal with qualitative and rich explicit constraints. These scenarios highlight some of the contributions of the dynamic semantic-based adaptation services assembling: a user's profile aggregates a very rich set of qualitative constraints, the customized quality includes both quality outputs and quality of adaptation; the semantic substitution of adaptation services on mobile devices depending on context variation represents key contribution of our approach. In this case, the service is replaced by another service or combination of services that provide the same functionality as the requested quality of service. The semantic information plays a central role in the discovery, dynamic selection, dynamic composition, and substitution of services.

## 8.5.2 Analysis and Discussion

### 8.5.2.1 Analysis of the Algorithm

In this section, we detail the complexity of the proposed algorithm.

**Theorem 1:** $O(T\ N)$ is the time complexity for service determination.

**Proof:** Consider T the total number of service roles to resolve the adaptation constraint and N the number of adaptation services for each service role. The analysis assumes that there is a constant number of adaptation services for each service role. The time complexity for start adaptation (e.g., some adaptation directives) is $O(T)$ and for accessing services is $O(N)$. Thus, $O(T\ N)$ is the time complexity for the service determination.

**Theorem 2:** $O(S\ N)$ is the time complexity for service substitution.

**Proof:** Consider S the total number of service roles needed to process a role substitution and N the number of adaptation services for each substitution role. The analysis assumes that there is a constant number of adaptation services for each substitution role. The time complexity for substitution is $O(S)$ and for accessing services is $O(N)$. Thus, $O(S\ N)$ is the time complexity for service determination.

### 8.5.2.2    Performance Results and Discussion

In this section, we evaluate the performance of Dynamic Semantic Adaptation Services Combination Algorithm (DSASCA) for dynamic semantic adaptation services, assembling into heterogeneous mobile applications. The experiments are conducted to compare our DSASCA algorithm with a QoS-based Dynamic Web Composition and Execution algorithm [509], named QoSDSC. This algorithm is a quality and distributed based services composition model for mobile-based applications. The reason for choosing [509] is that both our work and this reference provide a QoS mechanism based on a composition algorithm for a mobile heterogeneous environment.

The first experiment is to test the feasibility of our approach. We have evaluated our DSASCA, which includes a quality service semantic level algorithm and resources-awareness semantic level algorithm on an Android 3.2 Samsung Smartphone with 1GB of RAM and a double core Tegra 2 processor (1GHz). Figure 8.9 illustrates some results when considering almost 30 Java-based adaptation services. We have also measured the computation cost of the inferred adaptations' paths. In Fig. 8.9, when the adaptation service repository size is low (number of services = 4), the computation cost of DASCA is 8ms. When the service repository size increases to 6, the computation cost of DSASCA increases to 32ms. Naturally, when the adaptation service repository size rate increases quickly, the match of the services is increasing, and it increases the computation cost of adaptation paths.

The second experiment tests the performance of our DSASCA against QoS-DSC. Figure 8.10 shows the evaluation results, meaning that our approach turns out to be the best. Compared with QoSDSC, the response times of DSASCA increase more slowly than QoSDSC when the service repository increases. This result is practically significant, as well related to the two aspects of the QoS: one is the QoS for maximizing the benefits of output quality and the other is for optimizing the revenue of adaptation efforts as consequences of dynamic composition for environment evolution (e.g., network bandwidth) guided by the adaption directives and semantic relationships.

Based on these experiments and the validation given in the previous sections of this chapter, the herein proposed approach is usable and powerful. Other scenarios can be implemented and evaluated using our ontology. With 30 adaptation services, the dynamic semantic assembly using online semantic reasoning is very interesting with sufficient computation time, especially with situations of user mobility and context changes. From the above experiment results, we can draw some conclusions:

- The extension of existing profiles to include both low level and semantic fea-

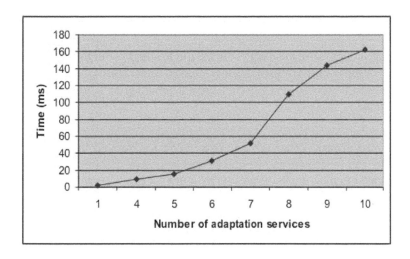

FIGURE 8.9: Computation time of adaptation paths.

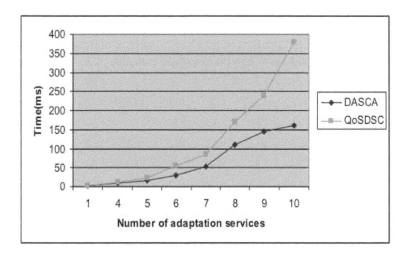

FIGURE 8.10: Response time under various adaptation services.

tures enriches profiles with explicit constraints. Moreover, the semantic rules the accurate generation of a dynamic quality composition of heterogeneous adaptation components.

- The exploitation of service semantic information and semantic services relationships (e.g., services with common quality and common types, annotated with concepts that are semantically close to each other) permits also to select the relevant adapted services that are not provided by traditional adaptation techniques.

- The automatic deduction of all the adaptation chains, that solve all the problems of adaptation through inference rules, are defined by service semantic relationships (dependency, equivalence, and substitution). These semantic relations are defined between two or more services to easily make the combination of adaptation services.

- The quality evaluation of an adaptation path was customized to make it more suitable for particular user needs and specific media types.

- The quick discovery of the relevant adaptation services, using the adaptation guides and adaptation rules detailed above, improve the composition speed (the dynamic service composition time decreased by 60% factor).

- The adaptation services that violate quality preferences, filtered during service selection from the list of candidate services, reduce their number and the size of the constraint satisfaction.

- The facility of implementation of our dynamic determination of adaptation service and quality combination algorithm. In the first step, we must easily resolve the adaptation constraint and for the remainder steps, we must check each service role and associated adaptation services.

## 8.6   Conclusion

The contribution of this chapter deals with an on-the-fly adaptation of multimedia documents based on semantic service information. We have detailed: (1) how a semantic profile analyzer converts the semantic profile constraints explicated in qualitative terms to quantitative values, (2) we have shown how to produce an adaptation guide that contains some adaptation directives for solving the constraints detected, and finally (3) we have shown how to use a generic ontology that exploits semantic rules allowing the automatic generation of a dynamic and a high quality composition of heterogeneous adaptation components. Our proposal is a fruitful idea for using service semantic information and explicit constraints for describing profiles. It enhances the dynamicity and agility of the services composition. Therefore, our semantic-based adaptation service assembling architecture provides great flexibility, adaptation of multimedia documents and customization of quality of service properties. More precisely, the proposed approach is based on three key elements: explicit qualitative constraints (suitable for understanding the variety semantics of users), semantic informations (to manage the service registry and service relationships), and semantic quality models (to enable an easy customization of user needs and selection criteria). Some aspects of the provided reflective facilities have been validated in the developed prototype upon the Java NetBeans platform toward a mobile environment. We have developed several real-world examples like e-SMS exchanges and on-the-fly multimedia traveling heterogeneous services in our prototype to validate our approach. Of future interest is the effort to refine the quality of service ontology to provide high level services to mobile clients. Investigation toward using cloud computing approaches is also in progress to attempt to improve the performance of the service identification and composition.

# Part III

# Multimedia Personalization

# 9

## When TV Meets the Web: Toward Personalized Digital Media

**Dorothea Tsatsou**

*Centre for Research and Technology Hellas, dorothea@iti.gr*

**Matei Mancas**

*University of Mons, Matei.MANCAS@umons.ac.be*

**Jaroslav Kuchař**

*Czech Technical University in Prague & University of Economics Prague, jaroslav.kuchar@fit.cvut.cz*

**Lyndon Nixon**

*MODUL University, lyndon.nixon@modul.ac.at*

**Miroslav Vacura**

*University of Economics Prague, vacuram@vse.cz*

**Julien Leroy**

*University of Mons, Julien.LEROY@umons.ac.be*

**François Rocca**

*University of Mons, Francois.ROCCA@umons.ac.be*

**Vasileios Mezaris**

*Centre for Research and Technology Hellas, bmezaris@iti.gr*

## CONTENTS

The rise of new paradigms in the field of television and digital media distribution (e.g., Smart TV, IPTV, Social TV) has opened a new digital world of data communication opportunities, but at the same time it has exacerbated the information overload problem for media consumers and providers. Therefore, the need for personalized content delivery has extended from the traditional web to the networked media domain. This chapter presents comprehensive research in the field of capturing and representing user preferences and context, and an overview of relevant digital media-specific personalized recommendation techniques. Subsequently, it describes the vision and first personalization approach adopted within the LinkedTV EU project, for profiling and contextualizing users, and providing targeted information and content in a linked media environment.

## 9.1   Introduction

The convergence of TV and the Web has brought a new perspective in personalized systems by offering the possibility to interlink user behavior with respect to multidisciplinary content [649] for consumer usage, whether that might be audio, video or text (transcripts, articles, tags, comments, etc), while tailoring it to the user's needs and concrete situation.

The Television Linked To The Web (LinkedTV)[1] project aims to provide a novel practical approach to future networked media, where audiovisual content is decoupled from place, device or source. Browsing TV and web content should be interrelated in such a way that eventually even "surfing the Web" or "watching TV" will become a meaningless distinction.

Accessing audiovisual content will be seamlessly interconnected and will take place not only on a TV set, but on any other digital medium, such as smartphones, tablets, or personal computing devices, regardless of whether it is coming from a traditional or new media broadcaster, a web video portal or a user-sourced media platform. To this end, within the scope of LinkedTV, triggerable spatio-temporal media fragments[2] will be connected to additional content and information, thus augmenting the features of the audiovisual content with more comprehensive descriptions.

Nevertheless, the general issue pertaining to advanced personalization in networked media platforms, such as LinkedTV, is capitalizing from the information available in user-consumed digital data by understanding and unifying information from various heterogeneous sources (i.e., video, audio, text and social media), and aiming this information effectively at the content consumers. Consequently, one of the key challenges in intelligent information systems is personalization.

Moreover, the preferences of a given user evolve over time and various states of the user model are applicable in different circumstances. Therefore, the context of the user at a given moment has a significant impact in efficient decision-making in personalized environments. To this end, LinkedTV aims to extract different user contexts based on two perspectives: a) on-platform interaction with the content, depending on the concrete conditions such as time, location, and actions and b) behavioral tracking in order to determine the physical state of the user (alone, with company) and the reaction to the content (mood, attention).

## 9.2 Background

In a personalized environment, information about the user can be either explicitly defined by the user or implicitly extracted through his interaction with the system and relevant resources. Nonetheless, due to the obtrusive and effort-consuming nature of explicit preference acquisition, recommendation systems cannot rely on the user to provide this information. In effect, the focus of the LinkedTV personalization task will lie on implicitly extracting user preferences.

---

[1]http://www.linkedtv.eu
[2]http://www.w3.org/TR/media-frags/

The sources of implicit transactional information pertaining to a user are three-faceted, involving learning predictive user models from a) data extracted from a user's the content consumption history (i.e., user-assigned ratings [64], data/text mining [300]), b) identifying similarities with peers in a social information exchange environment (i.e., collaborative filtering [633], social network mining [1016]) and c) taking into account predefined knowledge encompassed in ontological structures [845].

These aspects come with advantages and challenges. The content-based approach is the most straightforward and indicative of a user's individual behavior, while peer-based approaches may provide insight about foreseen preferences that a user has not yet expressed. Both of these methods though are hampered by the cold-start problem [673] when new users or new content are introduced on a platform, as well as by data sparcity and scalability problems [633]. Knowledge-based approaches resort on a finite (therefore arguably scalable), uniform vocabulary [922] that can augment user-induced information with domain knowledge. They are, however, hindered by the need for predefined, usually manually constructed knowledge bases (i.e., ontologies) and by the lack of mappings between free-form information and the ontological knowledge. A hybrid approach [183] taking into account both content and peer-based information while understanding and aligning them under uniform ontological conceptualisations, thus capitalizing on available knowledge, is deemed as the optimal trade-off.

### 9.2.1   Profiling Users in Networked Media Environments

In personalized digital media systems in particular, the metadata desultory and sparcity in heterogeneous multimedia content are the prevailing challenge in understanding and capturing user preferences. In the approach of [649] for a personalized TV program recommendation, the system relies heavily on explicitly provided user preferences, while avoiding to delve too deep into automatically extracting fundamental user knowledge.

In [928], the authors propose automatic metadata expansion (AME), a method used to enrich TV program metadata based on the electronic program guide (EPG) data and an associated concept dictionary (ACD). Transactional information such as program recording, program list browsing, program searching, and voting on viewed programs are processed by a rule-based engine automatically, in order to update the user interfaces. Indirect collaborative filtering (ICF) is used to trace peer induced user preferences. Typical users are clustered prior to the similarity calculations of CF.

The iEPG [401] "intelligent electronic programming guide" platform utilizes a graph-based organization of content metadata in the multidimensional information space of TV programs, used to structure user preferences and visualize them for the users to access, assess, and define nodes of interest, where "me" is the central node of the graph.

In [442], the authors underline the difference between playback duration and the actual viewing time when talking about TV consumption and further classifying user behavior in accordance with their purpose to consume, store, or approve what they consumed.

In the NoTube project[713][824], a user profile is generated automatically from the user's social web activities using the NoTube-developed Beancounter API[3], where the user can select which social web sources are to be used (e.g., Facebook, Twitter, GetGlue, last.fm), and can add, delete or alter the weightings of selected topics in the profile. The Beancounter monitors changes in the user's social web activities over time and evolves the user's profile. The user profile in NoTube extends the FOAF[4] vocabulary with weighted interests (adding a weight value on the FOAF interest property), published as FOAF-WI[5]. The interests are identified using categories from the DBPedia concept space.

## 9.2.2   Contextualization in Networked Media Environments

User context can be defined by manifold manifestations: recently browsed/rated items, social environment, physical environment, physical reaction to presented content, as well as location, time, external events, etc. The majority of existing contextualization approaches rely on segmenting user behavior or filtering content based on factors like the time, location, and company of other people [45].

Sieg et al. [844] illustrate the importance of distinguishing long-term preferences and short-term context-related preferences in a dynamic personalization environment, with the short-term preferences stemming from recent content consumption. The approach achieves a trade-off between accuracy-achieving factors, like continuously making "safe" recommendations of the most prominent long-term user preferences and diversity, the lack of which might significantly reduce recommendation performance.

Palmisano et al. [722] elicits hidden contextual information from data by identifying latent variables for specific sessions, used to identify patterns in the user behavior regarding specific topics, able to recognize contexts without having to map these contexts to a specific user in a multi-user environment.

In [649], implementing a personalized TV program recommendation system identifies the periodical habits of the user, so that the recommendations target the users' leisure time or time available for media consumption. Profile information is enhanced with supplementary information such as demographics and lifestyle, such as age, gender, profession, TV schedule, or leisure time.

In NoTube [824], the following context factors were taken into consideration: current location, time of the day (dinner time, evening), day of the week, time of the year (summer, winter), activity (traveling), device and multimodal capabilities, social settings, and also moods and feelings. They use them to

---

[3]http://notube.tv/category/beancounter/
[4]http://www.foaf-project.org/
[5]http://xmlns.notu.be/wi/

extract the user preferences that are relevant to the context from a user model. They use layers to describe different knowledge domains of a user model, i.e., temporal, spatial, geographic, music-specific, movie-specific. These are separated into hardly changing, slowly changing, and quickly changing (context) parts.

Songbo et al. [868] highlights that context related to IPTV services should be classified into user context information, such as the identity, location, preference, activity and time; the device context information, such as the screen, the supported content format and the terminal's location, the network context information, meaning the bandwidth and the traffic condition, and finally the service/content context information, which in interactive television is divided into the content description, the video objects and the program interaction.

The combination of user and technical context is also considered in mobile TV environments. Mobility allows the terminal to change location without service disruption, incorporating situational context information. CoMeR [1044], a context-aware media recommendation platform, underlines the importance of utilizing both facets of user-pertinent information: the long-term user preferences and the context. Given the challenges posed in the resource-constraint platforms they consider two distinct types of context: the user situation context, and the device capability context represented based on an ontological structure in OWL. The user's situation is defined by his location, activity, and time, and the media terminal's capability is defined by the device's operating capacity, its display characteristics, the network communication profile, and the supported media modality.

### 9.2.3 Implicit Preference and Context Capturing

The first step toward unobtrusively profiling users is implicit information tracking. Also called implicit feedback, this information is vital to understanding user interests and disinterests. Implicit feedback has the advantage that it can be collected unobtrusively, but it is potentially ambiguous and more difficult to interpret. There are many measures which can express this information based on the behavior of the users.

#### 9.2.3.1 Transactional Information Tracking

In transactional information tracking, three main sources of implicit feedback exist [101]: Attention time, Click through, and Mouse movements. Several other measures and behavior patterns related to the implicit feedback are proposed in [714], including actions that refer to the examination, retention, and reference to content. The information about users can be collected by various techniques. In general, there are many channels for collecting implicit information. This section will provide a summary of approaches related to browsing and activities on the web or media consumption.

Implicit feedback collection does not require intervention by the user. All information is gathered in the background, using agents that monitor user activity [507]. The basic task in the data acquisition (or in some sources data collection) phase is to record visitor actions such as content views, events, etc. Additional information may be gathered in order to identify user sessions and assign weights to the actions. The most prominent user information collection techniques are compared in [358].

Many studies are based on the use of the browser cache and proxy servers [261], [350]. In this case, only one setup at the beginning is needed and all the interactions of the user can be collected at the proxy server [422], desktop, or browser agents [1006]. All approaches, including hybrids, were compared, but there is no clear answer, as to which approach is more or less accurate [358]. In general, the approach that does not require any additional software on the client side is preferred.

The approach based on the browser cache [462] has no limitation on web sites, but it needs to send out cache information periodically. Using a proxy server, on the other hand, is a compromise solution [261], [350]. The drawback lies in the identification of users. A group of users can use the same proxy server, which gives rise to the problem of distinguishing them fom each other. This solution is very often limited on the use of a single computer. However, there are solutions that use login information and install proxies on various numbers of computers.

A third category is based on agents. Desktop agents are implemented as standalone applications, and browser agents are implemented as plug-ins to an existing browser [717]. The main disadvantage is that the user has to install additional software. The advantage is in the possibilities and intelligence of this solution. An agent has much more available information about user interactions and can provide better browsing assistance (fill out forms, highlight content, modify content, etc.).

The logs-related category (web and Search Logs) does not require additional software. Although it is prone to collect much less information than agents, it is one of the most used approaches. There are two main sources: browsing activity and search interactions. Search interactions can provide information about queries, and help to collect information about a user [516]. The drawback is in collecting information related only to search interactions.

User identification is a crucial ability for all the previously mentioned approaches. There are five basic approaches to user identification: software agents, logins, enhanced proxy servers, cookies, and session IDs [358]. Cookies are the least invasive technique and are widely used. Software agents, logins, and enhanced proxy servers are more accurate. The drawback of these approaches is the need for user participation. The general characteristic is the user register, and logging on to the system.

Cookies and session IDs are less invasive. This method is unobtrusive and there is no need for user participation. The drawback is in the different sessions across different clients for the same user, or one shared session for multiple users using the same client.

*Client-side vs. server-side tracking*

Data can be collected on the client machine (client-side), by the application server itself (server-side), or both. The quality of the collected data available is overlooked, but is a critical factor for determining the success of any further processing. Client-side monitoring is required to get a precise record about a user interaction with the website [1006]. Nonetheless, data acquisition for implicit information analysis is often taken as a synonym for processing server logs.

Client-side tracking is a process that is absolutely unobtrusive for the user, who does not know that he is being tracked and, particularly, does not need to install any tracking software. On the other hand, server-side monitoring is most commonly served by storing server log files. A server-side application may log any interaction (or its implications) with the visitor, which is propagated to the server. Some approaches use an application-level server-side tracking, meaning that the server application responsible for the generation of content is also in charge of tracking.

For example, when the server application generates the content, it can record which pieces of information it contains. An authoritative survey [317] marks application-level server side tracking as "probably the best approach for tracking web usage."

Custom server-side monitoring has several important virtues, when we consider its deployment in a dynamically built website:

1. Unobtrusiveness: no demands on the client, including scripting support.

2. Invisibility: the client can't know (in principle) whether he is being tracked and what data are collected.

3. Speed: the data are readily available on the server-side, no additional client-server communication needed.

4. Interlinked with content: all information in the underlying database, from which the content is generated, is available for tracking purposes.

Despite these advantages, custom server-side tracking has apparently never gained much popularity. It suffers from the same accuracy problems (that concerns the omission of hits) as log based solutions. Foremost, handling personal data on the server side seriously compromises user privacy, a fact that requires careful management of user data, including user anonymization and secure data transmission.

### 9.2.3.2    Behavioral Tracking

Behavioral tracking technologies generally assemble all means to observe, analyze and process reactive behavior of a user. This is a very large topic spanning from speech (audio) analysis and gesture understanding, to emotions reading, mouse clicks, or navigation history. The possibility of automatically understanding human behavior has already been vastly explored. In [956] for example, an extensive survey on social signal processing and behavior tracking for non-verbal communication can be found.

In this section, the focus is oriented toward audiovisual features and the behavioral tracking, which falls under the scope of computer vision technologies and particularly on TV or home related experiences applications where the state of the art is much less deployed. Moreover, this Section will be oriented toward implicit behavior analysis of the user's physical reaction to the presented content; therefore explicit gestures for interfaces' control are out of scope, even if the technology used for behavioral tracking can of course provide such information.

User behavior can be extracted in various ways from different kinds of input data. With the availability of cheap cameras which are able to acquire both the classical RGB data, but also the depth map representing the distance of each pixel from the camera, vision-based behavioral tracking has made a huge progress. Before those cameras were available, behavioral tracking was made by a few research groups using expensive cameras like time of flight cameras or 3D capture systems, implying several cameras and infra-red markers, like [954]. Other research groups worked on more classical sensors like simple cameras or stereo cameras, but those devices needed a lot of complex algorithms to provide less precise information, and produced ineffective results for real-life environment and applications.

The Microsoft Kinect sensor is a cheap and effective device which also gained popularity within a large public with its explicit gesture analysis for video games such games for Xbox[6]. Other 3D camera-based explicit interfaces developed for TVs and interactive ads [866], [905] are more and more popular and lead the public to be more aware about these technologies for home and TV-based applications. This trend has already pushed some TV manufacturers like Samsung to propose new cameras to be directly embedded into TVs [808]. Even if those efforts mainly intend to provide explicit control on TV interfaces, the same systems can also be used in implicit interfaces and behavioral tracking.

Within the work on implicit interfaces, Microsoft research has employed the Kinect to extract face emotions and provide them to avatars [671], as correspondingly also employed in [344] (cf. Figure 9.1). Figure 9.1 shows the face emotions which are extracted in real time (top) and then simulated on the left image avatar (bottom). Another work from Microsoft examined the representation of several avatars together mimicking real users' face emotions. Interfaces showing data (music data) use real-time context as social networks or related songs, to provide more context-aware information where developed in [670].

In the field of proxemic interaction, proxemic relationships between people,

---

[6]http://www.xbox.com

FIGURE 9.1: Face and emotion tracking. Extracted from [344].

objects, and digital devices are used together. The design intent is to leverage people's natural understanding of their proxemic relationships to manage the entities that surround them.

[379] identified five essential dimensions as a first-order approximation of key proxemics measures that should be considered:

- *Orientation*: the relative angles between entities; such as two people facing toward one another (interpersonal orientations).

- *Distance*: the distance between people, objects, and digital devices; such as the distance between a person and an interactive display (interpersonal distances).

- *Motion*: changes of distance and orientation over time (orientation and distance variations in time).

- *Identity*: knowledge about the identity of a person, or a particular device.

- *Location*: the setup of environmental features, such as the fixed-feature location of walls and doors, and the semi-fixed features including movable furniture (this includes 3D reconstruction of the scene).

Hello Wall [882] and Vogel's public ambient display [959] introduced the notion of "distance-dependent semantics," where the distance of a person from the display defined the possible interactions and the information shown on the display. The space around the display is separated into four discrete regions having different interaction properties. In [484], the authors developed a proxemic-aware office whiteboard which is able to switch between explicit (drawing on the whiteboard) and implicit (data display) interaction depending on the user's position.

Ballendat et al. [97] developed a system which activates when the first person enters, shows more content when the person is approaching and looking at the screen, switches to full screen view when a person sits down, and pauses the video when the person is distracted. This system was initially designed to be used with precise motion capture systems [954], but part of it works also using low-cost depth cameras like the Kinect sensor.

Consequently, systems applying implicit interaction and behavior tracking in TV setups are commercially existent, although limited in number. As the market (both for depth cameras and TV manufacturers) moves towards explicit and implicit interaction based on people's behavior, LinkedTV follows this emerging paradigm toward a system which is fully adapted to TV setups, and which integrates with other profiling technologies.

## 9.2.4 Semantic User Profiling

Expected information overload, instigated by the diversity and vastness of information in multimedia environments, can be efficiently managed by the representation of the information in compact, lightweight, ontological conceptualizations. These issues can be efficiently compensated for by extracting the semantics of user behavior with respect to the viewed and consumed content, and the user's interactions with peers, and express them in a uniform ontological vocabulary.

While the term "semantics" is loosely interpreted in personalization literature as the retrieval of meaningful relationships between content features or user attributes [1029], this subsection refers to paradigms that employ formal ontological knowledge as the background for building structured semantic user model descriptions. The use of ontological (or simply taxonomical) knowledge in order to improve recommendation accuracy and completeness has been explored widely in the past.

In [749] the authors use ontology-based profiles for personalized information retrieval, employing both a static facet and a dynamic facet for the profile, allowing them to capitalize on the expressivity and adaptability of formal semantic conceptualizations. However, no specific method for implicitly understanding the semantics of raw contextual information is defined.

The adaptability and flexibility potentials offered by the uniform descriptions in ontology-based profiling methodologies render them suitable candidates for personalization in context-specific systems, and even in resource-limited mobile environments [982].

Popular knowledge-based preference mining techniques map user interests onto the ontology (or taxonomy) itself, or an instance of the ontology, by activating concepts and assessing their impact based on the interaction of the users with content described by predefined concepts. For instance in [845], variants of the spreading activation algorithm are employed to propagate weighted interests up the hierarchy of a taxonomy per user. These ontology-based user profiles, however, lack the ability to express more complex, axiomatic semantic information about user behavioral patterns, and are further memory or server communication dependent since they require storage of the full ontology on the client in order to reflect the user profile.

Several profile learning techniques use a hybrid approach by combining the aforementioned implicit preference extraction methods (content transactions, social interactions, reference knowledge), which help compensate for the limitations of the individual systems [183].

In [845] emerging concepts-preferences stem from observation of the navigational behavior of the users on semantically pre-characterized content by taking into account interest factors such as "the frequency of visits to a page, the amount of time spent on the page, and other user actions such as bookmarking."

The authors of [271] capitalized on the synergy between user generated content encompassed in folksonomies and provided publisher information over the content to discover potential semantic interpretations of user interests. The semi-structured content information is analyzed by machine learning techniques to discover interesting concepts, and a probabilistic user model is formed via supervised learning of user-rated content.

Standard collaborative filtering (CF) preference learning techniques are extended in [844], in order to infer user similarities based on their interest scores across ontology concepts rather than explicit item ratings, thus significantly outperforming traditional techniques in prediction accuracy and coverage.

In [922] content-based and knowledge-based approaches are combined to tackle both the vocabulary impedance and cold-start problems by creating a semantic user profile through observation of the transactional history of the user. The authors perform semantic analysis of raw domain content by means of a lexical graph, used to automatically interpret and annotate consumed and provided content. This approach was extended in [923], in which the use of community detection for analyzing the domain is explored, thus rendering the system able to receive aggregated preference information from social networks.

LinkedTV aims to follow a hybrid approach, where background semantic knowledge will serve as the basis for understanding users based on their transactions, while imparting initial information that can leverage the lack of data about new users. It will also provide the means to unify, structure, and reduce the information load pertaining to the users.

## 9.2.5    Knowledge Bases for Personalization in Networked Media Environments

It is significant for the efficient elicitation of user preferences to have a holistic but somewhat lightweight vocabulary under which to classify this information. To this end, ontologies provide the needed expressivity and structure to represent all relevant domain and user-specific concepts, and provide uniform, compact conceptualizations for ambiguous, synonymous, and multilingual knowledge. Such an ontology can be used as the backbone for densely representing user preferences, as well as the inferential knowledge base (KB) for the production of targeted recommendations, e.g., to conduct profile-content matching.

A core ontology aiming to adequately describe knowledge relevant for a user in a heterogeneous hypermedia environment is expected to be rather broad. On the other hand, efficient handling of the immensity of information requires dense conceptualizations in a highly expressive, formal ontology for it to scale well and maintain the accuracy advantage of logical inference algorithms.

### 9.2.5.1 Representing Information in Digital Media

In the field of broadcasting, the RDF-based BBC, Programs Ontology[7] provides a descriptive vocabulary for TV programs, describing concepts such as broadcasting events, brands, episodes, etc. It is a lightweight ontology recording the broad spectrum of rather abstract broadcasting-relevant aspects. It notably provides semantics for media-related temporal concepts and objects, thus rendering a considerable basis for an upper level vocabulary in personalized TV environments.

For the purpose of personalized TV-content recommendations, an expressive OWL[8] ontology was developed within the AVATAR system [122]. This ontology consisted of a full hierarchy based on three levels of granularity of program-related categories and subcategories used to classify TV programs. It also comprised the properties that interrelate them and different important entities within the context (actors, directors, places, scriptwriters, etc.) [123].

The IPTC[9] news codes classification is widely used by news agents to categorize news content. It is merely a shallow set of categories and subcategories of news subjects, but it offers good coverage of topics in the media superdomain. The categories are available in a human-readable taxonomy by the WebTLab[10].

The Linked Open Data[11] (LOD) initiative attempts to provide structure to the vast information available online. Most current personalization approaches for networked media environments have turned toward employing such open linked vocabularies to efficiently describe and expand the diverse and continuously evolving information in digital media content, and in extension reflect and address the variety in user preferences.

DBPedia [86] stands out as the most prominent organization and concentration of knowledge in the current literature in LOD datasets. It is a shallow ontology that interlinks and semantically structures cross-domain information from Wikipedia[12]. It is released in a variety of languages, allowing for multilingual alignment. However, the broadness of this information restrains DBPedia to relatively low expressivity, corresponding to the $\mathcal{ALF}^{(\mathcal{D})}$ complexity of Description Logic (DL) [540].

Schema.org [813] is a collection of schemata, published through the collabo-

---

ration of three major search engines (Bing, Google, and Yahoo!). It introduces a set of vocabularies used to mark up data within web pages. These schemata span from vocabularies for representing audiovisual concepts to taxonomies for representing the metadata that describe web content. Schema.org provides a somewhat more structured representation environment than DBPedia, however sharing its expressivity shallowness.

Freebase [140] is a public collection of community-contributed interlinked data, or as the community itself describes it "an entity graph of people, places and things." The Freebase ontologies are again user-generated and edited, consisting of semi-structured information in the form of folksonomies. It was recently employed by the Google Knowledge Graph [375] to expand Google search results about such entities with related information.

YAGO [885] unifies WordNet [680] with Wikipedia, thus enhancing the semantic relations between entities and individuals of Wikipedia with more descriptive properties. It additionally offers correspondences between the entities and their lexical description (term) while taking into account synonymy and term ambiguity, thus allowing for advanced content classification. However, it is easily understandable that while such a vocabulary adds to the semantics of Wikipedia information, it also adds to the complexity of Wikipedia-based knowledge.

The NERD (named entity recongnition and disambiguity) ontology [784] provides a frame for mapping named-entities (NEs) described across several multi-discipline vocabularies on top of the NER named-entity extraction, classification, and disambiguation tool. This ontology can be used for extracting NEs within textual manifestations of digital media (audio transcripts, articles, etc.) and support semantic annotation of media content with coherent and interlinked instances belonging to popular LOD schemata, thus substantially enhancing semantic interpretation of diverse user-consumed content.

Furthermore, the LOD cloud encompasses many interconnected datasets of domain specific knowledge, a collection of which can be found in [436]. These KBs may offer richer information or/and deeper semantics pertaining to many aspects important to represent a user's preferences and context, as well as mappings to more general upper knowledge bases and other ontologies in the cloud, e.g., describing detailed geographical information (Geonames), music relevant semantics (MusicBrainz), etc. [436].

The wealth of information is the most significant advantage of the LOD cloud. In a vastly heterogeneous and broad environment such as networked media, the LOD datasets offer structure over the magnitude of data. This structure and information abundance is additionally augmented through the interconnectivity between different datasets within the LOD cloud.

Moreover, the knowledge encompassed in LOD datasets is not static, and does not require manual contribution from experts. Evolving knowledge is constantly updated, mostly through community contributed metadata. This process is further alleviated through the conformity of the knowledge bases to widely accepted and used standards (e.g., Dublin Core[13], SKOS[14], SIOC[15]).

### 9.2.5.2 Addressing User Needs

At the outset, LOD datasets provide structure and semantics to a large amount of entities but with a notably shallow structure [259], since they describe a vast amount of generic conceptualizations about the world that might concern different applications of different purposes. However, no complex information is available conveying the distinct axioms and specific relations between concepts that adequately describe the semantics prominent to a user or regarding the user's context across domains. Consequently, there are no semantic descriptions in such ontologies for specific people-related information that comes from general world perceptions. However, a personalization system should be able to specifically understand how the users generally perceive and interact with the world.

Furthermore, the extremely high dimensionality of several of these KBs [459], and the lack of consistency checking of community-contributed knowledge, raises serious knowledge management issues. Reference knowledge inconsistencies can throw the results of inference engines off track, while the problem for personalized services in particular is twofold: a) Recommendation services are server-bound even in resource-rich devices, since background knowledge is too large to be handled outside of a server at any instance. This renders constant client-server communication obligatory, thus giving rise to user privacy compromise problems; b) The volume of data and complexity in the KBs themselves and the additional information stemming from mappings across other vocabularies in the Linked Data clouds can prove to be unmanageable for inferencing services to handle.

Therefore, coherent conceptualizations and semantics pertinent to users of personalized platforms have also been proposed (e.g., skills, contextual situations, mood, etc). The Cognitive Characteristics Ontology [248], for instance provides a vocabulary for describing cognitive patterns for users within contexts, their temporal dynamics and their origins, on/for the Semantic Web.

In [113], the authors use the OCUM (Ontological Cognitive User Model) as an upper ontology that "encompasses the core human factors' elements for hypertext computer-mediated systems" and can be reused to provide enhanced user-centric mappings for personalization systems.

The NAZOU project [705] also specifies a dedicated user model ontology. In their approach, an ontology-based user model defines concepts representing user characteristics and identifies relationships between individual characteristics connected to a domain ontology.

The GUMO (General User Model Ontology) [407] and its descen-

---

[13]http://dublincore.org/
[14]http://www.w3.org/2004/02/skos/
[15]http://sioc-project.org/

dants/hybrids record general upper concepts in combination with characteristic attributes of a user. The ontology is very general and broad, consisting of hierarchical/categorical relationships only, but its user-related subsets include concepts describing user state and actions, e.g., personality, facial expression, motion, skills, location, social environment, etc, which are especially useful to semantically describe user context and the semantics of sensor extracted related information, i.e., the reactional behavior of a user.

## 9.3  Personalization and Contextualization in LinkedTV

Personalization and contextualization in LinkedTV focuses on implicitly learning user preferences, based on the semantic description of (seed and related) content, following the interpretation of raw audiovisual and textual data. The implicit profiling mechanism involves producing a semantic user model from a cold start and evolving it over time based on the age, the history of user transactions (what video the user watches, what media fragments s/he looks at, what concepts s/he chooses, what additional content s/he browses, the actions of the user on the media player and additional content), and user reactional behavior (engagement and attention to viewed/consumed content).

There are several issues that need to be dealt with in the context of implicitly capturing and representing a semantic user profile: the ontology that can provide a meaningful, uniform, and lightweight reference knowledge base potent enough to capture domain and user-pertinent semantics; the means to align this uniform knowledge base with the semantic information in the multimedia content; the means to unobtrusively capture the user's transactional and reactional behavior; the means to understand user behavior, i.e., map it to available knowledge and determine its impact; determining the most suitable representation schema of the user model in a manner that renders the synergy between the model, available knowledge and the inferencing mechanisms used for producing feasible recommendations.

### 9.3.1  Augmenting Semantic User Profiles with Aggregated Media Information

The ability of the semantic approach to user profiling to represent a wide range of content of multimedia information and to represent it in a meaningful way enables us also to augment these semantic profiles (originally based on implicit user generated information) using aggregated information available on the web.

Augmenting user profiles using semantic information from the Web has been in the context of LinkedTV being split into several separate tasks. The first aspect concerns techniques of web mining that should support both the

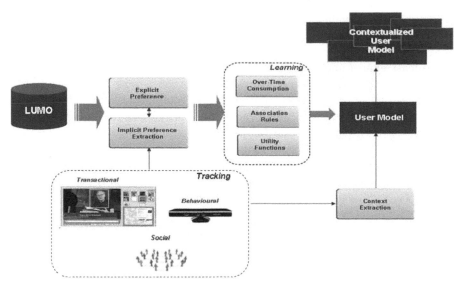

FIGURE 9.2: The user modeling workflow.

media annotation process and user profiling. This task employs existing tools (e.g OpenCalais[16], GATE[17], SProUT[18]) but also focuses on development of some new or integrating approaches on top of those already developed. An important aspect in this task is the problem of concept identification in hyper-video. Besides the development of tools for analysis and mining in structured data there is also research focusing on additional web mining techniques on completely unstructured data.

The second aspect involves development of a Linked Media Layer, composed of a fine-grained mechanism for addressing fragments of multimedia and annotation schemas. This mechanism is again based on a semantic approach for representing media data. The Linked Media Layer integrates the media fragment specification[19] and several metadata schemata together, those schemata belonging to the LOD cloud, such as DBPedia, schema.org, NERD, etc., with the annotations gathered from the web mining processes in order to provide a search and retrieval module.

A separate aspect also focuses on semantic techniques that, for example, take a concept of semantic user profile as the starting point and gather additional information around this concept based on semantically annotated information obtained from web-mining tools, and interlinked using the Linked Media Layer.

While the above described approach is very general and applicable in a

---

[16]http://www.opencalais.com/

[17]http://gate.ac.uk/

[18]http://sprout.dfki.de/

[19]http://www.w3.org/TR/media-frags/

wide range of domains, the focus of LinkedTV is the domain of broadcasting. Therefore it is possible to employ some domain specific enhancements on top of the described general techniques. Since there is a limited number of different broadcast genres and assuming that the viewers are likely to have different preferences and requirements for each of the genre, the project focuses on creating topic-specific information gathering templates. These templates can be integrated to semantic user profiles and also to media annotation tools. These templates then provide the necessary granularity and adaptability to accomodate the user's requests and interests.

Concerning ethical issues, the project takes special note of the fact that some online sources are not reliable. That is why the project's application partners have decided to use white-listing of online sources to ensure the required level of reliability of additional information obtained from web.

## 9.3.2    The Reference Knowledge Base: LUMO

Effective implicit profiling depends on representing implicit user feedback in a machine-understandable format appropriate for predictive inferencing of relevant concepts and content. While within LinkedTV media fragments, as mentioned in the previous section, will be semantically described based on LOD vocabularies, not all information is necessarily useful in reflecting user-pertinent semantics. The personalization task will attempt to bridge disparate information from the different LOD vocabularies to a more lightweight and meaningful (to the user) knowledge base in the interest of alleviating the processing and storage load. Minimizing this load is expected to minimize the need for server-client communication, thus venturing toward better safeguarding user privacy.

Such an ontological knowledge base should be able to a) support meaningful representation of world semantics under a single uniform vocabulary, b) encompass the minimum possible concept space among the ample information in the networked media domain with regard to addressing user needs and c) sustain abstract user- and context-specific conceptualizations such as user status, skill, and situation.

To this end, the first version of the LinkedTV User Model Ontology (LUMO)[20] has been composed, by adopting and extending existing broadcasting-related and user-related ontologies. LUMO brings together context-related (sensor data, user situation) and interest-related semantic information, as depicted in the top-level hierarchy snapshot of Figure 9.3.

LUMO is an OWL-DL ontology, with its expressivity, however, limited to the DLP fragment [382], in order to maintain the minimal complexity per the LinkedTV requirements, address scalability and user privacy issues, and at the same time render it able to encompass rules that can represent more complex

---

[20]http://data.linkedtv.eu/ontologies/lumo. See also http://mklab.iti.gr/project/lumo

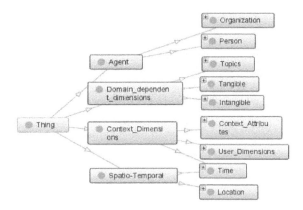

FIGURE 9.3: The top level concepts of the LUMO ontology.

user-specific knowledge. Should it be extended to OWL 2[21] semantics, its expressivity will correspondingly be restricted to the OWL 2 RL[22] fragment.

LUMO engineering aims to model the most relevant entities and semantics from open vocabularies and adapt them to the networked media domain and the needs of the LinkedTV users. This includes adopting, adding and appropriately discarding domain-extraneous information from relevant established vocabularies, as well as redefining semantics, with respect to leveraging LOD inconsistencies and enhancing the coherency and completeness of modeled semantics. The selection of the appropriate entities and semantics imported from source ontologies to the LinkedTV reference ontology will focus on:

- Addressing main conceptual features of the LinkedTV scenarios (news, art and culture with focus on antiques).

- Addressing the most relevant sensor information obtained through behavioral tracking.

- Addressing the specific contextual parameters pertaining to the users of the LinkedTV platform.

- Finally and foremost, minimizing the concept space in such a manner that will render it as lightweight as possible, though complete enough to make predictive inference based only within the LUMO concept space, with regard to scalability and user privacy safeguarding issues. This includes:

  - adapting relevant information based on existing vocabularies under a single, uniform and well-structured reference ontology, i.e., without importing related ontologies per se, and

---

[21] http://www.w3.org/TR/owl2-overview/
[22] http://www.w3.org/TR/owl2-profiles/#OWL_2_RL

- detaching the reference ontology to be used for modeling user interests and infering from its mappings to existing related vocabularies.

The major vocabularies that have inspired and guided the engineering of LUMO are:

*GUMO* has guided the design of the top level context and domain-dependent dimensions, while several GUMO contextual dimensions were adopted and adapted to LinkedTV's platform requirements (including camera-based sensor related concepts). Similarly, some LUMO "Topics" have been modelled to a great extent according to the corresponding GUMO "Topics."

The *IPTC news codes* were the main influence towards modeling the LUMO "Topics" hierarchy. Most of the upper topic categories were adopted per se, and subcategories and concepts related to them where revised and adapted to the topics' hierarchy. For instance, several additional subcategories were created (w.r.t. also to Wikipedia categories), while concepts in IPTC that were semantically direct descendants (with the "is-a" relationship) of the "Tangible" and "Intangible" LUMO categories were moved to that sub-hierarchy and related to "Topics" with the "hasTopic" and "hasSubtopic" object properties.

*Schema.org* influenced the modeling of the "Agent," "Location," "Intangible," and "Tangible" sub-hierarchies. These categories where also populated in correspondence to information from the *DBPedia schema* and the *NERD ontology*; to a smaller extent, they were influenced by several other open vocabularies. Modeling these aspects heavily relied on representing information stemming from content annotation within the LinkedTV platform.

### 9.3.2.1  Mappings

As a result of the principles described before, mappings to LOD vocabularies that semantically describe content are available, though detached from the actual reference ontology. Consequently, the current version of LUMO is accompanied by a separate mappings ontology[23] that maps LUMO to existing vocabularies.

More specifically, mappings in LinkedTV aim to align multilingual and cross-vocabulary information under the uniform LUMO vocabulary, and serve as the means to a) interpret content annotation and b) facilitate re-use of the ontology by the Semantic Web. To this end, a single, English conceptualization for each entity in the LUMO concept space is adopted and the separate mappings archive includes all the necessary information to reflect multilingual, multifaceted content annotations under these conceptualizations. Therefore, the user model can be represented in the minimum representative concept spaces and recommendation algorithms can avoid redundant steps in the inferencing process, such as dbpedia:Building ≡ schema.org:'Civic Struc-

---

[23]http://data.linkedtv.eu/ontologies/lumo_mappings

ture' ≡ lumo:Building or de.dbpedia:Fubal ≡ dbpedia:AssociationFootball ≡ Lumo:Football.

Mappings were generated automatically via the LogMap tool [472] and evaluated and revised manually. Currently, mappings are available to the main vocabularies that influenced the engineering of LUMO: DBPedia schema[24], schema.org[25], NERD ontology[26], IPTC news codes - as categorized by the WebTLab[27] and GUMO[28].

In future extensions, we consider indexing mappings under the annotation description of each class, through a dedicated annotation property. E.g., given a class lumo:Building, if it is found to be semantically equivalent with the class dbpedia:Building, then instead of adding the concept dbpedia:Building to a mappings ontology by an axiom dbpedia:Building ≡ lumo:Building, we can instead add a predefined descriptive property in the annotation of lumo:Building denoting that dbpedia:Building describes the class:

```
<owl:Class rdf:about="lumo:Building ">
     <isDescribedBy> dbpedia:Building</isDescribedBy>
</owl:Class>
```

The purpose of this considered approach is twofold. First, it will accommodate representing the (fuzzy) concept similarity between mapped concepts, assessed by some ontology alignment tool like LogMap, i.e., we can assign a degree of certainty that the mapped class describes the seed class. This degree will improve the quality of the classification as it can be used to modify the confidence degree by which the given concept participates in the content and to some extent, in the user profile. It is also expected that by using such indices the classification process can be conducted through simple search and retrieval modules rather than real-time inferencing in a mappings ontology, thus rendering it arguably faster and lighter.

The second justification is hypothetical and based on the expected outlook of the semantic interpretation of the additional content provided by the content analysis framework in the LinkedTV project (cf. Section 9.3.1). We might assume that real-time semantic classification of additional content might be incomplete, or at times infeasible, due to resource limitations and lack of information, but textual descriptions (tokens) might be easily extracted. Thus plain textual metadata might also need to be considered alongside available semantic metadata. Indexing prominent textual descriptions along with semantic descriptions at the schema level is expected to render the classification process more agile and efficient.

In essence, the proposed approach aims to aggregate all conceptual and lexical information that describe a concept under a single descriptive feature vector. In effect, these features will be used to classify the content based

---

[24]http://wiki.dbpedia.org/Ontology
[25]http://schema.org/docs/schemaorg.owl
[26]http://nerd.eurecom.fr/ontology/
[27]http://webtlab.it.uc3m.es/results/NEWS/subjectcodes.owl
[28]http://www.ubisworld.org/ubisworld/documents/gumo/2.0/gumo.owl

on its semantic annotation or textual metadata, under a uniform ontological vocabulary with reduced dimensionality.

### 9.3.3    User Interaction Listeners

This section describes the mechanisms that will be used to track user interactions within the LinkedTV platform and extract descriptive features (transaction statistics, behavioral analysis, and social activity mining) that will determine which actions are performed by the user in relation to consumed content. These features will subsequently help detecting which given content item/fragment/object is most interesting/uninteresting to the user, and thus determine exactly which concepts the user implicitly demonstrates a preference for. They will also indicate the nature of that preference (positive/negative) and the impact of the preference on the profile.

#### 9.3.3.1    Transaction Tracking

The General Analytics Interceptor (GAIN)[544] tool will be used in order to collect information about the users' transactional behavior. This module sends information about the user behavior as events (user paused, skipped video, viewed additional content, etc.), which occur during the interaction of user with the interface. Events are sent from the tracking module to the predefined tracking API for processing.

GAIN exposes a REST[29] API. The current first version 0.1 provides mainly services for tracking and services for retrieving simple aggregated statistics about interactions. The next version of the API will provide services to get the results of tracking analysis suitable for update of the user profiles.

The tracking API processes events from many interfaces and is designed as a scalable service. The API provides services that can:

1.  collect interactions  e.g., play, pause, skip, etc.

2.  retrieve aggregated statistical measurements about interactions, e.g., number of interactions per day, average number of interactions per video, etc.

3.  retrieve results of analysis for each user - e.g., interest level per shot, video, content item, etc.

The input of the tracking mechanism is derived from the tracking module integrated into the player (click-through behavior) and from the annotations of the media content. Such click-through behavior might incorporate interactions on the LinkedTV player (play, pause, skip, etc), clicking on a triggerable (semantically interpreted) object in the multimedia content, clicking on a recommended concept, or clicking on a recommended content item. User reaction

---

[29]http://www.ics.uci.edu/~fielding/pubs/dissertation/rest_arch_style.htm

information will also be incorporated into the GAIN tracker, so as to provide a uniform output to the profiling component.

The output results are in the form of aggregated stats about the user behavior (e.g., average number of interactions per video) and the levels of user interest (e.g., interest level per video shot). This information will be used as input for personalization and recommendation of media content and can be also used as updates of user profiles.

### 9.3.3.2 Reaction Tracking

A Kinect [672] sensor which is able to acquire both color images and a depth map is used. The audio capabilities of the Kinect are not considered at this point. The use of the Kinect for implicit behavior tracking is mainly focused on two aspects: inferring the user context and the user interest while viewing some content.

The first aspect involves determining the user context to adapt the user profile in contextual instances. For example, is the user alone or not, are there children in the room, is there any light in the room or not? A Kinect-based system which counts the number of people in the sensor range has already been developed. This information can be fused with other contextual information, e.g., about the viewing hour or of the day of the week.

The second aspect is about complementing information able to provide a degree of interest of the viewer with respect to his/her reaction while consuming a content item (seed video) and its enrichment (media linked within the seed video). Current development is focused on monitoring user behavior during the watching of the seed video. Interest detection at given moments means that the user might be also interested in enrichment of the seed video (i.e., related content) at that precise moment.

The Kinect is located on the top of the TV, with a field of view of about 6 meters long by 4 meters wide, which fits quite well with classical TV arrangements in homes. It observes the scene, and is able to extract the following features:

- *User face recognition.* The user goes to a specific place and looks at the camera (this place is closer to the Kinect, as face recognition needs to analyze face details). The program analyzes the face using state-of-the-art algorithms like PCA decomposition [932] and asks the user if he validates the recognized profile. The user can validate, perform the recognition again, or register as new profile.

- *Regions of interest.* The user can select the sofa, for example, by a simple click. This area is automatically segmented, and the system knows each time the user is in this area. The sofa can be seen as an area where the user might be interested in watching TV. The user behavior tracking is only initiated if the user enters this area (Figure 9.4).

- *User head localization.* The user's head is localized in 3D coordinates as

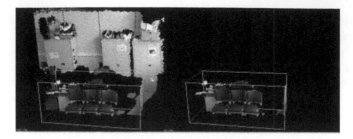

FIGURE 9.4: Left image: click on the sofa. Right image: the sofa is segmented.

a bounding cube centered on the skeleton head. At the same time, face detection is performed, to detect whether the face is looking toward the camera or not. This feature is not too sensitive as the face is well detected unless it has a direction very different from the one of the camera, but it is interesting to see if the user is really looking in a very different direction (something happens in the home, the user talks to someone else...). In this case we can infer that user attention is close to 0 during a period where his face is turned away from the TV set (Figure 9.5).

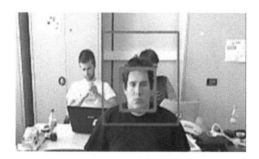

FIGURE 9.5: The face detector is present in a head area: the user is looking toward the camera axis.

- *User head direction.* After face detection, it is possible to extract the face direction. This direction needs the Kinect sensor to be closer to the face, but we can assume that the next generation of depth sensors will have a much higher resolution, and the same result can be achieved with a Kinect located above the TV. The result may be noisy, so a filtering post-processing is needed to detect the important head direction changes and especially the moments where the head points toward the TV for a period of time which is long enough to determine interest in the on-screen content.

- *User barycenter evolution.* The Kinect sensor can extract the user skeleton. This skeleton barycentre (which generally corresponds to the umbilical region of a person's body) is quite a stable feature compared to hands and feet. Thus, in a first approach, this feature shows the user excitement without explaining which body part is responsible for it. A change in excitement can be interpreted as a sign of interest in the content.

Regarding all these features, we must point out that they do not directly depict information on user behavior if taken globally. In effect, sudden changes relative to the user's short-term physical history and a combination of features will be more descriptive of the level of user engagement with the broadcasted content.

Indeed, if a user has been non-constant to his orientation toward the TV in his recent behavior but suddenly changes reactions and begins to look at the TV for a longer period of time, it is reasonable to infer that he might be more interested in the current content even if other aspects of his behavior (e.g., transactions with the platform) are neutral.

### 9.3.3.3   Social Transaction Tracking

Another possibility for generating and evolving a model of user interests is to use his activity in the Social Web as an indicator. Today's Internet users are sharing their interests using social platforms like Facebook[30] and Twitter[31], and a subject of research is whether past social web behavior can be a solution to the well known cold-start problem in recommendation systems. This was explored previously in the NoTube project, in a work that became known as the Beancounter[32].

The Beancounter is an online tool for generating a user model through semantic analysis of the user's posts/shares on specific social networks. For example, Twitter tweets are public and can be queried by an open API, while Facebook allows third party applications to access details of a user's timeline via its Facebook Connect API, in which the user logs in from the application to Facebook and expressly gives permissions for the application to read his or her posts. Many of the common social networks today support a specification called Activity Streams[33] which provides a structured model for the data being published over time (Facebook posts, Twitter tweets, etc). Activity Streams extend typical RSS/Atom feed structures (with a title, description, link, and some metadata potentially expressed for each posted item) with a verb and an object type to allow expressions of intent and meaning, as well as to give a means to syndicate user activities.

The Beancounter architecture makes use of components called Tubelets to interface between the Beancounter and different sources of social web data. The key goal of a tubelet is to be able to authenticate the application to the social web network's API and to retrieve via an API query a dump of the user's posts to that network over a period of passed time (the extent of the

---

[30]www.facebook.com/
[31]https://twitter.com/
[32]http://de.slideshare.net/dpalmisano/semtech2012-finalpdf
[33]http://activitystrea.ms/

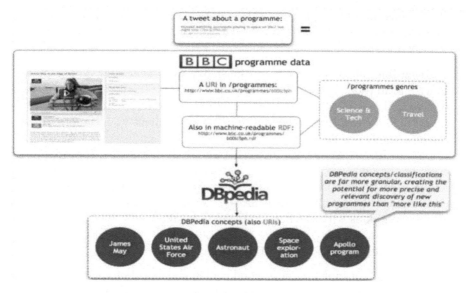

FIGURE 9.6: Illustration of Beancounter deriving topics from a tweet.

data returned varies from API to API). This dump, which is mainly textual (plus links to media), is subjected to entity extraction to derive a list of key "named entities" and number of occurrences from the user activity. For this, the Beancounter calls directly to the Alchemy API[34] to extract the named entities. Alchemy API supports the connection of named entities to concepts from the Linked Data cloud such as from DBPedia, Freebase, UMBEL[35], YAGO and OpenCyc[36]. This process is illustrated in Figure 9.6 with a tweet that mentions a BBC program. In this case, the RDF about the program is retrieved and an entity extraction is performed to derive from the tweet that the user has interests in the topics James May, US Air Force, Astronaut, Space Exploration and Apollo Program (all subjects of the tweeted program).

Hence for the user, Beancounter derives a weighted (by occurrence) list of Linked Data concepts. Also, each source can be given a different weighting, which is also applied to each concept in the list, based on from which source it was derived (for example in tests, public Twitter posts were weighted lower than Facebook posts shared only with friends, on the assumption private posts would more directly refer to an interest of the user than public tweets). As a result, an interest graph can be produced for a user made up of topics (using DBPedia URIs) each with a weighting value. Beancounter code is open source,

---

[34]http://www.alchemyapi.com/
[35]http://www.umbel.org/
[36]http://www.opencyc.org/

under the Apache 2.0 license[37]. It includes listeners for Facebook and Twitter, as well as a base implementation of topic weighting.

### 9.3.4 Capturing Implicit User Preferences

This section focuses on detecting user preferences based on the information tracked from the user's interaction with the platform. This includes understanding and representing the information extracted from the user's interaction with the seed media fragments within the LinkedTV platform, as well as with the provided enrichment (semantically related content). Effectively, it involves mapping the semantic description of the content to the lighweight and uniform chosen vocabulary (LUMO), determining the impact that a preference has to the user based on both its participation to the content and the behavior of the user towards it upon consumption, as well as the evolution of the preference over time and consumption. And finally, the serialization of the profile under a semantic description able to be used by the appropriate semantic inferencing engines for concept and content recommendation.

#### 9.3.4.1 Understanding Content

The first step of capturing user preferences involves translating the information about what the user has consumed into the lighweight and uniform vocabulary provided by the LUMO ontology. LUMO is designed to represent the backbone schema, i.e., concepts and properties, that describe the user preferences, and it cannot replicate the vastness of different specific individuals[38] who might belong to the networked media super-domain, but to a limited extent of few very concrete and finite individuals (e.g., days of the week).

To this end, a content annotation expressed at the schema level of an LOD vocabulary (referred to herein as class $D$) will be translated directly to a corresponding LUMO concept (referred to herein as class $C$), for which $C \equiv D$, expressed as an abstract concept which any individual might instantiate - in Prolog notation: $C(X)$, where $X$ is a variable that can be instantiated by any constant individual. Individuals in the annotation will be received directly from the LOD vocabulary, and instantiate the LUMO class that maps the more specific type of the individual subject in the corresponding LOD vocabulary, as a concept assertion of the form $\langle \alpha : C \rangle$, or more simply in Prolog notation $C(\alpha)$, where $\alpha$ is an LOD individual for which it applies $\alpha$ `rdf:type` $D$ and a mapping $C \equiv D$ applies.

For instance, if the semantic description of a content item that the user has consumed provides the more abstract annotation schema.org:Place which is mapped to lumo:Location, the preference *Location(X)* will be added to the

---

[37]https://github.com/dpalmisano/NoTube-Beancounter-2.0

[38]The term "individual" here is used in the same sense as in the OWL language, i.e., denotes a most specific entity (e.g., a particular person's or organization's name) which belongs to a more general class in an ontology's schema.

user profile[39]. Similarly, given a content annotation with the individual "White House" extracted from any given vocabulary, information about the entity will be received by the annotation process, letting us know that "White House" is of type schema.org:'Government Building'. The class schema.org:'Government Building maps to LUMO as a subclass of lumo: Building. Therefore the preference retrieved is an instance "White House" of the class Building, such as *Building(WhiteHouse)*.

In the LinkedTV content analysis context, the semantic annotation comes with a degree which represents the confidence with which the analysis process has identified the particular concept. Therefore, if a certain concept is present in the shot with a degree of 0.8, this degree also determines the primary impact of the concept-preference to the user before behavior analysis, such as (from the previous example): $0.8 \cdot Building(WhiteHouse)$.

**TABLE 9.1**
Direct Mappings: Mappings of Relevant schema.org Classes to LUMO Classes.

| schema.org | Mapping | LUMO |
|---|---|---|
| Government Building | $\equiv$ | Building |
| Place | $\equiv$ | Location |

**TABLE 9.2**
Direct Mappings: Mapping Entities to LUMO.

| Annotation (schema.org) | Annotation degree | LUMO preference |
|---|---|---|
| Government Building | 0.8 | $0.8 \cdot Building(White\ House)$ |
| Place | 0.3 | $0.3 \cdot Location(X)$ |

In the case of indexed mappings, the semantic similarity between two mapping concepts will appropriately influence the degree of the concept and to some extent its weight in the user's consumption history, e.g., the weight being the product of annotation degree of the LOD concept, and the semantic similarity of the LOD concept to a corresponding LUMO concept.

### 9.3.4.2 Understanding Preferences

The second step toward understanding what the user prefers is recognizing the nature (positive or negative) of the user interaction with the semantically interpreted content and the impact that this preference has on the user. The user can produce different kinds of implicit feedback while watching media content. There are three main categories of feedback:

---

[39]The "lumo" abbreviated URI is omitted herein in the Prolog notation for simplification purposes.

**TABLE 9.3**

Indexed mappings: Mappings of Relevant schema.org Classes to LUMO Classes.

| schema.org | Mapping | Semantic Similarity | LUMO |
|---|---|---|---|
| Government Building | isDescribedBy | 0.9 | Building |
| Place | isDescribedBy | 1.0 | Location |

**TABLE 9.4**

Indexed Mappings: Mapping Entities to LUMO.

| Annotation (schema.org) | Annotation degree | LUMO preference |
|---|---|---|
| Government Building | 0.8 | 0.72·Building(White House) |
| Place | 0.3 | 0.3·Location(X) |

- *Basic player interactions.* This involves capturing user interaction with the media player controls: Play, Pause, Skip, Fast forward, Rewind, Jump to.

- *Additional interactions.* This involves extended player controls and additional content interactions: Select object, Bookmark, View additional content, Percentage watched.

- *Physical interactions.* Actions based on tracking physical behavior: User leaves/comes to a region of interest; User sits down/stands; User is still/excited; User looks at TV/looks in another direction.

A combination of these interactions will express the level of user interest. Some of these interactions express positive interest, and some of them are negative. There are many approaches available, which can be used. The first approach is to manually define a set of heuristic rules to express preference impact per action and to apply learning using a genetic algorithm described in the following section.

*Weighting*

The impact of a preference will be depicted by a weight that will be assigned to each user preference. This weight will take into consideration a) the classification confidence of each consumed entity (cf Section 9.3.4.1) and b) a modification value based on the concrete user behavior towards that entity.

As mentioned, we have manually defined a set of heuristic rules that will define the level of user interest/disinterest based on the transactional and physical actions a user performs. Our idea is to transform a set of interactions on the fragment level into one number that will express the level of interest. Preferences may undulate in the [-1, 1] interval, with positive values ([0, 1]) representing interest and negative values ([-1, 0)) disinterest. Priorities per

performed action are also set. Only rules with the highest priority are activated. The number is normalized after the contribution of all activated rules
is summed up. If the sum is greater than 1 the result is 1, if the sum is smaller
than -1, the result is -1.

Table 9.5 lists an overview of the proposed heuristic transactional rules.
Stop and start times refer to the actual time, and not the playback time, to
detect the duration of the action performed.

**TABLE 9.5**

Heuristic Rules for Determining the Impact of an Action on the User
Preference.

| Action | Rule | Priority |
|---|---|---|
| Bookmark | Interest = 1 | 1 |
| View Additional Content | Interest = Interest + 0.2 | 2 |
| Skip | Interest = Interest − (time to end - current time)/total time | 3 |
| Play | Interest = Interest + 0.1 | 3 |
| Pause | Interest = Interest | 3 |
| Fast-forward | Interest = Interest − (stop time − start time)/total time | 3 |
| Rewind | Interest = Interest + (stop time − start time)/total time | 3 |
| Jump to | Interest = Interest - (stop time − start time)/total time | 3 |
| User situated in an area of interest | Interest = 1 | 1 |
| User sited | Interest = Interest + 0.5 | 3 |
| Sudden change in sitting: user standing | Interest = Interest 0.3 | 5 |
| User looking toward the screen | Interest = Interest + 0.2 | 6 |
| Sudden look toward the screen more than 5 seconds | Interest = Interest + 0.6 | 2 |
| Sudden change in body motion (barycenter) | Interest = Interest + 0.4 | 4 |

An example of how the interest values are computed based on the heuristic
impact estimation rules can be seen in Table 9.6. The interest level is computed
from actions listed in "User action subvector" using rules in Table 9.5. E.g.,
for the video "stadt" the interest is $0.6 = 3 \cdot 0.2$, based on the rule "View
Additional Content" $\implies$ Interest = Interest + 0.2.

**TABLE 9.6**

Preference Weighting.

| Video name | Semantic subvector | | | User action subvector | | | Interest level |
|---|---|---|---|---|---|---|---|
| | *Precise location* | *Sports* | *Archi-tecture* | *Book-mark* | *Additional content viewed* | *% skipped* | |
| prozess | Berlin | 0 | 0.8 | 1 | 3 | 0 | 1 |
| stadt | Berlin | 0 | 0 | 0 | 3 | 0 | 0.6 |
| infanticide | Postdam | 0 | 1.0 | 0 | 0 | 0 | 0 |
| ratze | Berlin | 1.0 | 0 | 0 | 0 | 1 | −1 |

*Weighting using a genetic algorithm from training data*

A second approach is based on learning the user interest level using a genetic algorithm (GA). The advantage of this approach is that it depicts the result as one value, which represents user interest. This approach is based on the application of symbolic regression on the training data. The training data are obtained during experiments, using questionnaires and values assigned by an expert. An example of training data can be seen in Table 9.7. Symbolic regression evolves an algebraic function, which will take into account available variables from the User Action Subvector (example in Table 9.7).

Symbolic regression is based on GA. GA has a set of equations (individuals) and combines these equations during the evolution process to find better equations. An example of such an equation is $Interest = 0.5 \cdot Bookmark + 0.1 \cdot additional\ content\ viewed - skip$. The results of this equation for each row in Table 9.7 represent the interest level (as seen in Table 9.8). The quality of the evolved equation is defined as the fitness of this equation. Fitness is actually the reciprocal of the total sum of the difference between the ground truth value and the computed value. Fitness is higher if the results of the equation are more similar to the ground truth derived from the training data. The result of symbolic regression is the best equation from a set of equations observed during the learning process, and this equation will finally be used as the model to compute interest level. Examples of differences between ground truth and the results of the learned model are seen in Table 9.8 . More details can be found in [543].

Fitness denotes the quality of the solution. In this case, fitness is in reciprocal proportion to the model error on the training data. Model error is computed as the absolute value of the difference between the estimate and the ground truth.

It should be noted that genetic algorithms generally do not guarantee to find the optimal solution. The quality of the model and its true error on unseen data can be verified using a separate validation dataset.

**TABLE 9.7**

The Values in the Last Column Were Assigned by an Expert.

| Video name | Semantic subvector | | | User action subvector | | | Ground truth conver- sion level |
|---|---|---|---|---|---|---|---|
| | *Precise location* | *Sports* | *Archi- tecture* | *Book- mark* | *Additional content viewed* | *% skipped* | |
| prozess | Berlin | 0 | 0.8 | 1 | 3 | 0 | 0.9 |
| stadt | Berlin | 0 | 0 | 0 | 3 | 0 | 0.5 |
| infanticide | Postdam | 0 | 1.0 | 0 | 0 | 0 | 0 |
| ratze | Berlin | 1.0 | 0 | 0 | 0 | 1 | −1 |

**TABLE 9.8**

Fitness Value Computation Example.

| Video name | User action subvector | | | Fitness computation | | Error |
|---|---|---|---|---|---|---|
| | *Bookmark* | *Addi- tional content viewed* | *% skipped* | *Ground truth conver- sion level* | *Computed with GA formula* | |
| prozess | 1 | 3 | 0 | 0.9 | 0.8 | 0.1 |
| stadt | 0 | 3 | 0 | 0.5 | 0.3 | 0.2 |
| infanticide | 0 | 0 | 0 | 0 | 0 | 0 |
| ratze | 0 | 0 | 1 | −1 | −1 | 0 |
| | **Total error (negative fitness)** | | | | | 0.3 |

### 9.3.4.3   The User Model

The user model will consist of concepts from LUMO *only*, or individuals instantiating again *only* LUMO concepts, but also of more complex preferences in the form of logical expressions, denoting associations of primitive concepts in the user's consumption history. Primitive concepts detected in the first transaction of the user with the platform will initialize the user model if no previous explicit information is provided.

From then on, implicitly learning user preferences involves the aggregation of the information in the user's consumption history over time. The learning mechanisms will have two main foci: a) adding new or updating preferred concepts, and their preference weights based on their frequency, age, and utility over consecutive consumptions and b) discovering persistent associating relations between concepts.

Whenever a new concept or instance appears in the user's consumption history, it is added to the user's preferences. This preference will carry a weight that expresses the level of interest or disinterest of the user to the concept/instance based on the three aforementioned facets: the participation of the concept/instance in the content item consumed, the transactional and the reactional behavior of the user toward the content item that is described by this entity.

The age of the preferences is decayed over time based on the recency factor introduced in [874] defined as $decay = 2^{-\lambda(t-t_{u,c})}$, where $\lambda \in [0, 0.1]$ is the decay rate. The higher the $\lambda$, the lower the importance of past preferences compared to more recent ones is [874]. Variable $t$ denotes the current time and $t_{u,c}$ denotes the last time the user $u$ has accessed the concept $c$, i.e., the last timestamp for concept $c$.

The preference weight is then updated by the formula:

$$\sum_i w \cdot \frac{f}{max(f) \cdot decay}, \tag{9.1}$$

where $w$ is the weight with which the preference appeared in the last transaction, $f$ is its frequency of appearance in the user's history, and $max(f)$ is the frequency of the preference that has appeared more frequently than all the others in the user's history.

Separate instances of the user profile will be stored for each different contextual situation of the user, thus each contextual profile will carry different information on frequency and time. In the case of contextualized user models we might consider modifying the variable $t$ to $t_{\text{last}}$, denoting the last time the user was in this contextual situation, and conversely use the relation between $t$ and $t_{\text{last}}$ to determine the importance of this contextual situation to the user.

This weighting scheme will update user preferences over time and content consumption. Further weight adaptation will be achieved via the estimation of utility functions [523] among the upper taxonomy concepts of the reference ontology. In addition, a pruning threshold will be statistically determined in order to delete the most obsolete preferences (based on their weight) and maintain a manageable profile size. It is also considered that a significance threshold will be similarly established in order to spare the filtering algorithms from excessive data overhead, so only the top-N semantically significant concepts might be passed along to the filterer.

The learning process will also take into account association rules following the approach of [966], that will enable complex preferences (logical expressions) that denote persistent associations between concepts to be included in the user model, such as simultaneous conditions (e.g., the user wants to view a concept *only in conjunction with* another concept, fine grained negation of a concept, or conditional associations (e.g., interest in a preference only if a specific (possibly contextual) condition applies).

Association rules will be conveyed in a machine-understandable semantic representation that will enable their use through semantic inference engines such as reasoners. The semantic interpretation would comprise translating the rule into a DL axiom, where the body contains the complex relationship and the head contains an interpretation of the rule's impact by attributing an impact weight derived from the rule learning process. For example, a complex preference *Rule1*, that represents the interest of the user to content about local sports clubs around the user's vicinity would read $SportsClub(X) \sqcap UserLocation(Bradenburg) \sqsubseteq 0.8 \cdot Rule1$.

*Modeling the user*

In order to support the information derived from the profile capturing and learning processes, the user profile (and the different contextual instances) will be expressed in logical axioms within the DLP [382] expressivity fragment:

(1) $\bigsqcup_n \exists hasPreference.Preference_{i \in n} \sqsubseteq Interests$,

(2) $\bigsqcup_m \exists hasPreference.Preference_{j \in m} \sqsubseteq Disinterests$,

where

(3) $Interests \sqcap Disinterests \sqsubseteq \bot$.

(1) denotes the disjunction of all preferences that are interesting to the user, (2) denotes the disjunction of all preferences that are uninteresting to the user (what the user rejects), and (3) denotes that interests and disinterest are disjoint. The concepts that are recognized in the preference extraction process as negative, and thus denoting disinterests, will be induced to the disinterests with an absolute weight, thus activating the detection of inconsistencies based on the formula (3).

DLP is the expressive intersection between Description Logics and Logic Programming. It provides the means to "build rules on top of ontologies" [382] and thus facilitates the inclusion of more complex preferences in the user profile such as logical expressions but still retaining the complexity of the model on a relatively low level, i.e., in a subfragment of the full DL expressivity. Such complex preferences might include association between concepts via Boolean constructors (and, or), association between concepts via disjointness but also restriction of a concept to a specific property and association between properties via inclusion, transitivity and symmetry.

The modeling schema and expressivity fragment are deemed as the most complete and at the same time most lightweight method to convey existential/universal restrictions and boolean constructors (e.g., derived from association rules), disjointness and negation (e.g., to express disinterests) and fuzzy logic (uncertainty in content annotation, i.e., degrees, preference weights and possibly thresholds derived from association rules).

The user model serialization will employ a variant of the KRSS2[40] repre-

---

[40]http://dl.kr.org/krss-spec.ps: the first version of the KRSS syntax. KRSS2 supports additional DL compliant semantics, but no documentation of the additional semantics is currently available online.

sentation language. The main advantage of the KRSS2 vocabulary is that it is significantly more lightweight than other notations. The lightweight representation aspires to enable future storage (and even filtering) on the client.

### 9.3.5 Semantic Filtering

More than one methodology will be considered for filtering within LinkedTV, in order to achieve fast and efficient delivery of personalized concepts and content to the user. This section will briefly delve in the employment of a fuzzy semantic reasoner that enables semantic matching between content and the user profile. The reasoner, namely f-PocketKRHyper is an adaptation of an existing implementation [922] to the LinkedTV needs.

It consists of an extension of the Pocket KRHyper [522] mobile reasoner, which allows for the inferencing and recommendation process to take place either on the server or on the end-device seamlessly and effectively, while providing slim and meaningful results. PocketKRHyper is a (crisp) first-order logic (FOL) theorem prover that borrows semantics from Logic Programming (LP), designed to be used in limited resource devices such as previous generation mobile phones. It implements LP-like semantics for the DLP fragment of FOL [382], while additionally providing an interface for transforming DLs to first order clausal logic (again with LP semantics, so for instance logical implication is not supported).

f-PocketKRHyper has extended the original implementation's DL interface to compensate for missing DLP semantics (disjointness, negation added), while providing additional support for uncertainty handling in both annotation and user preferences, based on Zadeh's fuzzy sets [1027] and Straccia's concept weight modifiers [129], respectively.

f-PocketKRHyper has proven to offer significant accuracy in semantic matchmaking between a given user profile to a set of content items [922]. Its accuracy nonetheless, as in any reasoning service, depends on the correctness and completeness of the input data, i.e., the reference knowledge base, the annotation, and the user profile.

---

## 9.4 Conclusions and Future Work

This chapter conducts an overview on different aspects, knowledge bases, and technologies involved in the process of personalizing and contextualizing user experience in networked media environments, and presents the personalization approach within the LinkedTV platform. The approach involves determining the appropriate background knowledge, semantically interpreting user transactions based on this knowledge base, interpreting the level of interest/disinterest for a given preference based on the user's transactional and reactional behavior and learning a user model through iterative adaptation of the user profile over time.

Future work will investigate appropriate probabilistic methods to infer the impact of the preferences in a given transaction of the user with the system based on his behavior to improve the current heuristic methodology. The modeling task will also effectively extend to determining the contextual parameters of the user and adapting the long-term user model to different contextual situations.

Contextualizing user preferences will also serve to minimize the volume of background knowledge that will be active at each inference session, since in a multidiscipline domain such as digital media real-time inferencing is still hampered due to the unnecessarily high complexity that the large concept space introduces, i.e., a large terminological box that has no relation to the domain or the situation of the user at a given session, however light and compact the reference knowledge might be. Therefore, we will further explore and foster manifold user knowledge acquisition and adaptation techniques, such as pulling subsets of the reference knowledge based on user context to reduce the dimensionality of background knowledge.

Finally, we will extend the inference engine to more functionalities, mainly by expanding supported expressivity to fully exploit the KRSS2 fragment, in direct correspondence with the OWL 2 RL fragment expressivity. Further extensions to address required fuzzy semantics stemming from preference learning algorithms are also considered. Such extensions include the introduction of weighted sum and threshold semantics [129], as well as the implementation of fuzzy role assertions.

# 10

# *Approach for Context-Aware Semantic Recommendations in FI*

**Yannick Naudet**

*Henri Tudor Public Research Centre, yannick.naudet@tudor.lu*

**Valentin Groués**

*Henri Tudor Public Research Centre, valentin.groues@tudor.lu*

**Sabrina Mignon**

*Henri Tudor Public Research Centre, sabrina.mignon@tudor.lu*

**Gérald Arnould**

*Henri Tudor Public Research Centre, gerald.arnould@tudor.lu*

**Muriel Foulonneau**

*Henri Tudor Public Research Centre, muriel.foulonneau@tudor.lu*

**Younes Djaghloul**

*Henri Tudor Public Research Centre, younes.djaghloul@tudor.lu*

**Djamel Khadraoui**

*Henri Tudor Public Research Centre, djamel.khadraoui@tudor.lu*

## CONTENTS

This chapter addresses personalization of multimedia content in the Future Internet (FI) from a recommendation perspective. It presents recent research results towards providing context-aware recommendations in the Future Internet, based on the semantic modeling of knowledge and recommendation approaches, mixing both semantic and fuzzy processing for better personalization. It focuses on a Hybrid Ad-Hoc Network environment, constituting a common communication infrastructure today, where people share and manipulate multimedia content on both fixed and mobile network nodes. Context-awareness in recommender systems is addressed together with the needs for semantics and fuzzy user preferences. A context-awareness platform based on the Content-Centric Network paradigm is presented, together with ontologies for context and situation, thanks to which contextual elements can be formalized. It feeds a knowledge-based context-aware recommender with contextual data. In the presented recommendation approach, Fuzzy theory is used in addition to semantic modeling and processing, to better represent user preferences and interests, and the weight of context regarding these interests. An ontology for handling fuzzy properties is given, as well as the extension of the semantic matching formula that computes recommendation scores. Simulated as well as user-based experimental results are provided.

## 10.1   Introduction

Today, the personalization of multimedia content is a significant concern. It enables people to not only navigate the Internet but also to experience what suits their needs and meets their preferences. Especially in the scope of the Future Internet, personalization as content filtering and recommendation as well as personalization as content adaptation, including context-awareness, must be handled seamlessly and intelligently to respond to people's needs independently from who or where they are, from the technology they use to navigate the Internet, and the time at which their need emerges.

Internet has become a common component in the everyday life of a majority of people in our modern societies. Emerging from computer networks, it is now accessible from more and more mobile devices, ranging from our smart phones and tablets to computers embedded in vehicles. A mobile Internet has been born, where fixed and mobile nodes communicate, running on various heterogeneous networks. The diversity of these nodes has moreover extended from classical server and client devices, to more dedicated ones like, e.g., sensors. New trends, needs, and usages have appeared, together with new related services and technical challenges to solve. All of this tends to one single communication paradigm, which is a main objective for the Future Internet: *access to the right content, anywhere, anytime.*

In the current visions on the Future Internet, content- and user- centricity, as well as context-awareness, are the main concerns [924]. Hence, the three pillars for the Future Internet are: *content, user, context.* More than information or data, content has become the main thing exchanged across networks. To ensure suitable exchanges, mobility implies taking into account contextual information concerning e.g., the user himself and his environment, the network's availability and physical constraints limitations, and environmental conditions. Generally, this means that awareness of both user and context has to be considered: knowledge of user preferences and interests allows personalizing services (e.g., content filtering, recommendations, content adaptation), while context knowledge allows service delivery in best conditions (by adaptation of, e.g., format, schedule, routing path, etc.). Users are put back at the center, and their related context, or situation, is taken into account to provide them content best fitting their needs and interests. However, despite extensive research on ubiquitous and pervasive systems, context aggregation and handling in today's communication networks remains a challenging problem, as the mobility of nodes and users makes information routing as well as context processing difficult [701].

User-centricity can often be materialized by personalized services, which can be split into two main categories: adaptation and filtering / recommendation. The latter has become a current trend in most of e-commerce web sites as well as on well known search engines, making more and more common the presence of scripts tracking the user and gathering data for different purposes. Context-aware recommendation is the next step currently targeted, especially because the user has become mobile and uses multiple devices. But user-centricity also implies that users create and exchange content not only through fixed network infrastructures, but also through direct ad-hoc connections.

In the remainder of this chapter, we introduce Hybrid Ad-Hoc Networks (HANETs), which are common in today's communication networks where nodes can be both fixed and mobile. HANET is an extension of the Mobile Ad-Hoc NETwork (MANET) architecture, where some (fixed) access points are available. They constitute a real issue for efficient context data retrieval and processing, as well as for data transmission. On the one hand, information is transmitted through heterogeneous networks, while on the other hand the user is mobile. At any time, networks and information (including context data) available to a user are different. Moreover, if the user moves during transmission the available routing paths and information sources change. Handling context-awareness and providing context-aware recommendations in this kind of network requires solutions that are flexible enough to abstract the heterogeneity of networks and the mobility of nodes. We present such a solution based on the Content-Centric Network (CCN) approach, grounded on semantic web ontologies. The resulting context-awareness platform feeds a knowledge-based recommender with contextual data used to weight and filter user interests.

After introducing context-aware semantic recommendation principles as well as the limitation and needs of current recommendation approaches, we propose tools to bring context-awareness into HANETs: ontologies for context and situation, and the Clairvoyant platform for context-awareness (CLV-CAP). The recommender system embedded in the CLV-CAP, namely F-Sphynx, is then presented, together with its user model, eFOAF, and its matchmaking algorithm. We then discuss the use of fuzzy user preferences in recommendations, which can be modeled using the FuSOR ontology and processed during the matchmaking of user interests for items to recommend. Then, the handling of context and the way its importance could be personalized in interest matchmaking is discussed. Finally experimental results involving real users or in simulation are provided, assessing respectively the semantic similarity measure used by the recommender, the whole context-aware recommendation using CLV-CAP, the usefulness of fuzzy preferences, and the weighting of situation and interests.

## 10.2 About Context-Aware Semantic Recommendations

### 10.2.1 Context Awareness

From a systemic perspective [963], context-awareness refers to the ability of a system to sense or be aware of contextual elements, and react accordingly. The following definition can be given [699]:

**Definition 3** *Context-awareness is the ability of a system, knowing its state and its environment, to sense, interpret, and react to changes in its state or in the environment it is situated in.*

This definition considers both what the system knows, and what it learns from observations of the world. The former can be technically assimilated to the system's knowledge base: it contains statements about the world that the system is aware of, including itself. The latter can be obtained through three phases: *sensing*, the gathering of contextual elements from some sensors; *interpretation*, where gathered data are fused, aggregated, and processed; and *reaction*, the actions taken by the system in response to the interpretation phase output.

Context is a controversial notion that has brought many discussions among researchers from different domains. As it can be interpreted differently depending on the domain or application, it should firstly be understood from a wide perspective. Dey and Abowd [282] have provided such a generic definition, which is still widely referred to:

**Definition 4** *Context is any information that can be used to characterize the situation of an entity.*

In this definition, *entity* has a very wide sense and refers to anything of interest.

The concept of context is highly connected to that of situation, and they are often used as synonyms. Without falling into semantic debates, we can draw on the latter definition of context to assume that context is more generic than situation, which is necessarily linked to some entity. Hence, the following definition can be given [699]:

**Definition 5** *An entity's situation is the entity's state and activity in a specific environment at a given point in space and time, inferred from the context elements.*

The completeness of a situation estimated by computation obviously depends on available knowledge and observables. Moreover, situations of interest vary according to the application domain. Hence an entity's situation is a subset of the domain or application-specific situations of interest, which is a set of partial situations related to this entity that are determined according to the available knowledge and observations. Situations are valid at a specific point in time and space. Then, they concern either (1) the state of an entity (e.g., the road is wet); (2) its activity (e.g., the user is sleeping); or (3) its environment. Assuming the environment is a set of entities in relation with the target entity, situations related to an entity's environment relate to the presence, the state or activity of other entities in this environment (e.g., the presence of snow on the road).

## 10.2.2 Recommender Systems

Initially used mainly for academic purposes, the Web has nowadays taken an important place in a very large number of people's life around the world, becoming every year more popular. Young people are now spending more time on the Internet than watching TV. Moreover, the advent of the Web 2.0 has completely changed our usage of the Web, turning passive actors into active contributors. For instance, more than 20 hours of video are uploaded each minute on Youtube, while Amazon sells hundreds of thousands of products around the world every day. This popular use of the Web as a global information system confronts us with an incredible amount of data and information. Hence, users are confronted with the same problem: how to filter and find the needed information in what seems to be tremendous chaos. From this situation, an important research field has emerged. Its objectives are to find some innovative and efficient techniques to access information, and to design automatic agents capable of intelligently assisting us in transforming this chaos into personalized and useful information. Recommender systems have appeared in this framework. Burke [182] describes the goal of recommender systems as guiding the user in a personalized manner to interesting items within a large space of possible options. Karypis [281] defines a recommender system as a personalized information filtering technology, used to either predict whether a particular user will like a particular item, or to identify a set of items that will be of interest to a certain user. According to Schafer [812], recommender systems are systems that provide users with an ordered list of items and information that help them decide which items to consider, or look at, based on the individual user preferences. Along the same line, Porcel [756] describes that recommender systems help online users in the effective identification of items suiting their wishes, needs, or preferences. They have the effect of guiding the user in a personalized way to relevant or useful objects in a large space of possible options. Generally, as stated already in 1997 by Resnick [781], a recommender system aims at providing personalized suggestions about items, actions or content considered of interest to the user.

Different approaches exist, taking into account either the user's own interests (content-based filtering approach to recommendation), or the neighborhood of content or users (collaborative filtering based on content consumption in a community of users, through content or users' similarity computations). For a few years now, some well known e-commerce web-sites propose such recommendations with good success, based on records of user actions, user ratings, or correlations between different users or consumed content. Different domains have been targeted: e.g., cultural and other kinds of product recommendations at Amazon.com [603], movie recommendations on MovieLens [677] or Netflix [1045]. In the multimedia domain, recommendation has multiple applications especially related to television, to filter out program guides, recommend what to watch, propose targeted advertisements or additional content to consume, both on fixed and mobile devices (see [704] for examples).

Within today's mobile networks, the situation of the targeted user, or more generally the context, becomes an important element that can greatly influence the relevance of personalization. Achieving pertinent recommendations in such an environment bears some issues, especially because of the nodes' mobility. The latter implies considering contextual information concerning the user himself and his environment, but also the network's availability and physical constraint limitations, as well as any useful environmental and situational information.

### 10.2.3 Limitations and Needs of Recommender Systems

In this section we stress the remaining drawbacks of the most adopted recommendation approach, namely collaborative filtering (CF), and highlight the need for semantic, fuzzy preferences and context-awareness toward the building of more powerful recommender systems, able to make suggestions closer to users' expectations.

#### 10.2.3.1 Collaborative Filtering

While Content-Based Filtering (CBF) was the first approach used by recommender systems, it has been criticized for some drawbacks: the necessity of suitable descriptions for items, overspecialization [95], and a high computational cost, which can be an issue for dynamic and large environments with the frequent addition of contents [44]. Currently, CF is the most explored and adopted technique because of its efficiency and ease of integration. Introduced by Tapestry [371], it was automated by GroupLens [780] and Ringo [838]. The principle of CF is to recommend to the user items that people with similar tastes or preferences liked. One of the success factors of CF techniques is that they do not require any description of the items, and can give good results using only user ratings. In a Web environment, gathering a large number of ratings is often not a problematic task. Among a large number of successful implementations of this method are MovieLens [677], Amazon.com [603] and NetFlix [1045].

Some weaknesses of CF techniques have nevertheless been identified [53]: the cold start problem (new user and new item problem), the sparsity of data, the scalability, and the lack of neighbor transitivity [161]. The sparsity problem occurs because most often each user rates only a small number of items with regard to the global number of available ones. In this case, weak recommendations may be obtained because of the difficulty in successfully finding near neighbors. The scalability problem is because the computational cost of CF is growing fast along with the number of users and items. To explain the loss in not taking into account the neighbor transitivity let's assume that we have three users $a$, $b$, and $c$. Let's also assume that $a$ and $b$ are very similar, while $b$ and $c$ are also very similar. We could expect that the system would be able to take into account some kind of transitivity to represent the probability that users $a$ and $c$ are therefore also correlated. Standard CF techniques do not capture this transitive relationship unless users $a$ and $c$ have rated many items in common [965]. Besides, CF techniques are not suitable in environments where there is only one single instance of an item (like an event) and which as such cannot be repetitively proposed [240]. In the multimedia industry, a newspaper article published in *The Economist* in 2009 [907] reports the natural tendency of people to consume blockbusters despite the large choice they have, which leads to a weak use of content providers' back catalog. This suggests that classical CF approaches would finally tend to reinforce this tendency because of the group effect, giving an additional argument in favor of content-based recommendation approaches.

### 10.2.3.2 Need for Semantics

For a few years now, the Semantic Web [567] and its associated technologies has been gaining interest, and has been in particular used for pushing personalization to a semantic level, using ontologies. Brought back from Artificial Intelligence by the Semantic Web, ontologies provide a formal representation of a knowledge domain, whose semantics is defined by its composing concepts and relationships between them. The resulting knowledge model provides not only an unambiguous representation of important concepts of the represented domain, but also allows inference-making on knowledge repositories. In personalization, some of the benefits that can be quoted are: the reduction of ambiguities between user profiles and data description; the exploitation of structure between concepts, and of the semantics associated to properties; and reasoning based on semantics.

Using ontological user profiles allows to perform inferences to discover interests that were not directly observed in the user's behavior, or explicitly expressed [674]. Ontologies also bring an answer to the semantic ambiguity problem [942]. In addition, during the last few years the Linked Data initiative has kept growing, and now, billions of triples are available on the Web, in enormous datasets such as DBpedia[1], Freebase[2] or Geonames[3].

Using ontologies for content description or indexing, and at the same time for user modeling, allows better matching of content with users in an information filtering process or in content-based recommendations. As concepts used in content and user profiles are formally defined by a common represen-

---

[1]http://dbpedia.org
[2]http://www.freebase.com/
[3]http://www.geonames.org/

tation framework provided by ontologies, ambiguity in terms matchmaking is removed. This leads to more accurate results and opens the door to richer personalization thanks to inference possibilities. It is of particular interest in the multimedia domain to relate people's interests not only to the content itself, but also to other elements like the content author, related topics, technical characteristics, etc. (see, e.g., [704]). Semantic representations of multimedia content can be built from existing standards like, e.g., MPEG7 or TV-Anytime for TV. Additionally, these descriptions can be enriched by exploiting the large amount of metadata available on the Web, on dedicated web-sites like, e.g., IMDB[4], or on collaborative web-sites (e.g., DBPedia), or finally, any tag or annotation posted by the crowd of web users. The benefit of using ontologies in multimedia content filtering and retrieval has been shown in recent studies, among which we can quote [80, 181, 188, 513, 700].

Recommender systems based on ontological descriptions and reasoning can also be named knowledge-based recommenders, following the classification provided by Burke [182]. They are known in particular to avoid the problem of the lack of ratings for new items, or when there are too many items in regards to the number of users. While the classic content-based recommenders rely only on implicit profiling, using the history of user ratings to guess what users will consume next, knowledge-based systems can use both ratings and any explicit knowledge of users, plus what can be inferred.

While the computational cost of reasoning on semantic data remains high, ontological knowledge-based recommender systems relying on ontological descriptions and reasoning are suitable candidates for using the recommendation approach in the Future Internet. Assuming that homogeneity of metadata models and the interoperability between applications on the Web is ensured by ontologies, such recommender systems can access a large distributed knowledge base where data on users, items to recommend, and context can be retrieved and inferred to provide richer recommendations better answering actual users' needs.

### 10.2.3.3 Need for Fuzzy Preferences

Implicit profiling can predict user interests to a certain extent, based on their behavior and consumption history. When this is not possible (because, e.g., of information lack) or when letting the user control his profile by expressing his interests and preferences, accurate profiling (this time explicit) is not always possible.

Human beings usually characterize things using specific terms rather than precise values. Such terms as "young," "hot," or "far," can be called *linguistic values*. These are imprecise notions conveniently used to evaluate corresponding linguistic variables (respectively, "age," "temperature," and "distance"). Linguistic values can be represented by fuzzy sets usually used to express imprecise notions such as an age, a temperature, or a distance. Recent research

---

[4]http://www.imdb.com/

works [651] [757] have shown their usefulness for modeling user preferences. Applied to recommender systems, they can simplify the expression of such preferences and allow one to more precisely formalize the system's behavior at the boundaries of an interest.

More generally, the interest in fuzzy logic for recommender systems has been illustrated in, e.g., [651], [694] and [980]. Combined with description logic, fuzzy sets can be used to represent the membership of an individual to a concept. However, having to define such a degree of membership may not make much sense for a user. For instance, what does being "young" at 80% mean, or being "tall" at 20%? Therefore, it seems sound to first define linguistic values such as "young" by defining their associated membership functions [881], and let the user use the linguistic values only. Applied to properties in ontologies, this means assigning them (as an object's statement) a linguistic value that is formally defined by a membership function.

Such an approach has the advantage of allowing a user to associate to a linguistic value its own understanding of it. For example, one can define the concept "young" as someone of an age between 0 and 15, while another person can define it as being between 0 and 40. With existing recommender systems, it is often not possible to express such complex preferences as "I usually prefer restaurants where a lunch costs no more than 20EUR, but I could accept up to 25EUR even if I would be less satisfied." This is illustrated by Figure 10.1, where the *johnCheap* concept is defined as a membership function specifying that the interest for restaurants is the highest (=1) until a price of 20, and decreases for higher prices until 25, where it becomes null (=0). Using membership functions thus allows defining how the interest evolves when the recommended content deviates from an ideal preference.

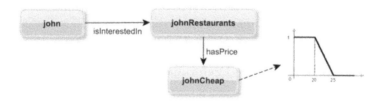

FIGURE 10.1: A user interest represented by a fuzzy set.

Finally, fuzzy logic can also bring some new possibilities [471][943]. It can, for example, help represent information provided by digital sensors. It would thus be possible to deal with imprecision, as in GPS positioning, environmental information like the temperature, or with not clearly quantifiable information like the user's state of mind. Users would be able to use linguistic values such as "near," "hot," or "sad" to express context-related information.

### 10.2.3.4  Need for Context-Awareness

The consumption of multimedia content today is highly context-aware. People manipulate such content using different devices, fixed or mobile, and thus in different locations and while they are in different environments. Content needs to be adapted to this changing context, and recommendation systems need additionally to deal with the changing user interests and needs.

Although context modeling and processing has been extensively studied for a decade in context-aware, ubiquitous, or pervasive computing research communities, it only started to be considered for personalization a few years ago. In particular, most of the existing recommender systems do not fully consider context information because of accuracy and reliability problems [1007]. According to a survey in 2009 [936], this was still an open issue a few years ago, although it is considered to be crucial for the success of personalization systems in real life. However, some works have shown the usefulness of context for automated personalization. In particular, information retrieval with context-awareness brings considerably improved results [188].

User preferences and interests are indeed context-dependent, and only few approaches try to take that into account, especially in an evolving context (see e.g., [703]). They can, however, be filtered according to a specific context or situation. Semantic models can help, by exploiting the semantic proximity between contextual data and the context of validity defined for user interests [704]. As context is time dependent, a personalization process taking into account an instantaneous context can also benefit from considering previous close in time contexts (see e.g., [693]). Finally, a recommendation requires finding the intersection of three elements: *user, things to recommend (e.g., content)*, and *context.*

## 10.3  Bringing Context Awareness in HANETs

Hybrid Ad-Hoc NETworks (HANETs) constitute one of the main communication network architectures supporting the Future Internet. In order to bring context-awareness into the Future Internet, it is necessary to create a common formalization of context data together with semantic interoperability among the context providers. In addition, HANETs require keeping network heterogeneity and mobility concerns away from the applications. Before presenting a complete platform for context awareness, this section details solutions to these issues: ontologies for context awareness and a network abstraction solution, Content Centric Networking.

## 10.3.1    Ontologies for Context Awareness

Multiple ontologies for context awareness have been created in the last decade. Depending on the application domains, they include different core concepts. For context-awareness in the Internet, the main core concepts are the user, the devices he uses, the content he manipulates, the network on which he communicates, and of course context and situation, which are completed by sensors and observations providing contextual elements. As stated in Section 10.2.1, context and situation are dual concepts. Although they might be specialized for each application domain, they can be generically represented by a set of core dimensions, which constitutes a common ground whatever the domain.

Building on the existing work, such generic ontologies for context and sensors-observations have been proposed in [699]. The context ontology (CO), whose core elements are depicted in Figure 10.2, is based on the generic con-

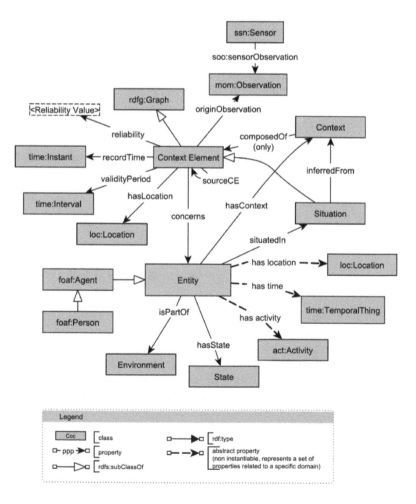

FIGURE 10.2: Context ontology core elements.

text definition of Dey [282]. The class *Context* is defined as a composition of *Context Element*, itself being formalized as a set of statements. The latter is defined as a subclass of a named RDF graph [194], to which a set of properties is added to characterize its origins, its validity and its reliability: *originObservation, sourceCE, hasLocation, recordTime, validityPeriod*, and *reliability*. The properties associated to *Entity* relative to fundamental context dimensions (see [699]) allow the expressing of statements about an *Entity* or its *Environment*, regarding *Environment, State, act:Activity, time:TemporalThing*, or *loc:Location*. The *Entity* related submodel in the ontology allows expressing such statements, which can be embedded into *Context Element* instances.

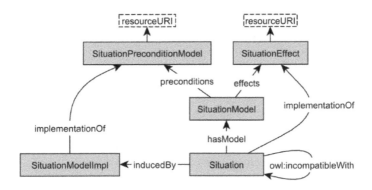

FIGURE 10.3: Situation ontology core elements.

Figure 10.3 illustrates the representation of the *Situation*. The *Situation* class is defined as a subclass of *Context Element*, considering that situations of entities are themselves context elements that are valuable parts of an entity's context, and can be used by applications to infer other situations or trigger actions. As an entity can indeed be affected by multiple situations, it is assumed that the global situation of an entity is the composition of all the situations related to it (through the *situatedIn* predicate). Situation instances that contain the instantiated situation effects are thus named RDF graphs, and have dedicated properties. Validity time, location and record time can be specified with the dedicated properties of the *Context Element*. The truth of a situation (i.e., its probability to be true) can be formalized using the *reliability* predicate. The subset of a context from which a situation is inferred can be directly formalized with the *inferredFrom* predicate, or indirectly using the *sourceCE* predicate attached to *Context Element*.

The *SituationModel* class represents the set of situation preconditions and effects. Preconditions, formalized by the *SituationPreconditionModel* class, allow representing a set of evidences that make a situation true. Hence, the occurrence of a situation is verified by searching for these evidences in the available knowledge base. Depending on the functionality and availability of the nodes, the latter could range from a small number of nodes sharing their data to the entire Internet. Effects, formalized by the *SituationEffect* class, will be used to define the effects of the situation, i.e., the state of the world after the situation has occurred. Multiple situation effects can be defined for the same preconditions, in order to represent the different effects a situation can produce, regarding the actual fulfilment of its preconditions. In addition, directly seeking facts related to situation effects in the knowledge base also gives insights on situations' truth.

### 10.3.2   Content-Centric Networking

The Internet and most of today's networks still rely on the client-server paradigm that was inherited from the origins of the computer networks. Peer-to-peer networks have started to change people's behavior with networked machines, and to drive their attention on the content they exchange. The presence of mobile devices facilitates the emergence of dynamic ad-hoc networks for short-life content exchanges. In these networks, the role and the behavior of the networks' actors has drifted from a simple producer-consumer dualism to a situation where each node may play several roles at a time, such as client, server, information relay, and data storage [260, 1030]. Finally, the recent interest in the Cloud infrastructure confirms that consumers are less, or no more, interested in the machines hosting the content than in the content itself. Machines tend to disappear behind services and content.

The classical client-server network model faces a growing number of caveats (see e.g., [701]). Node identifiers are strongly tied to the network topology, and thus can change with modifications of the latter. Another way is to move from a network of machines to a network of content, where the content becomes the addressable element instead of the machines, because it is the actual entity being manipulated by users. The concept of Content Centric Networks (CCNs) has gained interest in recent years [454], and in particular in the context of the Future Media topic of the European Future Internet Initiative (EFII). Content Centric Networks aim to focus on the delivery of content to users seamlessly, whatever the content and user location. The network itself is in charge of finding the content and delivering it from any number of locations where the content may be copied/cached. In CCNs, the fundamental unit is the data packet, and not the sender of this data. Multiple copies of a content item are stored in the network nodes, at router level, and delivered according to the needs, so that very popular contents can benefit from a large bandwidth, whereas contents with a limited audience do not need the same resources. Popular content is likely to be copied many times on the network, as opposed to scarcely used content that is only available from a few locations. Van Jacobson stated that "requests for content from a terminus do not have to go all the way to the host, only just as far as the first stored copy." [924].

To go further in the content-centric approach, the content itself can be represented as a Content Object, as suggested by [1030]. Such objects carry not only the content itself but also metadata, allowing applications to better manipulate and manage it. Although both concepts of Content Centric Network and Content Object have been proposed independently, they are complementary [343]. While the first one focuses on the network layer, and the transport and management of content, the second one concerns the modeling of the content itself, and its structure. However, CCNs do not specify the structure of content in their data packets and Content Objects do not specify the mechanisms for content delivery. Finally, they can be combined to form a COCN (Content Object-Centric Network). The latter is defined as a CCN where the content is represented in the form of a Content Object.

Context-awareness in HANETs can benefit from content-centricity and especially from the COCN approach. Context gathering from the network is made simpler by the CCN request mechanism, which can be used to ask for specific contextual elements as soon as they are recognized and identified as context-related content in the network. Context aggregation and reasoning is facilitated by the metadata bundled together with the contextual elements inside the content objects. The CLAIRVOYANT Context-Awareness Platform (CLV-CAP), which relies on these principles and technologies, constitutes a possible solution for bringing context-awareness to the Future Internet [701].

### 10.3.3 A Platform for Context Awareness in HANETs

The CLV-CAP is designed as a Multi-Agent System (MAS) middleware acting on top of the CCN and TCP/IP network layers. It provides a complete abstraction of the network layer, since applications using context data simply need to implement an agent able to communicate with the MAS of the CLV-CAP. The core agent set comprises, in particular, agents dedicated to contextual information gathering and aggregation, information filtering, data query, and sharing. In addition, user interfaces have been implemented for fixed and mobile devices, allowing the definition of agent behavior regardless of agents' implementation. The best possible information spreading and diffusion over both mobile and fixed networks infrastructures is ensured using the CCN communication protocol, at the agent level. CLV-CAP relies on ontologies for modeling all the elements of context-awareness (i.e., entity, context, situation, environment, device, network, user, etc.). The use of these ontologies ensures semantic interoperability and allows reasoning on knowledge available in the network, across the different layers of the context-aware system.

Figure 10.4 illustrates the CLV-CAP architecture, which is based on two important prerequisites. First, mobile infrastructures rely exclusively on CCN communications. Secondly, the fixed infrastructure supports both CCN and classical TCP/IP communications. Then, communication between nodes sup-

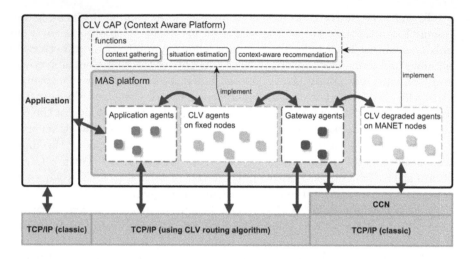

FIGURE 10.4: High-level overview of CLV Context-Awareness Platform.

porting different protocols is ensured by specific nodes acting as gateways. These gateway nodes have the capability to handle both TCP/IP and CCN protocols, and can dynamically switch from one to the other depending on the network needs. Each network node embeds an agent container belonging to one of the agent platforms deployed on the system. Handling of the context is performed both at the network and at the MAS levels. Specific agents, named context brokers, are in charge of retrieving context data from context sources on application request, and evaluating the situation of entities of interest. Context sources are interfaced with sensor gateway agents, retrieving observations from sensors or sensor networks, whatever their kind: physical, abstract (e.g., a web service) or logical (combination of sensor outputs), and making these available in the CLV-CAP. Context data spreading is ensured because nodes in the CCN maintain a cache of answers to each request they have answered. Moreover, context broker agents keep in their memory any context element they have retrieved or computed. With a context-awareness middleware such as the CLV-CAP, context-aware recommendations in complex network environments can be achieved by simply querying for context or situation data. Here, filtering and recommendation services are embedded in dedicated agents that can be spread across the network. Requests for context elements and situations will be handled by context brokers, which will then gather data from the network and their own knowledge base (memory) to provide an answer. The recommending engine can then process ontological user profile and content descriptions, and finally provide a list of weighted recommendations or filtered content that is valid regarding the context or situation aggregated by the context brokers.

## 10.4 F-Sphynx, a Semantic Knowledge-Based Recommender System

The recommending engine associated with the CLV-CAP described in the previous section is named F-Sphynx. It is a context-aware knowledge-based recommender system relying on the processing of ontological descriptions, including rules. This section presents the recommendation principles according to the approach followed by F-Sphynx, the user model on which its current version is based, namely eFoaf, and its matchmaking algorithm.

### 10.4.1 Principles

Following a content-filtering approach, F-Sphynx assumes that recommendation is a matter of finding items matching a user in a given context. The user profile, including interests and preferences, item model and context model, is formalized using dedicated ontologies such as the ones presented in Figures 10.2 and 10.3; and in Figure 10.5, presented in the upcoming Section 10.4.2.

Assuming the user has been profiled, items formalized, and context gathered and formalized, the recommendation algorithm comprises two phases: (1) inference and filtering; (2) matchmaking [700, 703]. Rules used during the first phase are of three kinds: a) inference rules, allowing to define the behavior linked to concepts defined in ontologies; b) filtering rules, allowing to exclude some contents from the processing; and c) interest creation rules, allowing to build new interests from the user profile. While the first kind is part of the ontologies, the others are specific to the recommendation mechanism. The inference phase is performed by applying the set of rules to the knowledge base formed by the user profile, items, and context descriptions, in a simple forward chaining inference process. This produces an inferred knowledge base on which the matchmaking is performed, to weight all the items in the knowledge base according to the user and the context, and thus computing recommendation scores (see [700] or [703] for details).

In this approach, the user profile can instantiate any characteristic providing any kind of information. While only the user interests expressed in the profile are used in the matchmaking phase, the other elements of the profile are used during rules processing (assuming they are concerned with at least one rule).

### 10.4.2 eFoaf, a Set of Connected Models for User Profiles

The Friend of a Friend ontology (FoaF) is one of the best known and used semantic user models. It aims at representing one person's profile, activities, and relationships with other persons. Numerous web-sites and software applications support FOAF information. For example, LiveJournal and the Dead Journal blogging web-sites can export the FOAF profiles of their members. FriendFeed or TypePad are other examples of services supporting FOAF. On the client-side, Safari and Semantic Radar extension for Firefox also support the format. A large number of FOAF descriptions is already available and interconnected. The foafPub dataset contains an increasing set of more than 200,000 triples describing user profiles. Some tools are available to assist the user in creating a user profile (FOAF-O-MATIC) or performing queries on this large quantity of information (foaf-seach).

However, FOAF alone is not enough when targeting context-aware recommendations, for instance to express contextualized preferences, or to allow documenting multiple contact points (phone number at work, at home, etc.). The eFoaf proposal [383] provides such an extension by coupling the original Foaf model with ontologies dedicated to, e.g., time, location, activities, abilities, interests, and preferences.

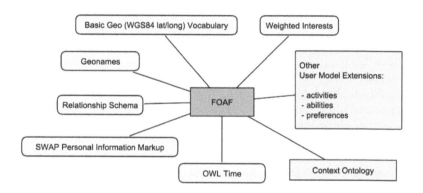

FIGURE 10.5: The eFoaf Semantic User Model.

We summarize below how existing schemas can be combined. We extend FOAF, to form a rich semantic framework to represent user profiles and their interests, as illustrated by Figure 10.5. The ontologies used are described in the following; we give the reason behind their choice, and their connections with FOAF:

- To allow a richer expression of the links between people, the RELATION-SHIP schema [266] is used. Examples of the relationships are *Child Of, Colleague Of, Employed By, Grandparent Of* or *Influenced By*. All are subproperties of *foaf:knows*, making the connection with the FOAF model straightforward.

- W3C hosts an RDF Schema vocabulary for contact information named SWAP Personal Information Markup [117]. It allows the expression of contextualized contact information. It is possible, for example, to express an address and phone number for work or home, etc.

- FOAF was not initially designed to represent contextualized and weighted interests. The Weighted Interests ontology [167] allows the expression of any kind of interest, not only documents, as FOAF does. It allows associating to each interest a context of validity as well as a level of interest.

- Geonames provides services to retrieve the components of a location (countries for a continent, administrative subdivisions for a country, etc.), the neighbors (neighboring countries), and nearby features (near the Eiffel Tower are Champ-de-Mars, Trocadero, etc.). This dataset contains information about most geographical locations.

- Basic Geo [166] is a simple RDF Schema to represent geographical coordinates. The three main properties are *lat*, *long*, and *alt*, standing for latitude, longitude and altitude, respectively; *foaf:Person* being a subclass of the main class of Basic Geo (*SpatialThing*), the connection with FOAF is direct.

- A *TemporalThing* class is introduced to link the OWL Time ontology [418] with the other models using a *hasTemporalEntity* property. Then, to express the fact that an entity can be associated to temporal information such as a date or duration, it only specifies that this entity is an instance of *TemporalThing*.

- To represent the user's language skills, a sight or hearing problem, or any other special ability, an Abilities model has been created [383].

- To represent biographical information such as a position held in a company during a particular period, or participation in a project and the role played, we added an Activity model [383].

As described, eFOAF can be used to specifically address the need for representing user profiles, including contextualized interests, so that they can be exploited by the personalization systems. Compared to other user semantic models, such as GUMO [407], our approach aims to ensure the maximal connectivity with existing semantic datasets available as linked data, making the model usable on the Web.

## 10.4.3  Matchmaking Algorithm

The matchmaking algorithm aims at computing a score, for each considered item, representing to what extent this particular item suits the user interests, in a given context/situation. This score is used to rank the items in order to provide a list of recommendations to the user. In F-Sphynx, the recommendation score of an item $I$ for a user $U$ is obtained by considering both the user's interests and non-interests targeting this kind of item, combined in a function that can be chosen according to the application objectives. While interests tend to increase the score, non-interest decreases it. The generic recommendation score of an item $IT$ for a user $U$ is given by the following function; more details about possible instantiations being given in previous articles [703, 700]:

$$RScore(U, IT) = \Gamma(\mathbf{M}_I(U, IT), \mathbf{M}_{NI}(U, IT)), \quad (10.1)$$

where $\mathbf{M}_I(U, IT) \in [0, 1]$ and $\mathbf{M}_{NI}(U, IT) \in [0, 1]$ are the vectors containing respectively the matching scores $MI(I, IT)$ and $MI(NI, IT)$ of interests and non-interests concerning the item $IT$; the $\Gamma()$ function being a vectorial function that can take different forms according to the application case.

Each interest concerns one or multiple topics $T_I$. According to the approach adopted in [703], topics of interest target either a category, an item, or any resource in the RDF sense. The matching $MI$ between an interest $I$ and an item $IT$ of the list of items to be recommended is expressed by:

$$MI(I, IT) = \frac{\alpha(I) * M_{CX}}{nb(T_I)} \sum_{i=1}^{nb(T_I)} M_{TI}(T_{I_i}, IT), \quad (10.2)$$

As multiple topics in an interest are considered as a conjunction of equally important preferences, $MI$ is calculated based on the average of topics matching scores $M_{TI}(T_{I_i}, IT) \in [0, 1]$. The weight of $I$ for $U$ (interest level) is $\alpha(I) \in [0, 1]$. $M_{CX} \in [0, 1]$ is the context similarity measure at the time the matching is computed. This weights the result according to the current situation matching it with the interest's validity context. Different strategies can be applied here. We will see in Section 10.6.1 that context should not necessarily be a discriminating factor.

Matching functions $M_{TI}$ dedicated to categories, item, or resource have been detailed in [700, 703]. The original function for items can be replaced by a semantic similarity measure, considering as much as possible all the complexity of semantic descriptions of user profiles and items.

There have been many propositions in the scientific literature for computing the similarity between classes of an ontology [292] or [786], but less attention has been given to the comparison of instances. Nevertheless, some interesting approaches use graph-based distance measures [962], while others use the number of direct or indirect relations between the compared instances [734]. However, it should be noted that attributes (properties having a literal value) are often not accounted for. One particular proposition by Maedche and Zacharias [638] tried to exploit most information available by differentiating three dimensions when comparing two instances $z_i$ and $z_j$:

- Taxonomy similarity, $TS(z_i, z_j)$: comparing the concepts (i.e., classes) to which both instances belong.

- Relation similarity, $RS(z_i, z_j)$: taking into account the relations that both instances have in common with other objects, including the incoming and outgoing links.

- Attribute similarity, $AS(z_i, z_j)$: comparing the literal values of the datatype properties of both instances.

These three components are aggregated to provide the final similarity measure between two semantic instances. The resulting MZS measure is defined as a weighted average of the three similarity metrics described above; see [638] for more detail on how the latter are computed:

$$MZS(z_i, z_j) = \frac{\alpha * TS(z_i, z_j) + \beta * RS(z_i, z_j) + \gamma * AS(z_i, z_j)}{\alpha + \beta + \gamma}. \qquad (10.3)$$

When the topic of an interest is an item, the MZS measure evaluates to what extent the considered item is similar to the one described in the interest. We have then $M_{TI}(T_{I_i}, IT) = MZS(T_{I_i}, IT)$ in Equation 10.2. When the topic of an interest is a resource different from an item, the approach described in [703] is still used. It consists of looking for resources of the same type connected to the item chosen to evaluate for recommendation. However, the matching function is also replaced by the MZS measure.

---

## 10.5    Recommendations with Fuzzy User Preferences

### 10.5.1    FuSOR, a Fuzzy Layer for Crisp Ontologies

Using linguistic values for recommender systems has been proposed in [651] and [757]. However, the integration with semantic knowledge-based recommender systems remains mostly unexplored. The FuSOR ontology [383, 702], depicted in Figure 10.6, makes this combination possible. It extends any existing ontology with a fuzzy layer for datatype properties.

A crisp value can then be replaced by a fuzzy set, usually corresponding to a linguistic value. By plugging this model into an existing OWL-DL ontology, any datatype property $p$ can be associated to a fuzzy property $f$ such as:

$$\left\{ \begin{array}{l} dom(f) = dom(p) \\ range(f) = FuzzySet \\ hasFuzzyVersion(p, f) \end{array} \right.$$

The fuzzy property $f$ has the class *FuzzySet* as its range and the same domain as $p$ (*range* and *domain* being the classical RDFS[5] properties). An annotation property[6] *hasFuzzyVersion* formalizes the link between the original property $p$ and its fuzzy version $f$.

Using eFOAF and FuSOR together allows the modeling of rich user profiles with interests and preferences, conveniently expressed with the user's own way of thinking, and also allows quantifying things, thanks to linguistic values and fuzzy sets. The RDF/Turtle sample below illustrates how

---

[5]see RDFS W3C Recommendation at `http://www.w3.org/TR/rdf-schema/`

[6]owl:annotation, see OWL W3C Recommendation at `http://www.w3.org/TR/2004/REC-owl-ref-20040210/#AnnotationProperty-def`

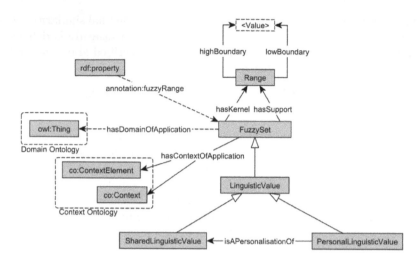

FIGURE 10.6: The FuSOR Model, associated with Context ontology.

the FuSOR model can be used to express a user's interest using a linguistic value. It formalizes the example given before in Figure 10.1, using the eFoaf ontology. According to the Turtle syntax, the example uses prefixes for URIs of concepts: `ex:` is the fictive URI of the example, `foaf:` is the URI of FOAF (`http://xmlns.com/foaf/0.1/`), and `wi:` is the URI of the Weighted Interest ontology (`http://purl.org/ontology/wo/core#`). See `http://www.w3.org/TeamSubmission/turtle/` for more information on the Turtle syntax.

```
ex:JohnDoe
   a foaf:Person ;
   foaf:name "John Doe";
   wi:preference [
      a wi:WeightedInterest;
      wi:topic  [
               a ex:Restaurant;
         ex:fuzzyCost ex:johnCheap;
               ] ;
      ] ;
```

The notion of *cheap* restaurant, according to John Doe's preferences, is formalized using a fuzzy property, *ex:fuzzyCost*, and a linguistic value, *ex:johnCheap*, defined below with the FuSOR ontology:

```
ex:Cost [
fusor:hasFuzzyVersion ex:fuzzyCost;
];
```

```
ex:johnCheap
  a fusor:LinguisticValue;
  fusor:hasSupport [
    a fusor:Range;
    fusor:hasLowBoundary -INF;
    fusor:hasHighBoundary 25;
  ];
  fusor:hasKernel [
    a fusor:Range;
    fusor:hasLowBoundary -INF;
    fusor:hasHighBoundary 20;
  ];
```

### 10.5.2 F-Sphynx and Fuzzy Preferences

#### 10.5.2.1 Recommendation Scores from Fuzzy Preferences

Topics of interest related to items or resources can contain references to properties evaluated using linguistic values. To compute the matching score, these values corresponding to different linguistic variables, must be aggregated with a specific function. Such an aggregation function takes into account the different degrees of membership of an item's characteristics to those user-defined linguistic values. Several aggregation operators have been proposed in the literature but their impact on the recommender systems is still to be investigated.

Let us consider an interest $I$, for which the user has expressed constraints or preferences using fuzzy values regarding a kind of item (e.g., restaurants in our previous example). Let $\{x_i, i = 1, .., N\}$ be the subset of characteristics of an item, concerned with elements of a set $FP_U$ of such user fuzzy preferences. $FP_U$ is formalized as a set of fuzzy sets $\{A_i, i = 1, .., N\}$ representing the preferences of the user $U$ for each respective characteristic $x_i$ of the items.

Consider two items $IT_j$ and $IT_k$ that are presented by the subset of their characteristics concerned by user fuzzy preferences: $IT_j = \{(x_i, \mu_{x_i}(IT_j)), i = 1, ..., N\}$ and $IT_k = \{(x_i, \mu_{x_i}(I_k)), i = 1, ..., N\}$. $\mu_{x_i}(IT_j)$ is the membership degree of the characteristic $x_i$ of an item $IT_j$ to the fuzzy set $A_i$ as defined here before. We write $R(FP_U, IT)$ the aggregation function computing a matching score for the item $IT$ regarding the user fuzzy preferences $FP_U$. The function $R$ has to comply with several heuristics. The matching score of an item $IT_j$ should be higher than the score of $IT_k$, i.e., $R(FP_U, IT_j) > R(FP_U, IT_k)$ if one of the following is verified:

1. $IT_j$ has a higher membership degree than $IT_k$ for all its characteristics:

$$\forall i, \mu_{x_i}(IT_j) > \mu_{x_i}(IT_k)$$

2. There are no characteristics of the item $IT_k$ having a membership

value higher than the corresponding one of $IT_j$ and at least one characteristic of $IT_j$ has a membership value higher than the corresponding one of $IT_k$:

$$(\nexists h, \mu_{x_h}(IT_k) > \mu_{x_h}(IT_j)) \wedge (\exists m, \mu_{x_m}(IT_j) > \mu_{x_m}(IT_k))$$

3. $IT_j$ and $IT_k$ have the same average of their characteristics membership values, and $IT_j$ has the highest minimum membership value:

$$(\overline{\mu_{x_i}(IT_j)} = \overline{\mu_{x_i}(IT_k)}) \wedge (\min_i(\mu_{x_i}(IT_j)) > \min_i(\mu_{x_i}(IT_k)))$$

A classical aggregation approach would consist of computing an average over the memberships. However, the last heuristic would not be verified. This function has important drawbacks when applied to recommendations, because it does not account for the diversity between the elements to aggregate. For instance, it would give the same matching scores for aggregating $(0; 1)$ as for $(0.5; 0.5)$, whereas in most situations a user could prefer a solution moderately satisfying each of his/her preferences rather than a solution where one of those preferences is not satisfied at all.

A way to obtain this kind of behavior from the aggregation function, while respecting the three heuristics, would be to use a combination of two common aggregators, the average and the minimum:

$$R_{am}(FP_U, IT) = \frac{1}{2}\left[min_i(\mu_{x_i}(IT)) + \frac{1}{N}\sum_{i=1}^{N}\mu_{x_i}(IT)\right]. \qquad (10.4)$$

Let us illustrate this aggregation function through a practical example, keeping our user *John* and his interest in a restaurant. We assume that when he looks for a place to eat, he prefers moderately priced restaurants that are close to his current location.

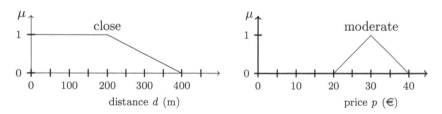

FIGURE 10.7: Examples of fuzzy membership functions for user preferences.

Figure 10.7 shows John's own representation of what a moderately priced restaurant is ($p$ being the average menu price) and what distance, $d$, he is willing to walk. Formally, those two linguistic values are defined by:

$$\mu_{price}(moderate) = \begin{cases} 0 \ \textbf{\textit{if}} \ p \leq 20 \\ \frac{1}{10} \times p - 2 \ \textbf{\textit{if}} \ 20 < p \leq 30 \\ \frac{-1}{10} \times p + 4 \ \textbf{\textit{if}} \ 30 < p \leq 40 \\ 0 \ \textbf{\textit{if}} \ p > 40 \end{cases}$$

$$\mu_{distance}(close) = \begin{cases} 1 \ \textbf{if} \ d \leq 200 \\ \frac{-1}{200} \times d + 2 \ \textbf{if} \ 200 < d \leq 400 \\ 0 \ \textbf{if} \ d > 400 \end{cases}$$

Applied to the two linguistic values described here, the aggregation function shown in Equation 10.4 gives the results illustrated in Figure 10.8. It appears clearly that $R_{am}$ rewards the satisfaction of both constraints together, but still rewards relatively well the fact that at least one constraint is satisfied.

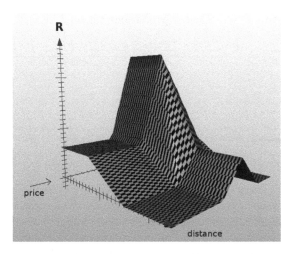

FIGURE 10.8: Recommendation score $R$ as a function of two user preferences defined by linguistic values.

### 10.5.2.2 Extension of the Matchmaking Algorithm

By extending the matchmaking algorithm described in Section 10.4.3 it is possible to take into account interests expressed with fuzzy sets. For each crisp datatype property, a corresponding fuzzy property will be assigned. For instance, a user will be able to express an interest for big cars by using a fuzzy set linked to the size property, or for environmental friendly cars by using a fuzzy set linked to the CO2 emission rate property.

During the matchmaking, fuzzy values are used as follows:

1. For each fuzzy set attached to an item characteristic in a user interest, we retrieve the corresponding non-fuzzy (crisp) property (e.g., *FuzzyHasSize* ⇒ *hasSize*).

2. For each instance in the available knowledge base having this crisp property instantiated (e.g., for each car having a value for the *hasSize* property), check if this value belongs to the *support* of the fuzzy set (i.e., it has a non null membership degree).

3. If the condition in step 2 is verified, compute the membership degree of this crisp value for the specified fuzzy set.

4. Once membership degrees are computed for each fuzzy preference in the interest, following step 1 to 3, compute the global matching score with $R_{am}$ (Equation 10.4).

The final recommendation score considering the fuzzy preferences assigned to topics of interests is obtained by taking into account the fuzzy matching score $R_{am}$ in the MZS measure (Equation 10.3), which becomes for an item $IT$, a user interest $I$ and a fuzzy preference set on $I$, $FP_U$:

$$FMZS(T_I, IT) = \frac{\alpha * TS(T_I, IT) + \beta * RS(T_I, IT) + \gamma * R_{am}(FP_U, IT)}{\alpha + \beta + \gamma}.$$

$$(10.5)$$

## 10.6   Context Dependent Recommendations

In the preceding sections, context was considered in the recommendation process as a weighting factor in the matching function between user interests and items to recommend. This approach, however, might not always be suitable. Consider a person that uses a recommender system for weekly food shopping. She expresses three preferences: 1) loves pizza; 2) Indian food, only when on weekend with family; 3) the day of an important football match, if vegetarian friends are present, likes to eat vegetable pizza or any vegetable meal.

This example contains three interests that are all differently dependent from contextual elements. In the first interest, there is no specific situation linked to pizza: it is always valid. For the second interest, the situation is extremely important: Indian food should be recommended only if the family is here during a weekend. The last interest is more flexible, regarding the meal to propose according to the situation.

If this were not already obvious, it becomes clear with this example that contextual elements have an influence on the users' interests, which is different for each user and interest. As seen in Section 10.4.2, user interests can be attached to a validity context or situation. The following first subsection explains how this is done in the F-Sphynx recommender supported by the CLV-CAP.

But more than having to consider both items to recommend, and context in their matchmaking process, recommenders should also consider the importance to give to interest-item matchmaking versus interest validity-situation [288]. This is the subject of the following second subsection, which explains how this could be handled with fuzzy logic.

## 10.6.1   Context Handling in Recommendations

Context-Awareness in recommendations almost always considers only the user's context. In a complex environment such as the environment created by HANETs, multiple entities evolve and influence each others. This is the kind of view taken by research works on ubiquitous or pervasive computing, also materialized in the Future Internet with the notion of the Internet of Things. In such an environment, context-aware recommendations should take into account the situation of any entity having an impact on the contexts of validity that are formalized in user interests and preferences.

The CLV-CAP presented in Section 10.3.3, supported by the context and situation ontologies (see Section 10.3.1), allows retrieving data related to any entity known by the system. For each user interest linked to a context element or situation, the recommender is then able to get related data from the CLV-CAP, estimate the current global situation, and compare it with the interests' context of validity in the computation of the final recommendation scores.

For a given user profile, the context of the validity of all interests is first retrieved. Using context and situation ontologies, these are instances of *Situation*, *Context*, or *ContextElement*. For all of them, the list of concerned entities is established (this is formalized by dedicated properties in the ontologies like, e.g., *situatedIn*, *hasContext*, *concerns*). The state of these entities (*hasState* property), as well as all the context elements and situations concerning them, are gathered in a virtual distributed knowledge base on which the recommender can reason to assess the validity of interests regarding the actual context. Each interest is then simply weighted in the computation of the recommendation score, depending on how much its context of validity is verified (Equation 10.2). This weight is directly proportional to the reliability of situations and other context elements concerned (*reliability* property in the Context ontology).

With the CLV-CAP middleware, agents keep in their memory any knowledge that they handle or meet in the network. Hence entity agents can provide an entity's state, context brokers build and keep the context elements, context, and situations, while sensor gateways handle knowledge related to sensors and observations. When assessing the validity of an interest context, the recommender thus sends queries on the network (CCN requests) for the entities' state and context elements. When dealing with situations, the preconditions formalized in the situation precondition model (see Section 10.3.1) are processed by a CLV-CAP context broker agent, which in turn queries the network for observations related to what is needed for preconditions processing. Once the occurrence of a situation has been evaluated, a *Situation* instance is sent back, together with a reliability value, which is the computed situation probability.

## 10.6.2    Interests and Context of Validity

As said previously in this chapter, context-aware recommendations consist of finding the intersection between a user, a set of items to recommend, and a context. Hence a user interest can be roughly reduced to a couple $I \approx < Topics, CX >$, where $Topics$ is the set of the interest topics, and $CX$ is the interest context of validity. In the matchmaking process, this leads to computing two things for each item: the matching between user interest topics and the item characteristics, $M_{TI}(T_{I_i}, IT)$, and the similarity between interests' context of validity and the context at the time the recommendation has to be done, $M_{CX}$. The final matching score of an item $I$ for a user is a function of both: $MI = f(M_{TI}, M_{CX})$, as is implemented in Equation 10.2.

Both functions are usually computed separately, and the final behavior of $MI$ depends on the chosen function. As there are no specific insights regarding the choice of this function, it can be simply a linear combination or a weighting of item matching by the context similarity. Alternatively the context similarity can be chosen to be discriminating: when under a given threshold, the score falls to zero. These functions, however, do not consider that the situation might be more or less important in certain cases, and that the way to combine both similarity metrics might depend on the user, as suggested by our example at the beginning of Section 10.6.

A way to address this kind of personalization of the matchmaking function, i.e., the weighting between item and situation influence, is to use a fuzzy model. This brings a flexible approach, allowing the user to define his own function or the function to self-adapt according to the user feedback or her/his consumption history [288]. The main idea is to propose a combination function based on fuzzy sets, and to use fuzzy inference to compute the recommendation score.

A user interest is then rewritten as a tuple $I \approx < Topics, CX, fzM >$, where $fzM$ is a fuzzy model attached to the interest. This model contains the definition of three fuzzy variables, `ItemMatching`, `ContextMatching`, and `MatchingScore`, and a set of fuzzy rules. Three linguistic values are associated to each of the three variables: `Low`, `Medium`, and `High`. Each fuzzy variable is defined by a fuzzy set composed of the membership functions for each of these linguistic values. The three fuzzy sets can be represented by the classical trapezoid function as shown in Figure 10.9.

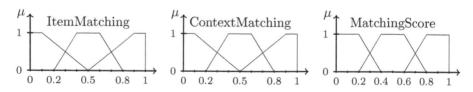

FIGURE 10.9: Fuzzy variables of the item–context weighting model.

Finally, nine fuzzy rules allowing one to compute the value of `MatchingScore` according to both `ItemMatching` and `ContextMatching` variables are defined as conditional statements, for example:

```
IF ContextMatching IS Low AND ItemMatching is High
   THEN MatchingScore IS Medium.
```

Weights are associated to each rule to allow adjusting their importance in the fuzzy inference process. Initially they can be all set to 1, forming a neutral model. Alternatively, rules can be designed in such a way that they favor either the item or the context matching. This prevents a cold start problem, but does not represent necessarily the behavior desired by the user.

Personalization of the fuzzy models can be achieved by adjusting the fuzzy sets and/or the rules. Since manual adaptation remains somehow a complicated and tedious task for users, automatic adaptation should be preferred. This can be done by exploiting the user feedback to construct a reference list of matchings and recommendation scores for a set of items, which will serve as a training set. Let $RefList$ be this reference list, defined as: $RefList = \{Mref_1, ..., Mref_n\}$, where each element is defined as a tuple $Mref_i = \langle IMref_i, CMref_i, MSref_i \rangle$. The three elements $IM$, $CM$, $MS$, stand respectively for `ItemMatching`, `ContextMatching`, and `MatchingScore` fuzzy variables.

One possible way to obtain this list is to show a list of items to the user, together with their recommendation score and matching values for respectively the user interests and situation, and to ask the user to give the recommendation score he would have given knowing this. As we speak here of linguistic values, this process is feasible. The way to consider multiple situations is trickier, since a list should be shown for different situations. However, if we consider an iterative process, where feedback is asked to the user each time she/he uses the recommendation system, then reference values can be exploited. As in any machine-learning approach, the fuzzy model attached to each interest could be self-adapted, and would converge to a suitable form, as the number of feedbacks related to each interest individually increases.

The fuzzy models attached to each interest can be optimized by minimizing the error between the current values and the training set after defuzzification. Three strategies can be followed: a) Variable optimization to adjust only the definition of membership functions; b) Rule optimization to modify the weight of rules; c) Full optimization to allow modifications on both the variables and the rules.

## 10.7 Evaluations and Experiments

Different variants of the F-Sphynx recommender have been applied to several cases in the interactive and mobile TV domains [700] [704], and for mobile e-commerce [704]. The matchmaking algorithm exploiting semantic descriptions of user interests and TV content descriptions performed well. From a functional point of view, the comparison of the recommender output with what should have been theoretically proposed shows very good Precision-Recall evaluations [700], P/R being the classical metric used to evaluate recommender systems. Context has been successfully used to provide geolocated targeted advertisements for products sold in a user's neighborhood shops, and to provide real-time additional multimedia content (here again, advertisements) to TV watchers according to what they were watching [704].

In order to push the approach forward and to expose it to the reality of users' behavior, we will present additional experiments conducted with actual users, with the extended algorithm, in Sections 10.4.3 and 10.5.2.2. First, an evaluation of the new semantic similarity measure (MZH) is proposed. Then, we report an experiment that uses the CLV-CAP platform to provide situated recommendations for car-sharing trips, based on both crisp and fuzzy user interests. Last, we propose theoretical results regarding the weighting of the situation versus the user interests introduced in Section 10.6.2.

### 10.7.1 Semantic Similarity Efficiency

The use of the MZS measure, as presented in Section 10.4.3, implies respecting some constraints regarding the knowledge that will be processed and the ontologies used to represent it [381]:

1. Instances are members of only one concept.

2. Ontologies are strictly hierarchical, so that each concept is subsumed by only one concept.

3. Each property has a defined domain and range.

4. No inconsistency is tolerated for instances, i.e., they must comply with the properties' domain and range.

5. Object and datatype properties are strictly separated.

If this is fine in theory, or in a closed world where everything is under control, large and popular linked data sources like DBPedia unfortunately tend not to comply with those constraints. Some solutions have been proposed to overcome those limitations [384]. To remove Constraints 1 and 2, the taxonomic similarity can be extended to compare the two sets of direct parents of both instances and all super classes of the classes of those two sets. Constraint 3 can be removed by conducting a usage analysis for each property whose range or domain is not defined to automatically detect them. To address Constraint 4, the inconsistent values can be ignored, and the matching score of the concerned instances can be lowered. To deal with Constraint 5, dual properties are introduced, i.e., properties that can take either resources or literals as a value.

Once these limitations of the MZS measure are addressed, it is possible to apply it to big repositories of semantic datasets such as DBPedia. It has been widely accepted [679] that a suitable criteria to evaluate the performances of semantic similarity measures between words is to compare their results to human evaluations. Following the same idea, we have designed an experiment for semantic resources [384]. It assesses the results of the modified MZS measure on the similarity between pairs of DBPedia movie instances.

**TABLE 10.1**

Human Evaluations of the Similarity between Movie Pairs.

| Movie Pair | Users | MZS |
|---|---|---|
| Inception – Manhattan | 0.22 | 0.31 |
| Inception – The Dark Knight | 0.62 | 0.61 |
| Beauty and the Beast – Eyes Wide Shut | 0.16 | 0.11 |
| Inglourious Basterds – Apocalypse now | 0.80 | 0.40 |
| Eyes Wide Shut – Apocalypse now | 0.10 | 0.30 |
| Inception – Apocalypse now | 0.46 | 0.35 |
| Beauty and the Beast – Dumbo | 0.63 | 0.40 |
| Manhattan – Dumbo | 0.07 | 0.31 |
| Manhattan – Eyes Wide Shut | 0.46 | 0.37 |
| The Dark Knight – Dumbo | 0.15 | 0.18 |
| The Dark Knight – Beauty and the Beast | 0.29 | 0.09 |
| Dumbo – Wall-E | 0.67 | 0.40 |
| Wall-E – Eyes Wide Shut | 0.12 | 0.29 |
| The Day After Tomorrow – Wall-E | 0.43 | 0.39 |
| The Day After Tomorrow – Vicky C. Barcelona | 0.16 | 0.19 |
| Pretty Woman – 4 Weddings and a Funeral | 0.65 | 0.34 |
| Pretty Woman – Manhattan | 0.64 | 0.38 |
| 4 Weddings and a Funeral – Beauty and the... | 0.35 | 0.37 |
| 4 Weddings and a Funeral – The Day After... | 0.11 | 0.11 |
| Vicky Cristina Barcelona – Manhattan | 0.72 | 0.57 |
| Vicky Cristina Barcelona – Inglorious Basterds | 0.18 | 0.38 |

Nineteen persons were asked to evaluate, on a scale of 1 to 7, to what extent pairs of movies are similar. A set of 21 pairs of movies was evaluated by each person. This was done through a dedicated Web interface showing the pairs, and allowing users to access movie descriptions. Those descriptions were dynamically queried from DBpedia. Additionally, users were able to browse DBpedia instances related to the movie, such as actors and directors, to learn more about the movie following its description. Our goal was to make sure that all the information used by the semantic similarity measure was available to humans as well. The user evaluations collected for each pair of movies after normalization of the ratings on a $[0, 1]$ scale and values computed with MZS are shown in Table 10.1. In this experiment, we obtained a Pearson correlation of $r = 0.69$ between the average human ratings and the results of the MZS measure. With $p$-value $< 0.001$, these are significant results showing the efficiency of the enhanced MZS measure. However, this remains to be assessed on a bigger experimental set.

## 10.7.2 Trip Recommendations in a Carpooling Scenario

Context-Aware fuzzy-semantic recommendations such as detailed in this chapter have been experimented with using a simulated carpooling scenario. In this scenario, our goal was to provide to a group of persons, the best possible car-sharing trips taking into account their interests and preferences. The latter included the user's own personal interests, as well as their preferences related to cars, drivers and passengers. When the trips are computed, the weather is taken into account to assess the pertinence of recommendations regarding user preferences related to snow tires.

Two sets of beta testers were explicitly profiled by filling out a questionnaire asking for some specific characteristics, such as their gender, age, or language level. They were asked about their interests related to generic topics presented in a hierarchical graph. Finally, they were also asked about their preferences regarding car sharing (age and gender of road companions, interests for activity, or topic sharing) and cars (model, size, speed, and pollution rate). The first group had to weight its preferences using crisp values, while the second group used fuzzy linguistic values. All these data have been integrated in individual user profiles using dedicated ontologies.

The users were each shown a list of car-sharing trips as computed by a carpooling algorithm, and were asked to rate them between 0 and 5, assuming they keep in mind the preferences they gave, 0 meaning that they would not like the proposed trip, and 5 that they would really like it. For the sake of simplicity, it was assumed that all the users have the same starting and arrival locations. The recommender system was then asked to rank the trips set for each user, taking into account the trip characteristics, user characteristics and preferences, and the situations potentially linked to certain preferences. The CLV-CAP was used to identify the situation of the road for each trip regarding the presence of snow, in order to check the behavior of the recommender regarding a preference for snow tires asked to testing users. Simulated sensors were used, giving observations that were gathered across the network using the CCN-based request mechanism. Following a data fusion process based on a Bayesian graph model of the *snow on the road* situation, context broker agents then provide the probability of occurrence, which is encoded as a *Situation* instance and sent back to the recommender system that asked for the situation check. Simulations have shown that the reliability of observations plays an important role in the final estimation of the situation. However, when taking it into account to weight the gathered observations in the fusion process, 5% of reliable observations are sufficient to converge to a reliable probability of the situation occurrence. In our experiment, observations were distributed so as to produce a high probability of having snow on the road, thus making the situation sought by the recommender to be true.

The comparison between the wishes of the users and the set of weighted trips proposed by the recommender gives good results on average in terms of Precision and Recall. Table 10.2 summarizes the results. The first column is

**TABLE 10.2**
Average Precision/Recall for Car-Sharing Trips
Recommendations.

|  | Crisp group | | Fuzzy group | |
|  | Recom. Set | First 5 | Recom. Set | First 5 |
|---|---|---|---|---|
| Precision | 0.6 | 0.71 | 0.74 | 0.72 |
| Recall | 0.66 | 0.6 | 0.77 | 0.61 |
| F-Measure | 0.63 | 0.65 | 0.75 | 0.66 |

obtained by normalizing both the user wishes and the recommender outputs
in $\{0; 1\}$ where 0 is assigned to trips scored below 0.5 on a $[0, 1]$ scale (bad
trips), and 1 is assigned to trips scored above the latter threshold (good trips).
The second column considers only the first 5 results, to identify to what extent
the first ranked recommendations contain the user most preferred trips. What
can be observed from this experiment is that:

1. The use of Fuzzy linguistic values increases recommendation quality
   by about 10%;

2. When considering the five first recommendations, results are similar.

Finally, the impact of the situation on recommendations has been observed
through the *snow on the road* situation, linked to the user preference for snow
tires on proposed cars. Actually, almost all of the users found this important,
and weighted the preference from 3 to 5, the majority being 5. Figure 10.10
shows for each user, the average weight computed by the recommender system
for trips where the car is equipped with snow tires, and respectively without.

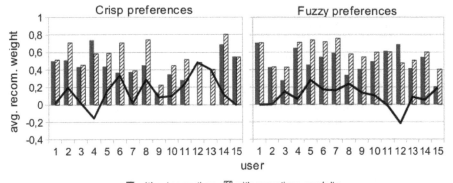

FIGURE 10.10: Situation impact on recommendations' weight, for user groups
using crisp or fuzzy preferences.

**TABLE 10.3**

Interest–Context Weighting Models RMSE for a Reference List of Item Matching Scores: Neutral and Optimized Fuzzy Model versus a Simple Weighting Function.

|  | N. Model | Opt. Model | Weight. Func. |
|---|---|---|---|
| RMSE | 0.071 | 0.010 | 0.279 |
| Max Error(%) | 53.343 | 18.781 | 99.872 |
| Min Error(%) | 0.000 | 0.003 | 0.066 |

For both user groups, the recommender clearly computes lower weight to trips without snow tires. Actually, the four exceptions came out to be errors by users who incoherently scored trips without snow tires regarding the high importance (5) they gave to the related preference. The results are coherent with the users' choices and the situation that was established from the gathered observations. Finally, this experiment provides a proof of the concept of the CLV-CAP for creating context-aware recommendations in Hybrid Ad-Hoc Networks.

### 10.7.3 Fuzzy Weighting of Context and Interest

The approach for the personalized weighting of items and the context of validity of user interests has been evaluated through simulation, using the jFuzzy-Logic library [232]. It has proved to be a promising approach compared to a weighting of the interests by a context of validity, which, however, still remains to be confronted by a real case.

With a reference list (*RefList*) of 1000 elements (see Section 10.6.2), created randomly and related to the same interest, convergence is obtained in 10 iterations for 200 elements with suitable settings of the optimization algorithm [288]. From the results summarized in Table 10.3, it can be seen that an initial neutral fuzzy model performs better than a classical weighting, and the optimized fuzzy model performs much better. Table 10.3 reports the Root Means Square Error (RMSE) for each model, as well as the maximum and minimum error, between the model and the reference. The optimized model (*Opt. model*) has the lowest RMSE, and the maximum error is 18.78%, which is the lowest, compared to 53.32% for the neutral model (*N. Model*) and 99% for the simple weighting function (*Weight. Func.*).

## 10.8 Conclusion and Perspectives

Context-Aware recommendations that correctly fit user needs according to their own preferences still remain an important research field. Especially on the Future Internet, where communication networks comprise both fixed and mobile nodes, as well as fixed and ad-hoc architectures. In such a context, this chapter presented a platform for context-awareness dedicated to Hybrid Ad-Hoc Networks, the Clairvoyant platform for context-awareness (CLV-CAP), together with its main ontologies, focusing on the context-aware recommendation solution adopted in this platform. This solution, the F-Sphynx semantic recommender system, is built as an extension of a content-based semantic recommender using semantic representations only, which has been exploited for different applications in the multimedia domain, especially for interactive or mobile TV as reported, e.g., in [704]. The specific content-centric approach implemented by the CLV-CAP, both at network and data levels, makes it particularly suitable for multimedia content transport and processing in today's communication networks and for the Future Internet.

We have explained how ontologies and fuzzy sets can be used in a semantic approach for content-based recommendations, also taking into account the context. Experimental results involving systematically real users have shown promising results both at the level of the semantic similarity measure efficiency and the usefulness of modeling user interests with fuzzy sets. The importance of context regarding user–item matching has been discussed and a solution using fuzzy logic to balance the weight of both context and user–item matching according to the user preferences has been assessed through simulation.

The system still ought to be tested with larger datasets and a more significant number of users in order to assess its performance. Numerous issues remain to be answered like, e.g., the real time gathering and processing of context data, reasoning with sparse sensor observations, distribution of knowledge processing in the recommendation process, and the prediction of users' fuzzy preferences from their behavior. However, it provides an insight on the various layers that need to be addressed on the Future Internet, to adapt mechanisms to multiple layers of the infrastructure, from network communication to user interactions with the recommender system.

## 10.9   Acknowledgment

The work presented here has been realized in the framework of the CLAIRVOYANT project (FNR C09/IS/12), supported by the National Research Fund, Luxembourg.

# 11

## Treating Collective Intelligence in Online Media

**Markos Avlonitis**

*Ionian University, avlon@ionio.gr*

**Ioannis Karydis**

*Ionian University, karydis@ionio.gr*

**Konstantinos Chorianopoulos**

*Ionian University, choko@ionio.gr*

**Spyros Sioutas**

*Ionian University, sioutas@ionio.gr*

## CONTENTS

In this chapter, we study the collective intelligence behavior of web users who share and watch video content. We discuss the aggregated users' video activity exhibiting characteristic patterns that may be used in order to infer important video scenes, thus leading to collective intelligence concerning the video content. Initially, we review earlier works that utilize a controlled user experiment with information-rich videos, for which users' interactions are collected in a testing platform and modeled by means of the corresponding probability distribution function. It is shown that the bell-shaped reference patterns are significantly correlated with the predefined scenes of interest for each video, as annotated by the users. In this way, the observed collective intelligence may be used to provide a video-segment detection tool that identifies the importance of video scenes. Accordingly, we discuss both a stochastic and a pattern matching approach on the users' interactions information, and report increased accuracy in identifying the areas indicated by users as having high importance information. Finally, in the last section, new insights in managing user interaction by means of a new stochastic algorithm are presented. In practice, the proposed techniques might improve navigation within videos on the Web and also have the potential to improve video search results with personalized video thumbnails.

## 11.1   Introduction

The Web has become a very popular medium for sharing and watching video content [199]. In particular, many individuals, organizations, and academic institutions are making lectures, documentaries, and how-to videos available online. Previous work on video retrieval has investigated the content of the video and has contributed a standard set of procedures, tools, and datasets for comparing the performance of video retrieval algorithms (e.g., TRECVID), but they have not considered the interactive behavior of the users as an integral part of the video retrieval process. Besides watching and browsing video content on the web, people also perform other "social metadata" tasks, such as sharing, commenting on videos, replying with other videos, or just expressing their preference/rating. Human-Computer Interaction (HCI) research has largely explored the association between commenting and micro-blogs, primarily tweets, or other text-based and explicitly user-generated content. Although there are various established information retrieval methods that collect and manipulate text, these could be considered burdensome for the users, in the context of video watching. In other cases, there is a lack of comment density when compared to the number of viewers of a video. All in all, there are few research efforts to understand user-based video retrieval without the use of social metadata [115].

In recent research [368], video consumption activity has been monitored in a well-instrumented environment that stores all the interactions with the player (e.g., play, pause, seek/scrub) for later research. Previous research [834, 684] suggested that implicit interactions between the people and the video-player can be of great importance to video summarization. To this end, in [368] a web–video interface was constructed, and a controlled user experiment was performed with the goal of analyzing aggregate users' interactions with the video, through their respective players.

## 11.2    User-Based and Content-Based Retrieval Approaches

### 11.2.1    Content-Based Semantics

Content-based information retrieval uses automated techniques to analyze actual video content. Accordingly, it uses images' colors, shapes, textures, sounds, motions, events, objects or any other information that can be derived from only the video itself. Existing techniques have combined the videos' metadata [893] with pictures [297], or sounds [584], while other researchers provide affective annotation [217], or navigation aids [512]. Even though content-based techniques have begun to emphasize the importance of users' content, still such approaches do not take into account peoples' browsing and sharing behavior. Moreover, low-level features (e.g., color, camera transitions) often fail to capture the high-level semantics (e.g., events, actors, objects) of the video content itself, yet such semantics often guide users, particularly non-specialist users, when navigating [244] within or between videos [512].

Since it is very difficult to detect scenes and extract meaning from videos, previous research has attempted to model video in terms of better-understood concepts, such as text and images [1010]. To evaluate methods for understanding video content, researchers and practitioners have been cooperating for more than a decade on large-scale video libraries, and tools for analyzing the content of video. The TRECVID workshop series provides a standard set of videos, tools, and benchmarks, which facilitate the incremental improvement of making sense from videos [865].

Thus, content-based techniques facilitate the discovery of a specific scene, the comprehension of a video in a limited time, and the navigation in multiple videos simultaneously. Again, the object of analysis remains the video content, rather than the metadata associated with people or how people manipulated and consumed the video. Accordingly, content-based techniques are not applicable to some types of web video, such as lectures and how-to instructions that present, respectively, a visually flat-structure or complex schematic information.

### 11.2.2    User-Based Semantics

In comparison to the more legacy content-based techniques, there are fewer works on user-based analysis of information retrieval for video content. One explanation for this imbalance is not the importance of the content, but the relatively newer interest in the social web, sharing, and the use of videos online. Nevertheless, there is growing research and interest in the user-based retrieval of video.

User interactions are one of the basic elements in user-based research. For this purpose, there is a need for detailed tracking of video browsing behavior. The authors of [892] developed a media-player-based learning system called the Media Miner. They tracked video browsing behavior, modeled users' states transition with Hidden Markov Model approaches, and generated fast video previews to satisfy the "interestingness" constraint of them. MediaMiner featured the common play, pause, and random seek into the video via a slider bar, fast/slow forward and fast/slow backward as well. Researchers tried to relate user activity to each user's browsing status, such as identifying whether the user was bored or interested.

Besides stand-alone videos, few works perform user-based information retrieval from videos on the Web. The principle example here is based on [834] and [1017], wherein the authors have highlighted the importance of implicit instrumentation and user-based semantic analysis of video on the Web. In the former work, the authors have proposed a shift from semantics to pragmatics, suggesting that content semantics follow the semantic utility of the interface. In the latter work, the authors have analyzed communicative and social contexts surrounding videos shared in synchronous environments as a means to determine a categorical genre, like Comedy, Music, etc., [1018] and video virality [835].

## 11.3   A Controlled Experiment on User-Interaction

### 11.3.1   Event Detection Systems

According to [368], several applications have been developed by the researchers, in order to evaluate novel event detection methods. Macromedia Director, a multimedia application platform, was used to develop SmartSkip [297]. The system used re-encoded videos in QuickTime format. Similarly, Emoplayer [217] was running locally on a laptop, and participants used a pointing device to interact with it. The system was developed with VC++ and DirectShow and the annotated video clips were stored in Extensible Markup Language (XML) files. In the case of [584] Microsoft Windows Media Player had been modified to develop the enhanced browser with its special features, because its default playback features are not sufficient for video navigation. The authors of [244] designed a system as a wrapper around an ActiveX control of Windows Media Player. A video recorder was used to collect video, at first, and then it was encoded in MPEG-1. The majority of these systems run locally, need special modification on software, and at the same time on video clips. Another important procedural parameter of the aforementioned experiments was that subjects had to be at a specific place where the experiment was conducted. Still, besides stand-alone applications, a number of web-based systems do exist. Hotstream [417] employed the Java 2 Enterprise Edition (J2EE) to develop a multi-tier web based architecture system (web-tier, middle-tier, backend database-tier, streaming platform) in order to deliver personalized video content from streaming video servers. In the same direction, the authors of [834] created different web-based platforms where the user can watch, browse, select, and annotate video material.

## 11.3.2 VideoSkip System Design

The VideoSkip player provides the main functionality of a typical Video Cassette Recorder (VCR) device [244]. The selection of the buttons was made to resemble the main playing/browsing controls of VCR devices because these are familiar to users. ReplayTV system and TiVo provide the ability to replay segments, or to jump forward in different speeds. In this way, the classic forward and backward buttons were modified to *GoForward* and *GoBackward*. The first one goes backward 30 seconds and its main purpose is to replay the last viewed seconds of the video, while the *GoForward* button jumps forward 30 seconds and its main purpose is to skip insignificant video segments. The thirty-second step is an average time-step used in previous research and commercial work due to the fact that it is the average duration of commercials. Next to the player's buttons, the current time of the video is shown followed by the total time of the video in seconds. A seek thumb is not available in order to avoid random guesses, as this would have made it difficult to analyze users' interactions. In [584] it was observed that when the seek thumb is used heavily, users have to make many attempts to find the desirable section of a video, and thus causing significant delays. VideoSkip [569] is a web video player developed with the Google App Engine and the YouTube API to gather interactions of the users while they watch a video. Based on these interactions, representative thumbnails of the video are generated. Users of VideoSkip should have a Google account in order to sign in and watch the uploaded videos. Thus, users' interactions are recorded and stored in Google's database with their Gmail addresses. The Google App Engine's database, the Datastore, is used to store users' interactions. Each time a user signs in to the web video player application, a new record is created, while whenever a button is pressed, an abbreviation of the button's name and the time it occurred are added to the Text variable. The time is stored within a second's accuracy.

## 11.3.3 User Heuristic for Event Detection

Every video is associated with an array of $k$ cells, where $k$ is the number of the duration of the video in seconds. The user activity heuristic consists of

**TABLE 11.1**

User Activity Heuristic Provides a Simple Mapping between User
Action and Value.

| User action | Play | *GoForward* (30 s) | *GoBackward* (30 s) | Pause |
|---|---|---|---|---|
| Heuristic | +2 | -2 | +2 | +2 |

three distinct stages. In the first stage, every cell is initialized to the number of
users who have watched the video. This initial value is used to avoid extremely
large negative values, to increase the viewing value of the whole video, and to
provide a balance for random interactions. In the second stage, the value for
each cell, that has been played by the user, is increased by two. Moreover, every
interaction means something for the event detection scheme. Each time a user
presses the *GoBackward* button, the cells' values matching the last 30 seconds
of the video, are incremented by two again. On the other hand, each time the
user presses the *GoForward* button, the cells' values matching the next 30
seconds of the video, are decreased by two. A set of different values has been
tested for interactions leading to the values of Table 11.1. For example, we used
for play, *GoForward*, *GoBackward* and pause "+1" or for play/pause "+1" and
for *GoForward*/*GoBackward* "+2." This combination was selected in order
to make the results distinguishable, while avoiding increased complexity. In
the third stage, the highest values of the array are considered, and also at
the same time, the number of values (interactions) that are gathered in a
specific cell area (i.e., the surface size). Moreover, a distance threshold of 30
seconds between the selected thumbnails was defined in order to avoid having
consecutive cells as a result. These specific scenes can be used as proposed
thumbnails to improve users' browsing experience. Each proposed thumbnail
begins at the first second of the selected area.

## 11.3.4   Experimental Methodology

### 11.3.4.1   Materials

One of the key points of the research in [368], was the exploration of methods
for event detection; accordingly, the selection of the suitable video content
is of high importance. The videos selected are as visually unstructured as
possible, because content-based algorithms have already been successful with
videos having visually structured scene changes. Another key factor considered was the length of a video. In general, the YouTube service allows video
uploading up to 15 minutes, while an option exists that allows one to request
an increase limit, leading to file uploads greater than 20GByte [1024]. Although there were videos that exceeded that limit, they decided not to use
them, because it would be tiresome for the majority of users. Indeed, some
early pilot user tests have revealed that user attention is reduced after they
have watched more than 3–4 videos of 10 minutes each. Narrative and en-

tertainment have been the most popular category, while according to [403] entertainment content is more likely to be watched in a leisurely manner, and costs so much to produce that it is reasonable to have a human produce previews and summaries. Moreover, lecture and how-to videos were selected, as users are actively watching them to retrieve information about a specific topic. Documentary videos could be categorized as video or audio-centric, lectures have an audio-centric content, and cooking videos have more video-centric features. The documentary video features a segment of a television program called "Protagonists" [1022]. The selected segment refers to the use of internet by young people. The lecture video is a paper presentation from a local workshop [1021] and the presentation's topic is "The acceptance of free laptops, that have been given to secondary education students." Finally, the how-to video is a segment of a cooking TV show for a souffle-cake [1023]. Each one lasts ten minutes and is available on YouTube.

### 11.3.4.2 Measurement

The measuring process adopted was based on the assumption found in [1025] that there are segments of a video clip that are commonly interesting to most users, and users might browse the respective parts of the video clip in searching for answers to some interesting questions. When enough user data is available, users' behavior will exhibit similar patterns, even if they are not explicitly asked to answer questions. In order to experimentally replicate user activity, a questionnaire was developed that corresponds to several segments of the video. Scene selection was based on a combination of audio and video factors. Thus, each question corresponds to a visual and/or a structural cue that can be used as a hint to find an answer. Furthermore, some irrelevant questions were included in order to check that the users are searching for the answers and do not attempt to guess the replies. The following parameters were under consideration: audio channel, speaker's channel, end-users' actions watching the talk to reveal the significant portions, and video channel used to select the thumbnails. Google Docs was used in order to create online forms for the users' questionnaires and was subsequently integrated in the user interface as presented in Figure 11.1.

### 11.3.4.3 Procedure

The goal of the user experiment was to collect activity data from the users, as well as to establish a flexible experimental procedure that can be replicated and validated by other researchers. There are several suggested approaches to the evaluation of interactive information retrieval systems [506]. Instead of mining real usage data, a controlled experiment provides a clean set of data that might be easier to analyze. The discussed experiment took place in a lab with an Internet connection, general-purpose computers, and headphones. Twenty-three university students, the characteristics of which are shown in Table 11.3, spent approximately 10 minutes watching each video, while the

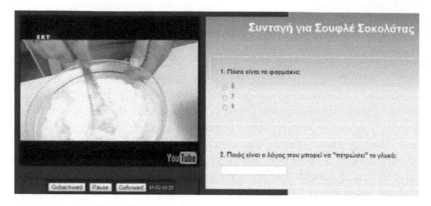

FIGURE 11.1: Screenshot of VideoSkip with the questionnaire (Greek language).

**TABLE 11.2**
Example Questions from Each Video.

| Video | Indicative questions |
|---|---|
| Lecture video | Which are the main research topics? |
| | What did the students not like? |
| | What time does the first part of the talk end? |
| Documentary video | What time do you see the message "coming next"? |
| | What is the purpose of hackers? |
| | What is the name of the girl in the video? |

buttons were disabled. All students had been attending Human-Computer Interaction courses at a post- or under-graduate level, and received course credit in the respective courses. Next, there was a time restriction of 5 minutes, in order to motivate the users to actively browse through the video, and answer the respective questions. Example questions for each video are shown in Table 11.2. Users were informed that the purpose of the study was to measure their performance in finding the answers to the questions within the time constraints.

Before the experimental procedure, participants were introduced to the user interface of the video player and the questionnaire. The experimental session for each video consisted of two parts. Initially, the users had to watch a video, and afterwards they had to answer the respective questionnaire. They could not see the questions from the beginning, and the video player's buttons were disabled during the first part of the procedure. Buttons were re-enabled for use in the second part, and participants could use them to browse video and search for the answer. Figure 11.1 portrays the second part of the experimental procedure. Furthermore, there was a time restriction of 5 minutes in this

**TABLE 11.3**

Summary of Users' Characteristics.

| Number | 23 |
|---|---|
| Age | 18-35 |
| Gender | 13 Female, 10 Male |
| Occupation | Studying informatics |
| Motivation | Course credit |

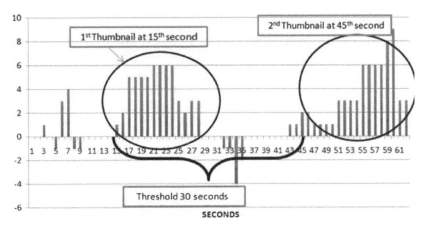

FIGURE 11.2: User activity graph with heuristic rules.

part, in order to motivate the users to actively browse through the video. The procedure was repeated in a random sequence for each video, in order to minimize possible learning effects. The result of this simple heuristic procedure is shown in Figure 11.2, where a coarse grained aggregation is evident.

## 11.4   Modeling User Interactions as Signals

The analysis to follow is based on the idea presented in [87]. Indeed, in order to extract pattern characteristics for each button distribution, i.e., scenes in which users exhibit high interaction with the video-player, three distinct stages (as shown in Table 11.4), are used.

In the first stage, a simple procedure is used in order to average out user activity noise in the corresponding distribution. In the context of probability theory, noise removal can be treated with the notion of the moving average [951]: from a curve $S^{exp}(t)$ a new smoother curve $S_T^{exp}(t)$ may be obtained as shown in Equation 11.1,

**TABLE 11.4**

Overview of the User Activity Modeling and Analysis.

| Stage | User activity signal processing |
|-------|--------------------------------|
| 1 | Smoothness procedure |
| 2 | Determination of users' activity aggregates |
| 3 | Estimation of pattern characteristics |

$$S_T^{exp}(t) = \frac{1}{T} \int_{t-T/2}^{t+T/2} S^{exp}(t')dt', \tag{11.1}$$

where $T$ denotes the averaging "window" in time. The larger the averaging window $T$, the smoother the curve will be. Schematically, the procedure is depicted in Figure 11.3. The procedure of noise removal of the experimental recording distribution is of crucial importance for the following reasons: first, in order to reveal patterns of the corresponding signals (regions of high users' activity), and second, in order to estimate local maxima of the corresponding patterns. It must be noted that the optimum size of the averaging window $T$ is entirely defined by the variability of the initial signal. Indeed, $T$ should be large enough in order to average out random fluctuations of the users' activities and small enough in order to avoid distortion of the bell-like localized shape of the users' signal which will in turn show the area of high user activity.

In the second stage, aggregates of users' activity are estimated by means of an arbitrary bell-like reference pattern. As a milestone of this chapter it is proposed that there is an aggregate of users' actions if within a specific time interval a bell-like shape of the distribution emerges. The bell-like shape implies that there is high probability that users' actions are concentrated at a specific time interval (the center of the bell) while this probability tends to zero quite symmetrically as we move away from this interval (Figure 11.4). Without loss of generality, the parameters of the width and height of the Gaussian function are set of the order of the averaging window and half of the number users' actions correspondingly. The same idea, in a rather premature form, was used in [498]. The notion of the bell–like characteristic pattern for users' aggregation is revisited and further detailed in Section 11.6.

The third step produces an estimation of the pattern characteristics, i.e., the number of users' aggregates for the specific signal and moreover their exact locations in time, by application of two different methodologies, a stochastic and a pattern matching.

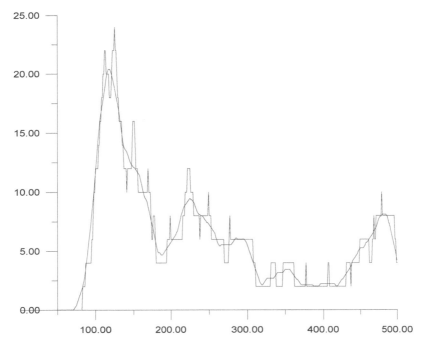

FIGURE 11.3: The user's activity signal is approximated with a smooth signal.

## 11.5 Treating User Signals

The aforementioned stochastic and pattern matching methodologies for estimation of the pattern characteristics are herein detailed.

- In the stochastic approach, the estimation of the exact locations can be done via the estimation of the generalized local maxima. The term generalized local maxima in this context refers to the center of the corresponding bell-like area of the average signal, as the nature of the original signal under examination may cause more than one peak at the top of the bell due to the micro-fluctuation. This is possible by estimation of the well known correlation coefficient $r(x, y)$ between the two signals (time series), that is, the average experimental signal and the introduced aforementioned reference bell-like time signal.

  It should be noted that while the height of the reference bell-like pattern does not affect the results, the width of the bell $D$ is a parameter that must be treated carefully. In particular, the variability of the average signal determines the order of the width $D$. Experimentation in [498] proposed that the bell width should be equal to the average half of the widths of the

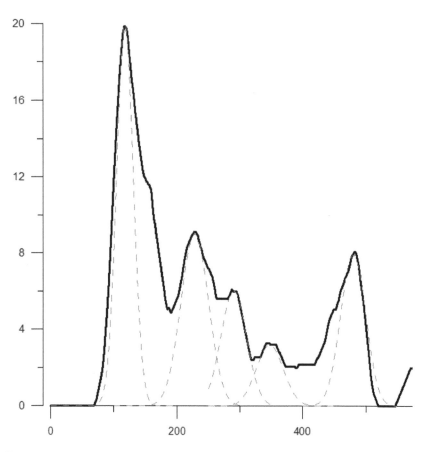

FIGURE 11.4: The users' activity signal is approximated with Gaussian bells in the neighborhood of user activity local maxima.

bell-like regions of the signals. This estimation was found optimum in order to avoid overlap between different aggregates.

- In the pattern matching approach, the distance of the reference bell-shaped pattern to the accumulated user interaction signal is measured using 3 different distance measures. Initially, a Scaling and Shifting (translation) invariant

Distance (SSD) measure (Equation 11.2), is adopted from [230]. Accordingly, for two time series, $x$ and $y$, the distance $\hat{d}_{SSD}(x,y)$ between the series is:

$$\hat{d}_{SSD}(x,y) = min_{a,q}\frac{\|x - \alpha y_{(q)}\|}{\|x\|}, \qquad (11.2)$$

where $y_{(q)}$ is the result of shifting the signal y by q time units, and $\| \cdot \|$ is the $l_2$ norm. In this context, and for simplicity, the shifting procedure is done by employing a window the size of which is empirically calculated to minimize the distance, while the scaling coefficient $\alpha$ is adjusted through the maximum signal value in the window context.

The second distance measure used is the Euclidean Distance (ED) measure (Equation 11.3) that has been shown to be highly effective [286] in many problems, despite its simplicity:

$$d_{ED}(x,y) = \sqrt{\sum_{i=1}^{n}(x_i - y_i)^2} \qquad (11.3)$$

Finally, the third distance measure utilized is a Complexity-Invariant Distance (CID) measure (Equation 11.4) for time series as discussed in [106]:

$$d_{CID}(x,y) = ED(x,y) \times CF(x,y), \qquad (11.4)$$

where the two time series $x$ and $y$ are of length $n$, $ED(x,y)$ is the Euclidean distance (Equation 11.3), and $CF(x,y)$ is the complexity correction factor defined in Equation 11.5:

$$CF(x,y) = \frac{max(CE(x), CE(y))}{min(CE(x), CE(y))}, \qquad (11.5)$$

and $CE(x)$ is a complexity estimate of a time series X, calculated as shown in Equation 11.6:

$$CE(x) = \sqrt{\sum_{i=1}^{n-1}(x_i - x_{i+1})^2}. \qquad (11.6)$$

The aforementioned distance measures produce another time series, *dist*, that describes the distance of the reference bell-shaped pattern to the accumulated user interaction signal, and thus requires the identification the locations of *dist* where its value is minimal, indicating a close match of the the reference bell-shaped pattern to the accumulated user interaction signal. To avoid using a simplistic global cut-off threshold a local minima peak detection methodology is employed, where a point in *dist* is considered a

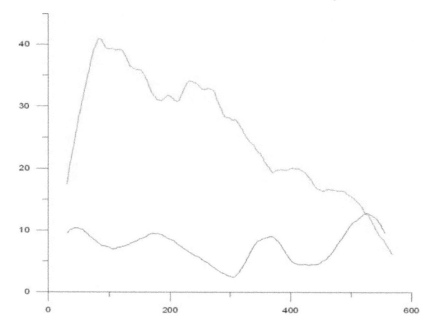

FIGURE 11.5: GoBackward signal (bottom) compared to *GoForward* signal (top), in order to understand which one is closer to the semantics of the video. The y-axis shows the measured activity of the user while the x-axis shows the time in sec.

minimum peak if it has the minimal value, and was exceeded, to the left of the signal, by a value greater by *DELTA*, the peak detection sensitivity value.

In the experimentation to follow, the focus has been on the analysis of the video seeking user behavior, such as *GoBackward* and *GoForward* after the previously described smoothing procedure. An exploratory analysis with time series probabilistic tools, such as variance and noise amplitude, verified what is visually depicted in Figure 11.5 concerning the lecture video. While the *GoBackward* button signal has a quite regular pattern with a small number of regions with high users' activity, the *GoForward* button signal is characterized by a large number of seemingly random and abnormal local maxima of users' activity. This is due to the experiment's design, where there was limited time for information gathering from the respective video and thus, usage of the *GoForward* shows users' tendency to rush through the video in order to remain within the time limit. We have also considered the use of the Play/Pause buttons, but for the current dataset, there were too few interactions. The following presents preliminary results demonstrating the results received from the aforementioned methodologies for detecting the patterns of users' activity.

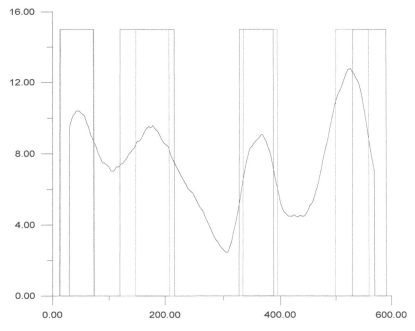

FIGURE 11.6: Lecture video: Cumulative users' interaction vs. time including results from stochastic approach. The y-axis shows the measured activity of the user while the x-axis shows the time in sec.

As far as the stochastic approach is concerned, the analysis of the users' activity distributions was based on an exploration of several alternative averaging window sizes. Results of the proposed modeling methodology for the lecture video are shown in Figure 11.6, and, in this case, the pulse width $D$ is 60 seconds and the smoothing window $T$ is 60 seconds. The results are depicted by means of pulses instead of the bell shapes, in order to compare them with the corresponding pulses of the ground-truth designated by the videos' authors. The mapping of between pulses and bells is based on the rule that the pulse width is equal to the width between the two points of the bell, where the second derivative changes sign. Similarly, results of the proposed modeling methodology for the documentary video are shown in Figure 11.7, while in this case, the pulse width $D$ is 50 seconds, and the smoothing window $T$ is 40 seconds. The smoothed signals are plotted with the solid black curve. Moreover, the pulse signals that were extracted from the corresponding local maxima indicating time intervals are depicted with the top line. Within the same figures, time intervals that were annotated as ground-truth by the author of the video to contain high semantic value information are also depicted with the bottom line.

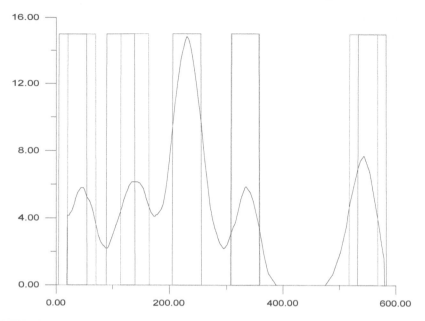

FIGURE 11.7: Documentary video: Cumulative users' interaction vs. time including results from stochastic approach. The y-axis shows the measured activity of the user while the x-axis shows the time in sec.

For the stochastic approach, the correlation of the estimated high-interest intervals and the ground-truth annotated by the author of the video, is visually evident. Cross correlation, between the two intervals, was calculated at 0.673 and 0.612 correspondingly, indicating strong correlation between the two pulses.

For the same two videos, the application of the pattern matching approach is examined for each distance measure using the F1 score and Matthews Correlation Coefficient (MCC) value for varying peak detection sensitivity values for each of the three distance measures, SSD, ED, and CID respectively. It should be noted that in the results to follow, the F1 score is linearly transposed from $[0, 1]$ to $[0, 100]$ in order to ensure ease of comparison.

**Lecture video** As shown in Figures 11.8 and 11.9, the SSD metric achieved an F1 score of 79 on a scale of $[0, 100]$, with 100 being the best value. Still, as the F1 score does not take the true negative rate into account, the MCC value has been computed leading to a 0.6 value on a scale of $[-1, 1]$, with 1 implying a perfect prediction. The claim of the ability of Euclidean Distance to be performing relatively high, despite its simplicity, is shown in this experiment where ED scored an F1 score of 72 and an MCC value of 0.42. Finally, the CID measure was outperformed by the other two measures having scored an F1 score of 70 and an MCC value of 0.39.

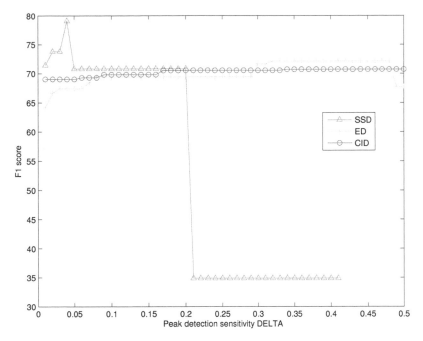

FIGURE 11.8: Lecture video, pattern matching approach, F1 score for SSD, ED, and CID metrics.

**Documentary video** As shown in Figures 11.10 and 11.11, the SSD metric achieved an F1 score of 75 and an MCC value of 0.56. The Euclidean Distance scored an F1 score of 66 and an MCC value of 0.34. Finally, the CID measure outperformed the ED measure, having scored an F1 score of 71 and an MCC value of 0.45.

## 11.6 New Insights in Managing User Interactions

The stochastic and pattern matching methodologies for estimation of the pattern characteristics discussed in this chapter were indeed tested on web videos under a controlled experiment, and were shown to present interesting results. Collective intelligence is attributed to the claim of being able to understand the importance of video content from users' interactions with the player. The results of this study can be used to understand and explore collective intelligence in general i.e., how to detect users' collective behavior, as well as how the collective behavior detected leads to judgment about the content from

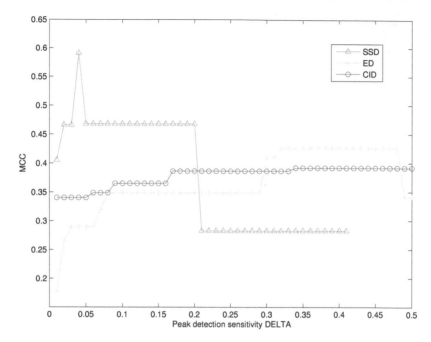

FIGURE 11.9: Lecture video, pattern matching approach, MCC for SSD, ED, and CID metrics.

which the users' activity was gathered. Moreover, collective intelligence may be used as a tool of user-based content analysis, having the benefits of continuously adapting to evolving users' preferences, as well as providing additional opportunities for the personalization of content. For example, users might be able to apply other personalization techniques, e.g., collaborative filtering, to the user activity data.

According to the definition provided for the two approaches for aggregates of users' activity estimation, it has been shown that the aggregate of users' actions locally coincides, to a large degree, with a bell-like shape of the corresponding distribution. The complete pattern of users' interactions is defined by the exact location of the center of bells of the total number of the bell-like patterns detected. In this way, one may map the different users' behavior to the different patterns observed. Moreover, these observed patterns of users' actions may reveal specific judgment about the content for which actions were collected, leading thus to collective intelligence. Indeed, for the case study presented herein, the exact locations of the bell-like patterns detected can be mapped to the most important parts of the content, as was shown by experimentation. On the other hand, collective intelligence could reveal new unexpected results, i.e., important intervals of users' behavior that were unexpected.

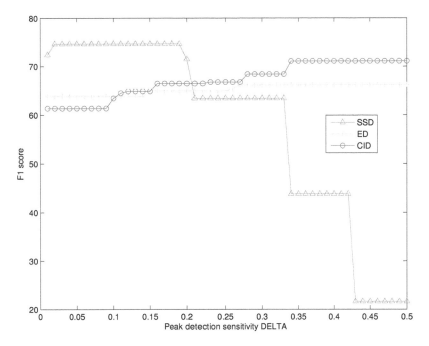

FIGURE 11.10: Documentary video, pattern matching approach, F1 score for SSD, ED, and CID metrics.

In a more general fashion, the methodology presented may treat general users' interactions for a specific (on line) content, by interpreting these interactions as an explicit time series. This could be a time series of clicks or plays of a video on YouTube, the number of times an article on a newspaper website was read, or even the number of times that a hash tag in Twitter was used.

Thus, this methodology can be applied for the detection of patterns emerging in the temporal variation of the corresponding time series, indicating the importance of a segment of content at a specific time interval of its duration. One may formally define this either as a problem of time series correlation, based on the correlation between the shape of the (experimentally collected) time series with the shape of a reference time series, indicating local maximization of users' activity, or as pattern matching of time-series wherein different similarity measures can be utilized for the detection of local minima in the distance. In both cases, a Gaussian function can be chosen as the optimum function for the reference time series.

Given that online content has a large variation during its duration, i.e., users' actions occur at arbitrary times, and with very different time intervals, a further extension of the proposed methodology is needed in order to adapt a time series metric that is invariant to scaling and shifting, i.e., to be able not

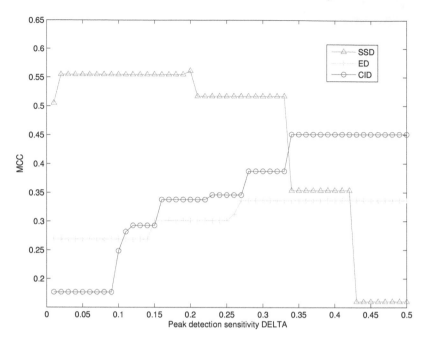

FIGURE 11.11: Documentary video, pattern matching approach, MCC for SSD, ED, and CID metrics.

only to detect the exact location of the local maxima of users' popularity, but also to estimate the corresponding absolute importance, as well as the corresponding time interval over which the specific piece of content was important enough.

To this end, it is possible, based on the aforementioned approach, to build a scale-free similarity metric introducing the notion of the aforementioned reference bell-like time signal. Indeed, the final result of this extended algorithm would be the estimation of the maximum correlation coefficient in terms of the optimum time moment and optimum bell width.

Accordingly, we may propose a two value correlation coefficient $r(t_c, w)$ where $t_c$ is the time center of the Gaussian bell and $w$ its width. In other words, we may construct a Gaussian time signal by shifting its center over the time domain of the experimental signal, and for each position we create a number of different Gaussian time signals gradually increasing its width $w$ (see Figure 11.12 from blue to red and to the green solid Gaussian bell). For each Gaussian reference signal of different width we then estimate the corresponding correlation coefficient with the experimental signal. In this way, a two dimension correlation coefficient is produced for each time location, and for different bell widths. We stated that whenever, for a specific time center of the Gaussian bell, a high correlation coefficient is identified during the time

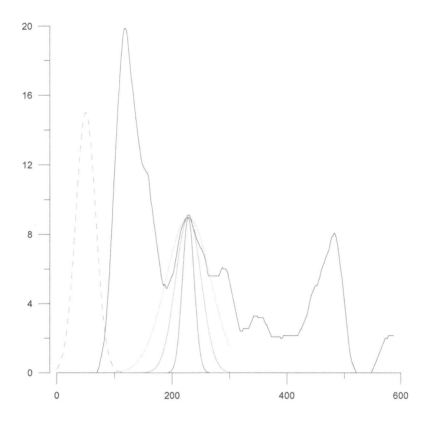

FIGURE 11.12: Gaussian bell is shifted over the time domain.

shifted process, a local maxima of the experimentally constructed time series is assumed. Indeed this can be seen in Figure 11.13 for an arbitrary time series (black solid line). With the red solid line, normalized to 10, the figure depicts the corresponding correlation coefficient between the arbitrary time series and the shifted Gaussian bell. Initially, we shall keep the width of the bell constant while a robust alternative measure for the initial width could be the variance of the smoothed experimental signal.

It is evident that there is a very clear maximum of the correlation coefficient exactly when the center of the Gaussian bell coincides with the maximum of the experimental series. As a result, the exact location of the experimental series is detected as the point of the local maximum of the corresponding correlation coefficient. Then, we relax the assumption of the constant bell width: keeping constant the center of the bell, we built Gaussian bells of different widths (as depicted in Figure 11.12). For each bell of variable width, a new correlation coefficient is computed. The maximum value of this second set of

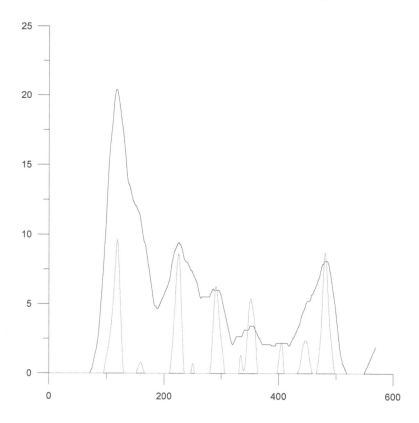

FIGURE 11.13: Local maxima of the correlation coefficient (bottom curve) coincide with local maxima of user's activity signal (top curve).

correlation coefficients is estimated, thus completing our process. The final result is the estimation of the maximum correlation coefficient in terms of the optimum time moment and optimum bell width. We argue that the optimum time moment coincides with the local maximum of the online media popularity, while the optimum Gaussian bell width coincides with the corresponding time interval over which popularity is important enough.

Summarizing, r-algorithm performs the following steps: it begins with an initial Gaussian bell, the center of which is located at the time origin of the content, and its width coinciding with the variance of the smoothed experimental signal. Then, a two step procedure follows, i.e., a detection step and a refinement or characterization step. Within the detection step, the bell is shifted along the time domain, computing the corresponding correlation coefficient between the Gaussian bell and the experimental signal. The local maxima of users' activity are identified as the time moments where the computed correlation coefficient reaches the local maximum, with the local maximum being above a specific threshold.

---

**Algorithm** The r-algorithm

---

**Require:** Experimental time series, upper part of Gaussian time series $g(ct, w)$ of center $ct$ and width $w$.

1: **for** $ct = 1$ **to** $L$ **do** {detection step}
2:    $r_c t$ {the correlation coefficient for different centers}
3:    **if** $r_c t > thress$ **then** {critical threshold of correlation}
4:       **for** $w = 1$ **to** $L/10$ **do** {characterization step}
5:          $r_c t_w$ {correlation coefficient for variable widths}
6:       **end for**
7:    **end if**
8: **end for**
9: **return** $r_c t_w$ {returns seconds of maximum user's activity and the corresponding time interval of popularity}

---

Within the characterization step, for each local maximum of the correlation coefficient, a series of Gaussian bells with variable widths is generated (beginning from a value of few seconds to a fraction of the overall duration of the content) and the corresponding correlation coefficients are computed again. The calculated optimal bell width gives an estimation of the time interval over which the content was important enough for the users.

---

## 11.7 Epilog

In this chapter, we attempted to present a method that detects collective behavior of users' activity via the detection of characteristic patterns in the corresponding signal monitoring users' activity. The methodology has been verified with web videos and user interaction data from a controlled experiment.

An algorithm for real time detection of collective activity was also presented herein, at the basis of which is the notion of a two parameter arbitrary Gaussian bell acting as a reference pattern for aggregation. Accordingly, the aggregation of users' actions coincides with the upper part of a bell-like shape of the corresponding distribution. The users' interaction pattern is defined by means of two parameters: the exact location of the center of the Gaussian bell, as well as the corresponding width. In this way, one may map different users' behavior to the different patterns observed.

Moreover, within the discussed methodology the exact height of each local maxima of users' activity can also be addressed. Indeed, as soon as the exact locations of each local maximum are estimated, then the corresponding heights coincide with the respective value for that time instance of the smoothed experimental signal. It should be noted that alternatively the use of a three-parameter value correlation coefficient could be used, i.e., for the determination of the users' maximum height, the corresponding computational cost is very high.

Further on, we need to stress that the robust determination of the relative heights of each maximum of collective activity is a very crucial parameter, since it scores the importance of each maxima. As a result, within the methodology presented, a ranking procedure can be built: the relative height of each maximum indicates the relative importance of each scene.

The results of this study could facilitate the understanding of collective intelligence in online media, i.e., how to detect collective behavior as well as how the detected collective behavior leads to judgment about the importance of fragments in time-based content. As a future work, the methods presented in this chapter may be improved, in order to capture not only quantitative measures of the Gaussian bells, but also qualitative features such as specific symmetries of the bells corresponding to the specific content over which actions are reported, thus leading to collective intelligence.

# 12

## Semantics-Based and Time-Aware Composition of Personalized e-Commerce Services on the Cloud

**Yolanda Blanco-Fernandez**

*University of Vigo, yolanda@det.uvigo.es*

**Martin Lopez-Nores**

*University of Vigo, mlnores@det.uvigo.es*

**Jose J. Pazos-Arias**

*University of Vigo, jose@det.uvigo.es*

## CONTENTS

Currently, most of the e-commerce recommender systems can adapt the selection of commercial items as the users' preferences evolve over time. However, the adaptation process commonly considers that a user's interest for a given type of product (or any of its features) always decreases with time from the moment of the last purchase, when it is true that certain products may indeed become more interesting or necessary over time. Some authors have addressed this issue by attaching temporal information to the items' metadata, but missing the point that the influence of time can be very different for different users. In this chapter, we present a time-aware recommendation strategy that solves these problems by linking an ontology of commercial products to parameterized time functions, whose values are adjusted per the users' membership in consumption stereotypes. This strategy plays a key role in a process of automatic composition of interactive services, grounded on a cloud-based personalization engine, which adopts semantic reasoning and rule-driven programming to engineer e-commerce personalized services ready to run on different consumer device platforms. This way, our approach goes further, assembling a list of items, for example, to select the most convenient offers from among various providers, to look for the pieces of information that describe each item in the most complete way for the user in question, and to arrange the most suitable interfaces through whichever device is being used.

## 12.1 Introduction

Discovering products that meet the needs of the consumers is crucial in such competitive environments as online shopping. Recommender systems provide assistance by automatically selecting the most appropriate items for each user according to his/her personal interests and preferences [46]. Obviously, keeping the users' satisfaction high requires a means to adapt the selection of items, as their preferences for products evolve over time. Product perception and popularity are constantly changing as new products emerge. Similarly, customers' inclinations are evolving, leading them to even redefine their interests. Thus, modeling temporal dynamics is essential for designing filtering strategies for recommender systems or general customer preference models. For many years, in most of the existing filtering strategies, data collection about the users' interests was regarded as a static process, weighing equally the ratings given by the users at different times. Later, some researchers proposed time-aware approaches that made the latest observations more significant than the older ones, which means assuming that a user's interest in a product always decreases from the moment of the last purchase (see examples in [631, 830, 299, 287, 565]). This may be true in certain areas of application, such as personalized programming guides that recommend TV programs to the users. Notwithstanding, the interest in (or the need for) commercial products in general may actually increase or vary in diverse forms over time. For instance, whereas car tires typically have a lifetime of 6 years for average drivers, it is expectable that taxi drivers or users interested in car tuning and motor sports need more frequent replacements (say, every 6 months). Analogously, it makes sense not to recommend dolls for some time after an average user has bought one, but the same is not true for doll collectors.

Bearing these conditions in mind, our chapter proposes a **time-aware recommendation strategy** that selects commercial products for each user, by considering the evolution of his/her particular needs and interests over time. The basic assumption is that the influence of time can be radically different, not only for different types of items as explained above, but also for different users. For that reason, our new approach makes tailor-made selections of items by exploiting the semantics formalized in an ontology to link items (and their features) to time functions, whose shapes are corrected by considering the preferences of like-minded individuals (modeled in consumption stereotypes) and the effects of time in their purchasing behaviors.

Instead of just providing the users with lists of the items that best match their preferences, one could expect time-aware recommender systems in e-commerce to select the best offers from among various providers, to look for the pieces of information that describe each item in the most complete or accessible way, to arrange the most suitable interfaces to place an order, etc. In other words, we believe the recommender should be responsible for building the shop in which the user will feel most comfortable to browse the suggested items. Obviously, this is not a task for human developers, because no workforce would suffice to provide specific services for all the different users, items, and devices in all possible contexts.

In this chapter, we explore the possibilities of the Semantic Web technologies for **generating automatically interactive e-commerce services** that provide the users with personalized commercial functionalities related to the selected items. Specifically, the procedure behind the generation of interactive services involves solutions from the field of *web services* and the development of *web mashups*, lodging our time-aware recommendation strategy, semantic reasoning and rule-driven programming to fit the particular interests and needs of each user. The goal is therefore that the resulting time-aware e-commerce recommender system goes further than assembling a list of items, by selecting for example the most convenient offers from among various providers and arranging the most suitable interfaces for whichever device. In this regard, bearing in mind the common adoption of multiple devices by the users, our approach aims for a unified user experience by redesigning the personalization logic of our time-aware recommender system. The idea is moving from a mixed client-based and server-based architecture to another one, following the *cloud computing* paradigm to ensure continuity of the personalized experiences over multiple devices and from different places.

The chapter is organized as follows. First, Sect. 12.2 includes a review of the (traditional and semantics-driven) recommender systems literature. Next,

Sect. 12.3 details the main parts of our personalization cloud, including the ontology, user profiles, and consumption stereotypes, along with the time dependence curves adopted in our time-aware approach. After focusing in Sect. 12.4 on the algorithmic internals of our time-aware filtering strategy, Sect. 12.5 describes the elements and the workflow of our automatic composition of personalized e-commerce services. Finally, Sect. 12.6 summarizes the main conclusions from our research and motivates our ongoing work in the area.

## 12.2 Background on Recommender Systems

The research community has proposed two main filtering paradigms for recommender systems, namely *content-based* and *collaborative filtering*, which are commonly combined in *hybrid approaches*. The main difference between these strategies lies within the kind of information they use during the personalization process: whereas content-based filtering considers the descriptive *features* of the items, collaborative proposals use the *ratings* that users assign to these items in their profiles.

### 12.2.1 Content-Based Filtering

*Content-based filtering* suggests to a user items which are similar to those he/she liked in the past, by matching their respective content descriptions (i.e., the features defined in the user's profile and the attributes of the available items). Even though there exist plenty of different similarity metrics for this purpose, they all miss much knowledge during the personalization process, because they are unable to reason about the meaning of the attributes (for example, it is not possible to link items about *"Golden Retriever"* with items about *"Boxer,"* because the two words are dissimilar). Actually, the most basic approaches establish simple syntactic comparisons among the attributes of the considered items, whereas more advanced metrics rely on the predictive capabilities of automatic classifiers — e.g., Bayesian networks, decision trees or neural networks — to decide about the relevance of an item for a given user (see examples in [141, 105, 632, 1050]). All these approaches have a common weakness in their syntactic nature, which only permits them to detect similarity among items sharing *the same* attributes. This results in *overspecialized* recommendations which only include items that are excessively similar to those the user already knows [46].

### 12.2.2 Collaborative Filtering

*Collaborative filtering* recommends to a user items which have been appealing to others with similar preferences (henceforth, his/her *neighbors*). First, the user's neighborhood is formed; next, his/her levels of interest in the items defined in the neighbors' profiles are predicted, so that the products with the highest ratings are finally suggested.

Traditional collaborative approaches compare the users' preferences by considering only the levels of interest defined in their respective profiles. Specifically, these approaches create a vector for each user containing ratings for each item available in the recommender system (a rating of 0 is assumed in case the user has not rated an item yet). Next, the correlation between the considered user's vector and the vectors of the remaining users is computed. Finally, the $N$ highest correlation measures corresponding to the $N$ nearest neighbors to the user are selected. From this way of doing things, it follows that traditional collaborative approaches only detect that two users share preferences in case there exists an overlap between the items defined in their respective profiles.

In order to predict the interest of the user in the items contained in his/her neighbors' profiles, collaborative approaches compute a weighted average of their ratings in these items, using as weights the correlation measures between their respective preferences.

Collaborative recommender systems suffer from some severe limitations. One of the most critical ones is the so-called *sparsity problem*, whose effects are apparent as the number of available items increases. In this case, it is unlikely that two users have rated the *same* items in their profiles, thus hampering the formation of the user's neighbors. Collaborative systems are also limited due to scalability-related concerns: as the number of available items gets higher, the users' rating vectors also increase in size. Consequently, the creation of neighborhoods becomes very demanding in computational terms. Last, since collaborative approaches only suggest to the user items defined in their neighbors' profiles, new items available in the system cannot be recommended before a significant number of users have rated them in their profiles.

## 12.2.3 Hybrid Approaches

In view of the limitations of content-based and collaborative systems, researchers have commonly opted for *hybrid approaches* to exploit the advantages of both filtering strategies, and to mitigate their deficiencies. Since *hybrid approaches* unify collaborative and content-based filtering under a single framework, they compute the similarity between two users' preferences by considering both the content descriptions and their respective ratings. This way, hybrid systems fight the *sparsity problem* by detecting that two users have common preferences even when there there is no overlap between the items contained in their profiles. However, in order to measure similarity in this case, it is necessary that these items have common attributes, and so traditional hybrid approaches are still limited by the syntactic metrics used in the traditional content-based techniques.

## 12.3   The Personalization Cloud

In our previous works with e-commerce we had used personalization engines built according to the traditional client-server paradigm [1009, 1008]. Now, since we are moving to scenarios of multiple devices converging into a unified user experience, we have designed a new personalization engine following the guidelines of the *cloud computing* paradigm [721]. The most significant difference from the users' point of view has to do with ensuring continuity of the personalized experiences over different devices, and from different places. Internally, by relying on a suitable IaaS (*Infrastructure as a Service*) provider, we also get significant advantages in terms of scalability, availability, performance, and flexibility for changes in comparison with our previous client-server designs.

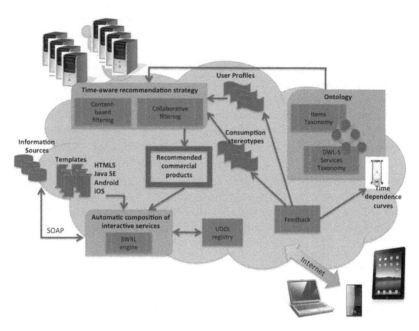

FIGURE 12.1: The elements and processes of our personalization cloud.

In this section, we will detail the main elements and processes of our personalization cloud (depicted in Fig. 12.1) which include the following aspects:

- The knowledge bases that support the personalization processes, namely an *ontology* of semantic concepts and their relationships, along with a a collection of *user profiles* and *stereotypes* that capture information about the users.

- The *time-aware recommendation strategy* that selects the most suitable products for each user, by (i) reasoning about their semantic descriptions and the preferences and interests stored in their profile, and (ii) considering the time dependence curves associated to the ontology concepts.

- The *automatic composition of interactive e-commerce services*, which is grounded on a *rule-driven engine* which is in charge of: (i) identifying the most appropriate *services* for each user, (ii) populating *templates* with contents retrieved from multiple sources, and (iii) translating those templates into a running application for the specific platforms of the *users' devices*.

- The *feedback* procedures that are used to update the knowledge bases and the functions adopted by our time-aware recommendation strategy, by monitoring the users' consumptions behaviors and their interactions with the automatically composed tailor-made services.

## 12.3.1  Domain Ontology

An ontology represents the knowledge of a domain by a language equipped with a formal semantics, by characterizing the semantics in terms of *concepts* and their *relationships* (represented by *classes* and *properties*, respectively). Both entities are hierarchically organized in the conceptualization, which is populated by including specific instances of both classes and properties. As an example, Fig. 12.2 shows a small excerpt from the ontology adopted in our approach where classes, items, and attributes are denoted by gray ellipses, white squares, and white ellipses, respectively. In this figure, the ontology organizes diverse commercial products and defines features like unique identifiers, the topics of a book, the actors in a movie, the materials of clothes, or the technical specifications of an appliance. Besides, there are labeled properties joining each item to its attributes and *seeAlso* properties to link strongly related items.

## 12.3.2  User Profiles and Stereotypes

Up to this point, everything is independent of the individual preferences and needs of any user (as it was in [1008]), so we need additional artifacts to incorporate the users' personal interests into the filtering process. To this aim, we do not proceed individually, but rather with groups of users who may be clustered together per some of their preferences.

In our work, a user's profile stores various data, including a record of the items bought in the past, the classes and attributes that describe those items in the ontology, and the time of the last purchase. Furthermore, each item is linked to a number on a scale from $-1$ to $1$ that measures the degree of interest (hereafter DOI) of the user in it ($-1$ represents the greatest disliking, $1$ the greatest liking). The numbers may be given explicitly by the user, or

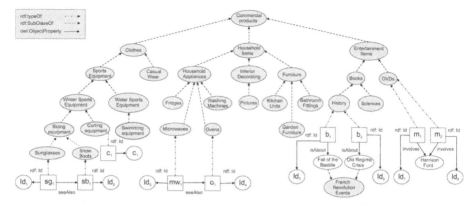

FIGURE 12.2: A micro-excerpt from our ontology.

inferred indirectly by monitoring their interaction with the recommendations (more details in [617]).

The stereotypes take the same form as the individual profiles, though they are completely void of information that might serve to identify individual users. In other words, a stereotype is an excerpt from the ontology with attached DOIs. These can be updated dynamically from *feedback messages* issued by the users, indicating only their degree of membership to the stereotype (a number whose computation will be explained later), the rating given to an item, and the current time instant.

### 12.3.3   Time Functions and Group Corrections

Our approach to time-aware filtering starts out by associating parameterized time functions to item classes and attributes, in order to model the variation of the potential interest of each type of product or any of its features with regard to absolute dates or purchase times. Even though it is possible to adjust values per marketing criteria, the specific time function associated to a given item hinges commonly on its nature/type (e.g., a book and a dishwasher are not purchased by a user with the same frequency, hence the fact that both items need to be associated to different time patterns or functions). We handle functions with diverse shapes, including combinations of constant, linear, exponential, sinusoidal, parabolic, hyperbolic, and elliptic segments, with values between 0 and 1. As a rule of thumb, low values are intended to prevent the recommendation of the items, whereas high values are intended to promote them. We also require valid functions to take the maximum value (1) at some point, corresponding to the time an item or an attribute is potentially most interesting for a general audience.

Next, we shall focus on the few functions represented in Table 12.1, which suffice to explain how the time-aware filtering works:

**TABLE 12.1**
A Few Examples of Our Parameterized Time Functions.

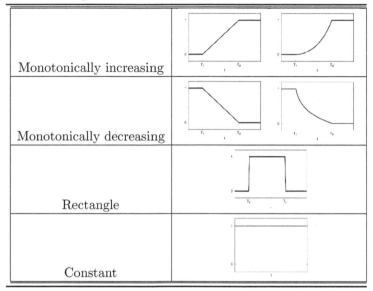

| | |
|---|---|
| Monotonically increasing | |
| Monotonically decreasing | |
| Rectangle | |
| Constant | |

- **Monotonically increasing function.** First, we consider items that are purchased sporadically due to their long average lifetime (e.g., consumer electronics, vehicles, and household appliances). The interest for such products can be modeled by a function that grows (linearly or exponentially) from the time of purchase. The zero value of the function coincides with the instant when the user bought the product (denoted by $T_1$ in Table 12.1), whereas the maximum value 1 is reached once its lifetime has expired ($T_2$). For instance, this function can prevent recommending a washing machine to a user who has just bought another one, to give way after a few years.

- **Monotonically decreasing function.** There are many products that are useful during a limited period and whose utility decreases over time, as it happens with seasonal items (e.g., swimming pool supplies and winter sports equipment). The temporal dependence here can be modeled with a function that takes the maximum value up to the beginning of the season (instant $T_1$ in Table 12.1) and decreases monotonically (linearly or exponentially) afterwards. The zero value is reached once the seasonal period has ended (instant $T_2$). For example, this function prevents recommending skiing equipment in summer.

- **Rectangle function.** A rectangle function (see Table 12.1) can be bound to products that may be repeatedly purchased during a given period of time. This is the case, for example, of food like nougat bars, which are only available around Christmas — outside of those months, the zero value of the function prevents such products from appearing in any recommendation.

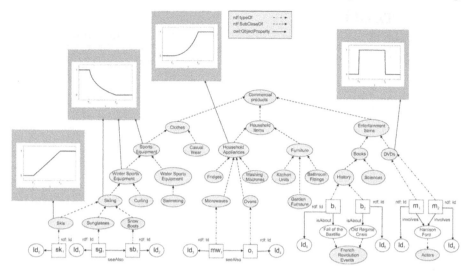

FIGURE 12.3: Some sample examples of time functions associated with classes of our ontology.

Likewise, Elvis Presley merchandising could be linked to a rectangle between July and September, surrounding the date of his death.

- **Constant function.** A constant (time-unaware) function can be linked to products that the user could purchase daily, such as books or personal hygiene items.

As we shall explain later, the time function for a specific item is computed from those of the classes it belongs to and those of its attributes. By default, attributes are associated to a constant function of value 1, which can be modified per marketing criteria/requirements (remember examples given in the description of the rectangle function), while each class inherits the temporal dependence from its immediate superclass (for example, in Figure 12.3, *"Sports Equipment"* is associated to a monotonically decreasing function, which is transmitted to *"Winter Sports Equipment"* and *"Skiing equipment"*).

In any case, it is also possible to disregard the inheritance process by assigning specific functions to the classes and attributes of specific items. For example, *"Skis"* appears linked to a monotonically increasing function in Figure 12.3, whereas its superclass defines a monotonically decreasing one.

The time function computed for an item from its classes and attributes is not used directly in the time-aware filtering, because in this way we would be assuming the same temporal dependence for all users. Instead, we modify the shape of that function by means of a group correction that is built from the feedback messages received from the users. Specifically, we use the following

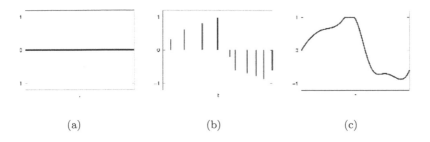

(a)                                    (b)                                    (c)

FIGURE 12.4: Computation of group correction: (a) the starting zero function; (b) a sample pulse train; (c) the resulting group correction.

procedure to compute one group correction $gc(\mathcal{C}_m, \mathcal{S}_j, t)$ for each class $\mathcal{C}_m$ of a stereotype $\mathcal{S}_j$ as a function of time:

- First, we record the ratings received for items belonging to the class $\mathcal{C}_m$ in the different time instants.

- Second, we build a pulse train by averaging the ratings of each instant, weighed by the degrees of membership to the stereotype $\mathcal{S}_j$ of the users who provided them.

- Finally, we approximate the pulse train by a natural smoothing spline [764], so that each piece of feedback has an effect over a period of time, and not only at the specific time instant for which it was issued. The resulting curve is trimmed between $-1$ and $1$.

This procedure is depicted in Figure 12.4, starting with a zero function, which is assumed by default (i.e., in the absence of feedback).

We devised group corrections as an additive adjustment of the items' time functions, so the result of the sum — trimmed between 0 and 1 and normalized to take the maximum value 1 — yields another time function. As shown in Figure 12.5, the corrections can completely modify the shape of the temporal dependence curves.

## 12.4    Our Time-Aware Recommendation Strategy

Having described the elements involved, we can now explain our new semantics-based time-aware recommendation strategy, which follows a four-step process (see Figure 12.6):

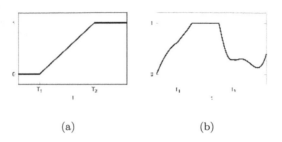

(a)                                        (b)

FIGURE 12.5: Adding a group correction to a time function: (a) a sample time function; (b) the function of Figure 12.5(a) corrected by Figure 12.4(c).

- **Step 1: Stereotype-driven pre-filtering**. In order to reduce the computational cost of making recommendations, initially, we perform an offline pre-filtering process driven by the available stereotypes, using a semantics-based *similarity metric* to sort out the different items by their potential

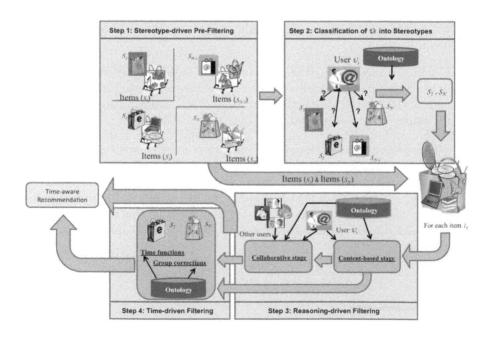

FIGURE 12.6: The four steps of our semantics-based, time-aware recommendation strategy.

interest for different groups of users. Our metric differs from existing similarity approaches due to the richness of its inferential capabilities: while other metrics only exploit the information explicitly represented in the domain ontology, our metric discovers hidden semantic relationships from the hierarchical links and properties formalized in the ontological model.

- **Step 2: Classification of the user into stereotypes.** Both to refine the pre-selection and to obtain the temporal dependence curves that affect a given user $\mathcal{U}_i$, it is necessary to identify the stereotypes in which the user fits best ($\mathcal{S}_2$ and $\mathcal{S}_N$ in Figure 12.6). To this aim, existing stereotype-based classification mechanisms are not appropriate because they typically classify users on the basis of answers they gave during first use of the system, which can be "confusing" and "irritating" [782, 354]. For that reason, we propose a taxonomy-driven classification method, which harnesses the hierarchical structure of the ontology when comparing the user's preferences and the available stereotypes. Different from other existing semantics-devoid mechanisms [957, 836, 800], our approach brings flexibility to the classification process allowing us to compare preferences and stereotypes by considering the interest in *classes/types* of products instead of the interest in *specific* products.

- **Step 3: Reasoning-driven filtering.** Having classified the user $\mathcal{U}_i$, we match the pre-selected items against the preferences captured in his/her profile, applying a hybrid recommendation strategy, which improves the traditional content-based and collaborative filtering approaches by semantic reasoning mechanisms to fight their limitations (recall Sect. 12.2).

- **Step 4: Time-driven filtering.** Finally, we assess the current interest of the user $\mathcal{U}_i$ in the items selected in step 3 by combining our contributions in terms of time issues in personalized e-commerce: first, the *time functions* of the classes and attributes of the selected items, and second, the *group corrections* computed for the stereotypes in which the user fits best.

## 12.4.1   Step 1: Stereotype-Driven Pre-Filtering

The pre-filtering consists of selecting the items that are most appealing to each available stereotype. The benefit of the pre-filtering process has to do with the computational cost of the recommendation strategy adopted in the filtering phase (step 3). In absence of pre-filtering, the hybrid strategy should match *all* the items available in the ontology against the user's preferences, which would be too time-consuming due to the complexity of the reasonings adopted in the filtering phase, as we will explain in Section 12.4.3. On the contrary, if we make a previous offline pre-filtering, the strategy will have to reason just about a subset of items potentially interesting for each stereotype, thus making the personalization process computationally feasible.

Specifically, the pre-filtering consists of computing a *matching level* between each item $\mathcal{I}_k$ in the ontology, and each one of the stereotypes, $\mathcal{S}_j$. To this aim, we need a similarity metric to compare $\mathcal{I}_k$ with each item $\mathcal{I}_r$ rated in $\mathcal{S}_j$. Briefly, $\mathcal{I}_k$ is marked as a potentially interesting item for users that fit in $\mathcal{S}_j$ when the *matching level* exceeds a configurable threshold $\alpha_1$.

In literature, it is possible to find numerous similarity metrics that measure resemblance by looking at the strength of the *hierarchical relationships* established in a taxonomy (see [249, 768, 1001]). According to these proposals, the notion of hierarchical similarity between two items depends on the existence and position of a common ancestor in the hierarchy. Specifically, the value of the hierarchical similarity between two items grows with the depth of their *lowest common ancestor* (LCA), and also with its proximity to both in the hierarchy. The depth of a node is given by the number of hierarchical links traversed to reach the node from the root of the hierarchy; thereby, the hierarchical similarity between two nodes is 0 if they do not have a common ancestor other than the root class. For example, the similarity value between the sunglasses $sg_1$ and the snow boots $sb_1$ in Fig. 12.2 is higher than the similarity value between the items $sg_1$ and $c_1$, because the first $LCA$ (*"Skiing equipment"*) is one degree more specific than the second one (*"Winter Sports equipment"*).

As Ganesan et al. explained in [720], taxonomy-based approaches do not accurately capture similarity in certain domains, such as when the data is sparse or when there are known relationships between the compared items. For that reason, we extend the existing similarity metrics by mixing the hierarchical relationships included in the domain ontology with other semantic associations hidden behind the properties defined in it, thus achieving richer inferential capabilities than those offered in approaches such as [801, 739, 771]. Specifically, our metric measures similarity by looking at relationships between the semantic attributes of the items compared, so that two items are considered similar if (i) they have common attributes, or (ii) they have sibling attributes, i.e., attributes belonging to the same class in some hierarchy. For example, the movies $m_1$ and $m_2$ in Fig. 12.2 share the attribute of having Harrison Ford as an actor, whereas the books $b_1$ and $b_2$ have sibling attributes in that they deal with events bound to the French Revolution.

Specifically, the similarity value measured by our metric grows with: (i) the specificity of the $LCA$ existing between the compared items in the ontology hierarchy, (ii) the number of their shared attributes, and (iii) the interest of the user in those attributes. Consequently, the matching between an item $\mathcal{I}_k$ and a stereotype $\mathcal{S}_j$ is high when $\mathcal{I}_k$ is very similar to items that appear with high DOIs in $\mathcal{S}_j$.

## 12.4.2    Step 2: Classification of the User into Stereotypes

The goal in this phase is to select the stereotypes where the user $\mathcal{U}_i$'s preferences fit best, from which we build adaptive temporal shapes to correct the time functions associated to the items in the ontology. This phase is crucial, due to the fact that stereotypes prevent us from working with individual corrections, which would be unfeasible to gather sufficient information from every single user to accurately characterize their potential interest in all item classes over time. Instead, it makes more sense to consider the success or failure of the recommendations made to like-minded individuals, by considering their (stereotypical) consumption histories.

In order to measure up to what point the user $\mathcal{U}_i$ is represented by a stereotype $\mathcal{S}_j$, we compute the *degree of membership*. To this aim, we need to match $\mathcal{U}_i$'s preferences against the consumption histories represented in $\mathcal{S}_j$. In the absence of semantics, this comparison should be tackled in terms of *items*, so that $\mathcal{U}_i$ would be only assigned to $\mathcal{S}_j$ if the items this user has bought are exactly those included in that stereotype. However, the product hierarchy defined in our ontology (see Fig. 12.2) makes it possible to match $\mathcal{U}_i$'s preferences and $\mathcal{S}_j$ in a more flexible way. Specifically, we propose a taxonomy-driven classification approach that infers relationships (i.e., similarity) between items from the hierarchical structure of the ontology, being able to detect that $\mathcal{U}_i$'s profile fits in $\mathcal{S}_j$ when the user has purchased either (i) the same type of product as those included in the stereotype, or (ii) products strongly related to them. This requires comparing the user's profile and stereotype by considering their classes (instead of their items), as shown next:

- First, we create a rating vector for $\mathcal{U}_i$ — denoted by $\mathcal{V}(\mathcal{U}_i)$ — including the DOI indexes of the most significant classes in this user profile (i.e., the ones with DOIs close to 1 or $-1$, representing the items that are most appealing or unappealing to the user).

- Second, we create a rating vector for each stereotype $\mathcal{S}_j$ — denoted by $\mathcal{V}(\mathcal{S}_j)$ — including the DOIs it assigns to the classes of $\mathcal{V}(\mathcal{U}_i)$.

- Third, the degree of membership of $\mathcal{U}_i$ to $\mathcal{S}_j$ — denoted by $DOM(\mathcal{U}_i, \mathcal{S}_j)$ — is computed as the correlation between the two vectors $\mathcal{V}(\mathcal{U}_i)$ and $\mathcal{V}(\mathcal{S}_j)$.

- Last, we consider that the user $\mathcal{U}_i$ is represented by the stereotype $\mathcal{S}_j$ if $DOM(\mathcal{U}_i, \mathcal{S}_j)$ exceeds a configurable threshold $\alpha_2$.

### 12.4.3 Step 3: Reasoning-Driven Filtering

Having classified the user, our recommendation strategy focuses on the items that were identified in step 1 as suitable for the stereotypes that represent him/her (identified in step 2). For each one of those items, $\mathcal{I}_k$, we compute a time-unaware recommendation value for the user $\mathcal{U}_i$ by content-based and collaborative criteria. In order to avoid the limitations of traditional approaches to content-based and collaborative filtering, we enhance them with our semantic reasoning mechanisms, resulting in the two phases we explain next.

#### 12.4.3.1   Our Enhanced Content-Based Phase

Our reasoning-driven content-based phase leans on the same semantic similarity metric explained above, which allows us to compute the *matching level* between $\mathcal{I}_k$ and the items rated in $\mathcal{U}_i$'s profile (denoted by $matching(\mathcal{I}_k, \mathcal{U}_i)$). Thanks to the semantic reasoning, our enhanced content-based phase alleviates the *overspecialized* nature of the traditional recommendations (recall Section 12.2.1), by detecting that two items are similar if they are semantically associated, even when their respective attributes are different. This way, $matching(\mathcal{I}_k, \mathcal{U}_i)$ is high when $\mathcal{I}_k$ is semantically related to items that were very appealing to $\mathcal{U}_i$. If the resulting value is greater than a configurable threshold $\alpha_3$, $\mathcal{I}_k$ is selected for the final filtering step. Otherwise, as shown in Figure 6, it is reconsidered from a collaborative filtering perspective.

#### 12.4.3.2   Our Enhanced Collaborative Phase

The collaborative phase attempts to predict $\mathcal{U}_i$'s rating for an item $\mathcal{I}_k$ by considering the preferences of individuals with similar interests (neighbors). The identification of like-minded users is driven by the same procedure as the taxonomy-driven classification mechanisms used to associate a user with a stereotype: roughly, we create and correlate the rating vectors of $\mathcal{U}_i$ and the other users, selecting as neighbors the ones who yield the $\mathcal{M}$ greatest correlation values. Then, $\mathcal{U}_i$'s predicted rating for $\mathcal{I}_k$ is computed from the interest of $\mathcal{U}_i$'s neighbors in $\mathcal{I}_k$. Again, $\mathcal{I}_k$ is selected for the final filtering step if the resulting value exceeds a configurable threshold, $\alpha_4$.

Our collaborative phase overcomes the *sparsity problem* of the traditional approaches (remember Section 12.2.2), because it allows us to detect that the preferences of two users are similar even when their respective profiles do not contain identical items or attributes. In our approach, it is only necessary that the classes of the considered items share a common ancestor in the hierarchy defined in the ontology. Furthermore, our process of neighbors' rating estimation enables us to suggest (without unnecessary delays) items which are completely novel for all the users because when $\mathcal{U}_i$'s rating in $\mathcal{I}_k$ is unknown, this value is predicted as the matching value between their preferences and that item. Therefore, we alleviate the *latency problem* of the traditional collaborative approaches, in which an item must be rated by many users before it can be suggested.

Last, after content-based and collaborative phases, we compute the time-unaware recommendation value of the item $\mathcal{I}_k$ for the user $\mathcal{U}_i$ we mentioned at the beginning of the section. This value is computed by taking either the *matching level* between $\mathcal{I}_k$ and the items rated in $\mathcal{U}_i$'s profile (if $\mathcal{I}_k$ were selected by content-based filtering), or the $\mathcal{U}_i$'s predicted rating for $\mathcal{I}_k$ (if it were selected by collaborative filtering).

## 12.4.4 Step 4: Time-Driven Filtering

The final step of our filtering strategy consists of assessing the current interest of the user in the items selected by semantic reasoning. To do this, we multiply the time-unaware recommendation value resulting from the previous step by a factor $\mathcal{CI}(\mathcal{I}_k, \mathcal{U}_i, t_0)$ that results from valuating the item's time function — modified by the pertinent group corrections — at the current instant $t_0$. This way, we punish items that reasoning and time criteria have detected as not appropriate for the user $\mathcal{U}_i$.

With this computation, we corrected the time functions associated to the classes and attributes of the item $\mathcal{I}_k$ in the ontology by using the group corrections built from the stereotypes where the user $\mathcal{U}_i$ fits best. Specifically, this process is organized as follows:

- First, we consider the time functions associated with the classes and attributes of the item $\mathcal{I}_k$ in the ontology. In the computation of the (uncorrected) time function for the item $\mathcal{I}_k$, we average the functions of its classes and reshape the resulting curve, multiplying by the time functions of its attributes.

- Second, we take into account the group corrections corresponding to the classes of $\mathcal{I}_k$ per the stereotypes in which the user $\mathcal{U}_i$ fits best. The influence of the group correction corresponding to those classes in each stereotype $\mathcal{S}_j$ is weighed by the degree of membership of the user $\mathcal{U}_i$ to it.

- Finally, we truncate values to fit the range $[0, 1]$, and normalize to have the maximum value 1 in some point, because this refers to the time an item or an attribute is potentially most appealing to a general audience.

An item $\mathcal{I}_k$ is finally recommended to the user $\mathcal{U}_i$ if the time-aware recommendation value exceeds a configurable threshold $\alpha_5$. Regarding the values of our $\alpha_i$ thresholds (with $i = \{1, 2, 3, 4, 5\}$), note that they should be set and tuned empirically, depending on what policies are considered most suitable when it comes to deciding what items are relevant for each stereotype (in the case of $\alpha_1$), what users belong to each stereotype ($\alpha_2$), and what items are relevant to each user ($\alpha_3$, $\alpha_4$ and $\alpha_5$). If the granularity of the items' and users' characterization is high enough, we can be picky, and choose values close to 1; if the characterizations are coarse (as it commonly happens when there are few items or users), we have to be more permissive, and choose lower values.

Once our time-aware recommendation strategy has selected the most appropriate item for each user, it is necessary to compose interactive tailor-made services about those products, by considering the personal preferences individually when choosing the functionalities provided by the composed services (e.g., purchase or only informative), the kind of elements most suitable to the performance of the user's device (e.g., maps, videos, or only text descriptions...), the sources to retrieve information... This automatic composition process is detailed in the next section.

## 12.5 Automatic Composition of Personalized e-Commerce Services

We conceive our interactive e-commerce services as mashups that are assembled by discovering, invoking, and automatically composing web services. Our approach personalizes this composition process by considering the preferences and context information of the users to whom the service is offered. As a novelty, this personalization process is enhanced by semantic reasoning techniques that are able to select the most appealing items for each user, starting from her preferences and semantic annotations of available items. Our approach composes the interactive service in a dynamic and personalized way by including multiple contents related to each selected item (e.g., photos, videos, maps, text descriptions, etc), which are located and retrieved by web service technologies. The next subsections include details about the elements involved in the composition process (Sec. 12.5.1) and its workflow (Sect. 12.5.2).

### 12.5.1 Necessary Ingredients

Our service composition process is grounded on the following elements:

- **Functional elements.** Our interactive personalized services are assembled by combining multiple *functional elements* that are characterized as web services' *semantic web services* described by machine-processable annotations compliant with OWL-S [648], which includes three interrelated subontologies, known as the *Profile, Process Model*, and *Grounding*. In brief, the profile expresses what a service does for purposes of advertising, constructing service requests, and matchmaking; the process model describes how it works and enables invocation and composition of web services; and the grounding maps the constructs of the process model into detailed specifications of message formats and protocols (normally expressed in WSDL).

  OWL-S provides an extensive ontology of functions where each class in the ontology corresponds to a class of homogeneous functionalities. Using such an ontology, web services may be defined as instances of classes that represent their capabilities in a *Service Taxonomy*. In order to assess the adequacy between a type of OWL-S web service and the user preferences, our domain ontology includes both the *Items Taxonomy* and the *Services Taxonomy*, so that the semantic reasoning can relate the available OWL-S *Service Profiles* to the items stored in the user profiles.

  A micro-excerpt from our *Services Taxonomy* is depicted in Fig. 12.7, where we see that each item is linked to different types of OWL-S web services by labeled properties (*hasTypeofService*). For example, the guidebook about *Modena* and the sports car are associated to two kinds of OWL-S services. First the guide is joined to a service where this item can be bought, and

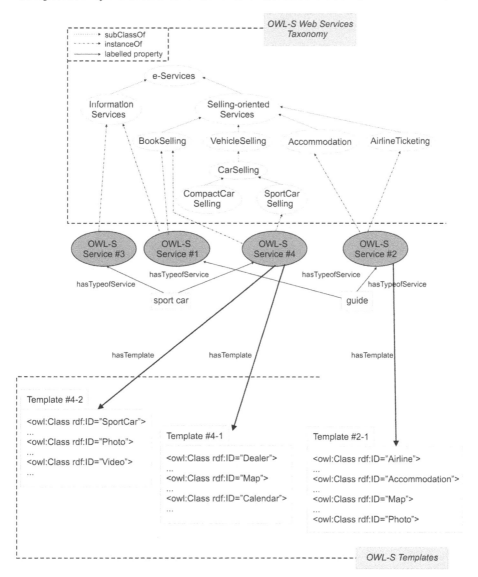

FIGURE 12.7: A brief excerpt from the taxonomy of our OWL-S semantic web services and templates to offer them.

also to where information about the main tourist attractions of *Modena* is provided (denoted by *OWL-S Service #1* in Fig. 12.7). The second service of the *guide* offers a travelogue of this Italian region with the option of booking accommodations and flights (*OWL-S Service #2*). Regarding the sports car, the first of the services (*OWL-S Service #3*) simply describes its features, while the second one (*OWL-S Service #4*) is oriented to the selling of the vehicle, hence this service offers the chance to arrange an appointment with a dealer.

- **Templates**. Our tailor-made services include several interactive elements (e.g., photos, text, audiovisual contents, etc) that provide information about the item being advertised. In order to select the elements to be lodged in an interactive personalized services, we have defined templates where the Simple Object Access Protocol (SOAP) logic necessary to retrieve them (by the corresponding OWL-S web services) is programmed. Actually, the same service can be offered through multiple templates that differentiate from each other in the type of interactive elements shown to the user. This fact is depicted in Fig. 12.7, where the elements of each template are identified by semantic concepts formalized in our OWL-S ontology (e.g., *Map*, *Dealer*, or *Calendar* for checking of availability in case of a booking).

- **Semantics-Enhanced UDDI registry**. Finding the elements of a given template requires that the OWL-S services used to retrieve them advertise their capabilities with a UDDI registry, which matches the requested services and the advertisements stored. We adopt a semantics-enhanced registry that combines OWL-S and UDDI so that the latter provides a worldwide distributed registry, whereas the former supplies the information required for capability matching. Specifically, the OWL-S/UDDI matchmaker integrates OWL-S capability matching in the UDDI registry, leading to automated service discovery that enables us to distinguish between services that have different names and equivalent functionalities, or services with identical names and totally unrelated operations.

### 12.5.2   How to Cook the Recipe

The composition of our interactive e-commerce services is triggered when the user shows interest in one of the items suggested by our time-aware recommendation strategy. In this process it is necessary to choose: (i) which type of OWL-S web service best matches the user's preferences, and (ii) which is the most convenient template to use to assemble the service. Finally, the template selected must be populated by retrieving the contents included in each one of its interactive elements, as we detail through this section.

In order to select the kind of OWL-S service, we adopt an approach similar to that presented in [750], where Dirk Plas et al. use SWRL rules to relate user preferences and context information in a personalization system. Being inspired by this work, we defined a set of SWRL rules to associate personal information of the users (e.g., hobbies and income) with the kind of services classified into the *OWL-S Web Services Taxonomy* shown in Fig. 12.7. In other words, the user preferences are antecedents to our SWRL rules, while the kind of services locate in their consequents. These rules make it possible to infer, for example, that a user (denoted by U) who is fond of traveling may appreciate services (denoted by WS) that permit one to book a tour (as stated by Rule #1 below) instead of services that only describe the tourist attractions of a given destination. Likewise (per Rule #2) it is possible to infer that a user with a high income will more likely be interested in services for the buying of luxury items than in services that only provide information about them.

| **SWRL Rule #1** |
| --- |
| user (?U) AND swrlb:notEqual (income(?U), "LOW") AND swrlb:equal(hobby (?U), "TRAVEL")) ⇒ appealingTypeOfService (?U,?WS) AND TourSellingService (?WS) |

| **SWRL Rule #2** |
| --- |
| user (?U) AND swrlb:notEqual (income(?U), "HIGH") ⇒ appealingTypeOfService (?U,?WS) AND LuxuryItemsSellingService (?WS) |

This kind of inference is carried out by a Description Logic (DL) reasoner that matches the user preferences against the available SWRL rules, by assigning a relevance factor to each kind of OWL-S web service associated to the item chosen by the user. Obviously, the factor of a kind of service is high when it appears in the consequent of a rule whose antecedent identifies the user preferences.

Having selected a service type, the next step is to discover in the UDDI registry OWL-S services fulfilling two conditions: first, they must be categorized in the *Services Taxonomy* under the previously selected type; and second, these services must permit us to find all the elements included in the selected template. With this information, the *OWL-S/UDDI matchmaker* maps the OWL-S Service Profile into the corresponding WSDL representations of the services advertised in our semantic registry. This matching process is reduced to subsumption between the capabilities categorized in the *Service Taxonomy*. As a result, the registry provides a set of web services to retrieve the contents to be lodged in the template that shapes the interactive service. For example, assembling personalized interactive services via *Template #2-1* shown in Fig. 12.7 requires us to discover selling services that offer information about airlines and hotel providers, as well as maps and photos of rooms.

To conclude the process of assembly, it is necessary to invoke the discovered OWL-S services by exploiting the WSDL descriptions of message formats and in/out arguments provided by the *Service Groundings*. By this invocation, our composition logic retrieves contents that are pre-filtered considering the user preferences. The resulting contents are finally lodged in the OWL-S template to shape the interactive personalized service.

## 12.6 Conclusions and Further Work

In this chapter, we have proposed a personalization strategy that combines semantic reasoning techniques with time-driven filtering. The aim was to show that the influence of time on the potential interest of an item can be radically different for different users, whereas previous works merely considered time on an item-dependent or class-dependent basis. Our approach starts out from the identification of consumption stereotypes, which make it possible to process the users' feedback so as to dynamically compute group corrections and reshape the default temporal dependence curves. The new strategy should be easy to incorporate in semantics-based recommender systems, because the time-driven filtering only depends on attaching information to the nodes of an ontology.

Our time-aware recommendation strategy plays a key role in a process of automatic composition of personalized e-commerce services, which can be considered as the second contribution of our research work. Specifically, instead of merely providing a list of potentially interesting items, the approach explored in this chapter takes advantage of an ontology-driven reasoning about items and user profiles to automatically compose interactive services providing personalized commercial functionalities. Our proposal paves the road for generation and provision of interactive capabilities which have been noticeable for their absence, despite being one of the major assets of the new technologies. The contributions in this regard have to do with the adoption of semantic reasoning processes in the automatic composition of tailor-made interactive services, and with the promotion of the convergence among web services, Semantic Web technologies, and networks.

Finally, note that our approach to personalized e-commerce has been grounded on a personalization engine following the guidelines of cloud computing, to fulfill a unified user experience through different devices and from anywhere. This engine could be adapted to work in other domains of personalized applications, as part of some SaaS (*Software as a Service*) offerings.

Our approach is flexible enough to be used outside of the e-commerce scope. Specifically, our reasoning-based recommendation strategy, along with the time-driven filtering process, can be easily adapted for new multimedia-related application domains. As examples, note the possibility of developing recommender systems for the creation of tailor-made TV/radio channels, or even virtual learning tools based on the expertise and knowledge of each pupil. In this regard, our recommendation strategy is able to pre-select a set of potentially interesting multimedia items (e.g., radio and TV contents, educational material), whereas the time-driven filtering process refines the previous selection by considering the users' preferences extracted from stereotypical consumption historials coming from like-minded individuals.

As further work, we are currently working on mechanisms to reduce the overhead due to group corrections in the ontologies. Specifically, if we manage to recognize cases in which the corrected curves resemble some of our time functions, we could save space by storing the corresponding parameters, instead of all the feedback messages that make up the correction. This idea should work both to adjust the parameters of the default curves and to replace them to match dissimilar consumption patterns.

Also, we are interested in developing mechanisms to develop analysis and visualization tools to evolve stereotypes from the relevant feedback gathered from individual users, supporting decisions on when to introduce a new stereotype, when to merge existing ones, or when to split one stereotype into several specialized versions.

# 13

## Authoring of Multimedia Content: A Survey of 20 Years of Research

**Ansgar Scherp**

*University of Mannheim, ansgar@informatik.uni-mannheim.de*

## CONTENTS

Multimedia has been one of the buzzwords in the mid nineties. Since then, it has not decreased in interest but rather has become a ubiquitous part of our environment. Thus today, high quality playback of images, audio, and video has entered numerous application domains and is available on various devices including advertisement screens at train stations, infoscreens in elevators, and mobile devices like cell phones. Authoring of multimedia content, i. e., creating content, has been investigated for about two decades now. In this chapter, we investigate the different authoring support that has been developed in the past. Based on an earlier study [815], we conduct an extensive analysis and provide a classification and comparison of the existing authoring support. This survey helps not only in better understanding the existing support and approaches for authoring personalized multimedia content, but also enables assessing future work.

## 13.1   Introduction and Overview

The notion of multimedia is ambiguous. Basically, multimedia is considered the composition of the words "multi" (multiple) and "media." This means that multimedia is the combination and usage of multiple media. A medium can be either discrete or continuous, determined by its medium type. While multimedia content represents the composition of different media assets into a coherent multimedia presentation, multimedia content authoring is the process in which this presentation is actually created. Today's multimedia applications need to provide personalized content that actually meets the individual user's needs and requirements. This means that the multimedia content must reflect the user's situations, interests, and preferences, as well as the heterogeneous network infrastructure and (mobile) end device settings. Consequently, personalization is considered as a shift from a one-size-fits-all to a very individual and personal provision of content by the application to the end users. The term personalization often appears together with the term of customization. However, the distinction between them is not always clear, often intermixed, and sometimes considered equal or interchangeable. Therefore, we must clearly distinguish between the notion of personalization and customization, and will show that this distinction is not a merely academic one, but provides for defining two different application families: the customized applications and the personalized applications. This is due to the different requirements customized and personalized applications have to implement their functionality. Finally, authoring is the process of selecting and composing media assets into multimedia content, i. e., into a coherent, continuous multimedia presentation that best reflects the needs, requirements, and system environment of the individual user. Typically, the result is a multimedia presentation targeted at a certain user group in a specific technical or social context.

This chapter provides an extensive review of today's support for authoring and personalizing multimedia content. We first present our notion of media and multimedia in Section 13.2. In Section 13.3, the notions of multimedia documents, multimedia document models, and multimedia formats are introduced. In addition, the central modeling characteristics of multimedia content are identified and presented, namely time, space, and interaction. These definitions lay the foundations for understanding the notion of multimedia content authoring introduced in Section 13.5. In Section 13.6, the different aspects of personalization are presented and our understanding of authoring personalized multimedia content is introduced. Subsequently, we review the state-of-the-art in the field of multimedia content authoring and personalizing multimedia content in Section13.8. To this end, we systematically present and analyze the existing approaches and systems for multimedia authoring. Finally, we categorize and compare the different authoring support and authoring approaches for multimedia content, before we conclude this chapter.

## 13.2   Notion of Media and Multimedia

Multimedia is the combination of the terms "multi" and "media" [876, 520]. It is the combination and usage of multiple media [664]. A medium can be either *discrete* or *continuous*, depending on its medium type (or medium form [408, 531]). Examples of media types for discrete media are text and image, e. g., computer graphics and pictures taken from a digital camera. They do not change in time. Discrete media types are also called time-independent [94, 876], time-invariant [376], and non-temporal [366], respectively. On the contrary, continuous media objects naturally change in time, like the media types audio, video, and animation (cf. [311]). These media objects are time-dependent [94, 876], time-variant [376], and temporal [366], respectively. An instance of such a medium type is called a medium asset.

Summarizing the discussion above, multimedia can be seen as the interactive conveyance of information that includes (a seamless integration of) at least two media assets that are of two media types [408, 901]. A frequently cited, extended definition of multimedia by Steinmetz et al. [876, 411] requires the use of at least one continuous medium type and one discrete medium type. The definition also considers further aspects of multimedia with respect to storage and communication. This leads to a definition of multimedia that is characterized by the computer-controlled, integrated creation, manipulation (i. e., interaction of the user with the media), presentation, storage, and communication of independent information [876]. This independent (multimedia) information is encoded at least through one continuous medium type and one discrete medium type [876].

## 13.3   Multimedia Document Models and Formats

The composition of different media assets such as images, text, audio, and video in an interactive, coherent multimedia presentation is the multimedia content or multimedia document (also called multimedia object [74]). Features of such a multimedia document are the temporal arrangement of its media assets in a temporal course, the spatial arrangement of the assets, and the definition of its interaction features. A multimedia document is an instantiation of a multimedia document model. A document model provides the primitives to capture the aspects of a multimedia document as sketched above. A multimedia document that is composed in advance of its rendering is called pre-orchestrated, in contrast to compositions that take place just before rendering that are called live or on-the-fly. A context-aware multimodal document model is presented by Celentano and Gaggi [198]. It allows for a rule-based approach for specifying the different ways that multimodal information is presented to a user in a specific context.

A multimedia format defines the syntax for representing a multimedia document for the purpose of exchange and rendering. Since every multimedia format implicitly or explicitly follows a multimedia document model, it can also be seen as a proper means to "serialize" the multimedia document's representation for the purpose of exchange. Examples of multimedia formats are the World Wide Web Consortium (W3C) standards SMIL [173, 177], SVG [890], and HTML 5 [970], the International Organization for Standardization (ISO) standard Lightweight Applications Scene Representation (LASeR) [448, 449], and proprietary multimedia formats such as the wide spread Flash format [40]. Finally, a multimedia presentation is the rendering of a multimedia document to an end user. For a more detailed discussion and definition of the terminology, we refer to the literature such as [815].

## 13.4   Central Aspects of Multimedia Documents: Time, Space, and Interaction

The expressiveness of a multimedia document model, i. e., the primitives it defines, determines the degree of functionality the multimedia documents can provide. These central features or central aspects [818] are the temporal course, spatial layout, and interaction possibilities of a multimedia presentation, i. e., how users can interact with the document [136, 139, 444, 424, 480, 797, 530]. We present an overview of these central aspects. For further discussions, we refer the reader to [818, 136, 139].

- Temporal course: A temporal model describes the temporal arrangement of media assets defined in a multimedia document [139, 138, 480, 349, 607]. With the temporal model, the temporal course such as the parallel presentation of two videos or the ending of a video presentation on a mouse-click event can be described. One can find five types of temporal models: point-based temporal models, interval-based temporal models [608, 62], enhanced interval-based temporal models that can handle time intervals of unknown duration [298, 416, 971], event-based temporal models [88], and script-based implementations of temporal relations [890]. The multimedia formats we find today implement different temporal models, e. g., SMIL 1.0 [172] and Flash provide an interval-based temporal model only, while SMIL 2.0 [88] also supports an event-based time model.

- Spatial layout: Not only the temporal synchronization of the media assets is of interest in a multimedia document but also the spatial arrangement of the assets on the presentation canvas [73]. The positioning of visual media assets in the multimedia document can be expressed by the use of a spatial model. It defines the spatial organization, i. e., the spatial positioning of the visual assets [139, 138, 480]. For example, one can place an image above a caption or define the overlapping of two visual media assets. Besides the arrangement of media assets in the presentation, the spatial layout is also defined in the document. In general, three approaches to spatial models can be distinguished: absolute positioning, directional relations [730, 728], and topological relations [306]. The absolute positioning of media assets with respect to the origin of the coordinate system can be found, e. g., with Flash [40] and SMIL 2.0 in the Basic Language Profile (BLP) profile [88], while relative positioning is provided, e. g., by SMIL 2.0 [88] and SVG 1.2 [890].

- Interaction possibilities: The third central aspect of a multimedia document model is the ability to specify user interactions. The interaction model allows the users to choose between different presentation paths [138]. Multimedia documents without user interaction are not very interesting, as the course of their presentation is exactly known in advance and, hence, could be recorded as movies. With interaction models a user can, e. g., select or repeat parts of presentations, speed up a movie presentation, or change the visual appearance. For modeling user interaction, one can identify at least three basic types of interaction [138]: navigational interactions, scaling interactions, and movie interactions. Navigational interaction provides for control of the flow of a multimedia presentation. It allows the selection of one out of many presentation paths and is supported by all multimedia document models (cf. hyperlink in [480]). Scaling interaction and movie interaction allow the users to interactively manipulate the visible and audible layout of a presentation [136, 139]. For example, one can define if a user is allowed to change the presentation's volume or spatial dimensions. Scaling interaction and movie interaction are rarely used, or not defined within today's multimedia documents. Typically, such types of interaction rely on the functionality offered by the actual multimedia player used for playback of the presentation.

Looking at the existing multimedia document models, both in industry and research, one can see that these aspects of multimedia content are implemented in two ways: The standardized formats and research models typically implement time, space, and interaction in different variants in a structured fashion as can be found with SMIL 2.0, HTML+TIME [823], SVG 1.2, Madeus [482, 483], and Z $y$ X [136] employing the Extensible Markup Language (XML) [160]. Proprietary approaches represent or program these aspects in an internal model such as the Flash format [40]. Examples of (abstract) multimedia document models in the research are Madeus [482, 483], Amsterdam Hypermedia Model [397, 399], CMIF [178], Z $y$ X [136, 135], and MM4U [817, 818], which is based on Z $y$ X.

## 13.5 Authoring of Multimedia Content

While a multimedia document represents the composition of different media assets into a coherent multimedia presentation, multimedia content authoring is the process in which the multimedia document is actually created. The process of multimedia content authoring involves parties from different fields including domain experts, media designers, and multimedia authors. Domain experts provide their specific knowledge in a field such as biology or sociology. The input from the domain expert is used by the media designers to create a storyboard of the intended multimedia document or set of multimedia documents, e. g., in form of an interactive multimedia application. Figure 13.1 depicts an example storyboard for the highly-interactive multimedia-based e-learning tool GenLab [814]. The virtual laboratory GenLab[1] allows students of genetics engineering to conduct experiments without risk, using the computer and preparing themselves for a real laboratory work.

Besides creating a storyboard of a multimedia document together with the domain experts, the media designers also create, process, and edit the media assets required for the multimedia document. To this end, the storyboard is used as the basis to create a list of required media assets and to plan the implementation of the multimedia content. Finally, multimedia authors compose and assemble the preprocessed and prepared media assets into the final multimedia document. This composition and assembly task is typically supported by professional multimedia development programs, so-called authoring software or authoring tools (see Section 13.8.1). Such tools allow for the manual (possibly assisted or wizard-based) composition and assembly of the media assets into an interactive multimedia document via a graphical user interface. If the multimedia document needs to be programmed using authoring tools, the media authors are often computer scientists. The implementation of the

---

[1]`http://virtual-labs.org`, last accessed: 20/1/2013

| Application: | Learning unit: | Page number: 3 |
|---|---|---|
| GenLab | gel electrophoresis | Originator: Chris Red<br>Coordinator: John Blue<br>Date: 7 December 2001 |

Sketch of the
lab procedure:

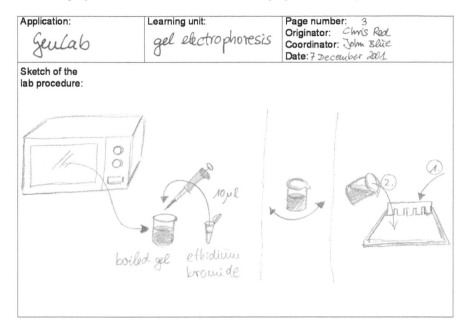

FIGURE 13.1: Example of the storyboard of the highly-interactive multimedia application GenLab (taken from [814]).

storyboard of the virtual laboratory GenLab is shown in the screenshot of Figure 13.2.

Even though we have described the authoring of multimedia content as a sequential process, it typically includes cycles. In addition, the expertise of some of the different roles involved in the process can be provided by a single person.

## 13.6 Personalization vs. Customization

The concept of personalization means different things to different people [783]. It is often intermixed and sometimes considered equal to or interchangeable with customization [406, 944]. In this work, we clearly distinguish between the notion of personalization and customization as done by [61, 709]. This distinction is not a mere academic one, but provides for defining two different application families, the customized applications and the personalized applications.

Customization is an activity that is conducted by and under direct control of the user [406, 709, 61]. This means that a customized application is actively

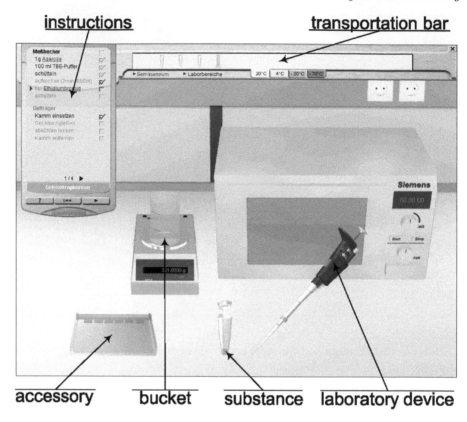

FIGURE 13.2: Screenshot of the final interactive multimedia presentation resulting from the storyboard depicted in Figure 13.1 (also taken from [814]).

adapted by its users to meet their individual needs and requirements. However, the customized application does not adapt itself to the users. For example, users actively customize their cell phone's user interface by selecting individual content such as apps, ring tones, screensavers, and wallpapers. However, the cell phone itself does not adapt to the user (although of course some apps installed on the phone do). As customized applications do not adapt to the needs and preferences of the users, there is no need for them to provide a user model, i.e., to gather and maintain information or assumptions about the users' needs and preferences. All customization activities are carried out by the users themselves.

In contrast, personalization is considered a process driven by the application [158, 709, 61]. Here, the content provided to the individual users is adapted by the personalized application itself. Thereby, the personalized applications take, e.g., the interests, preferences, and background knowledge of the users into account [333, 61]. For example, a personalized web application aims at providing web pages to the users based on information and assumptions of his or her information needs [709]. Consequently, a personalized application must provide a user model, i. e., it must gather and manage some information or assumptions about the user's needs and requirements. The personalized application is able to adapt itself to the needs and requirements of the user, i. e., to individualize [688, 311, 665] or tailor [688, 529] its content according to the information stored in the user model.

We support our decision to clearly distinguish between customized applications and personalized applications by the traditional distinction between adaptable systems and adaptive systems [176, 334]. While an adaptable system allows the user to change certain system parameters and adapt its behavior accordingly, an adaptive system can automatically change its behavior according to the information and assumptions about its users [176, 334]. Consequently, a personalized application is sometimes also called a user-adaptive application.

On the basis of a discussion about the notions of personalization and customization, we can now provide our definition of personalization. Personalization is defined as the offer and opportunity for special treatment in the form of information, services, and products provided by an application according to the interests, background, role, facts, requirements, needs, and any other information and assumptions about the individual. Personalization is conducted proactively by the personalized application, and is typically carried out to the user in an iterative process.

## 13.7 Authoring of Personalized Multimedia Content

Based on the definition of personalization and multimedia content, we consider personalized multimedia content as multimedia content targeted at a specific user or user group. It is able to adapt itself to the individual user's or user group's needs, background, interests, and knowledge, as well as the heterogeneous infrastructure of the (mobile) end device to which the content is delivered, and on which it is presented. Consequently, the authoring of personalized multimedia content is considered to be the process of selecting and composing media assets into personalized multimedia content, i. e., into a coherent, continuous multimedia presentation that best reflects the needs, requirements, and system environment of the individual user. A recent survey on multimedia personalization was conducted by Lu et al. [621].

The authoring of multimedia documents described in Section 13.5 represents a manual creation of such content, often involving much effort and high cost. Typically, the result is a multimedia document targeted at a certain user group in a specific technical context. However, this one-size-fits-all fashion of the multimedia document does not necessarily satisfy the different users' information needs. Different users may have different preferences concerning the content, and also may access the content in networks on different (mobile) end devices. Consequently, the authoring of personalized multimedia documents raises new requirements. For a wider applicability, the authored content needs to "carry" some alternatives or variants that can be exploited to adapt the multimedia presentation to the specific preferences of the users and their technical settings. This approach has been investigated in the past in projects such as aceMedia[2].

## 13.8    Survey of Multimedia Authoring Support

We have introduced and defined the central terms in the area of multimedia content authoring and personalization. These definitions lay the foundation for the subsequent analysis of the existing authoring support for (personalized) multimedia documents and the classification of this support. With respect to authoring personalized multimedia content, there has been research conducted for almost two decades. Proof of these achievements is, among others, the well-known Cuypers Multimedia Transformation Engine [361, 948], the Semi-automatic Multimedia Presentation Generation Environment [319, 320], and the Standard Reference Model for Intelligent Multimedia Presentation Systems [870, 318].

In the following, we provide an extensive overview in the field of multimedia content authoring, and analyze existing approaches and systems for multimedia content adaptation and multimedia content personalization. We present today's support for personalized multimedia authoring from different points of view and aspects. Following this overview and analysis of today's support for personalized multimedia content, the considered approaches and systems, families of systems, and research directions are categorized in Section 13.9.

### 13.8.1    Generic Authoring Tools

Multimedia authoring tools allow for the manual composition and assembly of media assets into an interactive multimedia presentation via a graphical user interface. For creating the multimedia content, the authoring tools follow different design philosophies and metaphors, respectively. Traditionally, these metaphors are roughly categorized into script-based, card/page-based, icon-based, timeline-based, and object-based authoring [765]. Examples of multimedia authoring tools are Adobe's Authorware [41], Director [42], Flash Professional [43], Toolbook [886], and the Edge Tools and Services [39] that allows for the authoring of Flash-like multimedia documents using HTML 5 [970].

---

[2]http://cordis.europa.eu/ist/kct/acemedia_synopsis.htm, last accessed: 20/1/2013

Also Tumult's Hype [930] allows for an easy and interactive generation of Flash-like multimedia documents in HTML 5. These domain-independent tools let the authors create very sophisticated multimedia presentations, typically in a proprietary format, or in HTML 5 in the case of Edge and Hype. In addition, the general purpose authoring tools typically require high expertise in using them and do not provide explicit support for personalizing the authored multimedia documents. Everything "personalizable" needs to be programmed or scripted within the tool's programming language. Consequently, the multimedia authors need programming skills and thus some experience in software engineering.

Adobe's icon-based authoring tool Authorware provides some support for creating personalized multimedia content [41]. It allows for a flow-chart oriented creation of multimedia documents, as shown in Figure 13.3, that supports defining the control of the presentation's flow with Decision Icons. A Decision Icon calculates the current value of a variable or expression that is attached to it and determines by this means which path of the decision structure is followed [41]. Thus, the Decision Icon can be used in principle to "personalize" the flow chart of a multimedia application developed with Au-

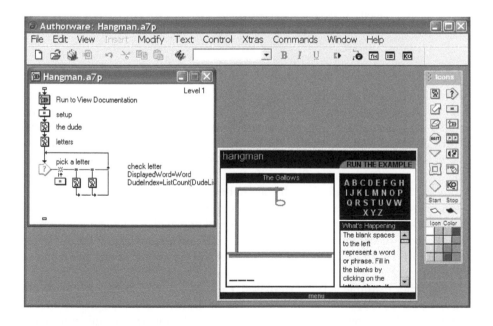

FIGURE 13.3: Screenshot of Adobe Authorware for a flow-chart oriented creation of multimedia documents (taken from a test installation of [41]). It shows the flow-chart of a simple Hangman game (top left) with the decision icons when the user provides correct and incorrect answers, and the actual rendering of the game (bottom right).

thorware. However, enhanced support for developing personalized applications is not provided.

In the field of adaptive multimedia models there are authoring tools such as the SMIL Builder [156] that allows for an incremental authoring of SMIL documents while verifying the temporal validity of the documents at any step. To this end, the authoring tool makes use of a temporal extension of petri nets. Another tool that provides for the authoring of personalized multimedia presentations is the Madeus authoring environment [482, 483]. Here, constraints are exploited to compose and assemble adaptable multimedia presentations. However, these constraints provide for personalization support only within a limited range, and do not support exchange of presentation fragments as it is supported, e.g., by SMIL's switch element (cf. Section 13.8.3). In addition, the generalized authoring tools are still tedious to handle and not practical for the domain experts, and following Bulterman in [175]: "Unfortunately, we have not seen the hoped-for uptake of authoring systems for SMIL or for any other format."

In order to provide multimedia authors with comprehensive support for personalized multimedia content, the multi-purpose authoring tools need to offer an editor that explicitly allows to define (abstract) user profiles. These user profiles need to be matched with, e.g., Authorware's Decision Icon, in order to select those paths in the flow chart that best fit the user's needs.

## 13.8.2   Domain-Specific Authoring Tools

In order to allow domain experts to author personalized multimedia documents, domain-specific authoring tools hide as much of the technical content authoring details from the authors as possible. They let them concentrate on the actual creation of the multimedia documents. The tools we find here [1002, 353, 551, 545, 342, 816, 137] are typically very specialized and target a specific domain. They are often organized in a wizard-like fashion that guides the experts through the authoring process. An example of such a domain-specific authoring tool is the page-based Cardio-OP Authoring Wizard [521]. The wizard supports the creation of personalized multimedia content in the field of cardiac surgery [521, 380, 137]. It guides the domain experts through the authoring steps in the form of a digital multimedia book on cardiac surgery.

Another example is the page-based *Context-aware Smart Multimedia Authoring Tool* (*xSMART*) [816]. It is based on the MM4U document model [817] and provides different domain-specific wizards for the (semi-)automatic, context-driven generation of multimedia documents, such as multimedia-based photo books [810]. During the different steps of creating the multimedia document, xSMART exploits contextual information as depicted in Figure 13.4 to guide the author through the content authoring process. For example, in the context of authoring multimedia-based photo books, this tool takes into account among other things, the quality of the photos, limitations such as the

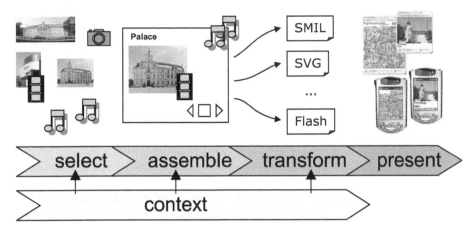

FIGURE 13.4: Depiction of the generic process for a context-driven generation of multimedia documents in xSMART (taken from [816]).

maximum number of pages, and targeted audiences like personal memories, for family like grandparents and friends, or for professional use such as architects and exhibitors. The authoring tool xSMART is designed so that it can be extended and customized (see Section 13.6) to the requirements of a specific domain by application-specific wizards. These domain-specific wizards can be developed so that they best meet the domain-specific requirements, and effectively support the domain experts in authoring the personalized multimedia content, while at the same time fully exploiting the generic infrastructure the authoring tool provides.

Similar to xSMART, the user-centric authoring tool by Kuijk et al. [545] allows for creating stories from photos. Another photo-driven authoring tool and layout system is by Xiao et al. [1002] for creating photo collages. The recent development of domain-specific authoring tools also takes into account the social web, such as the video authoring tool by Laiola and Guimaraes [551] that is aware of the users' social network.

### 13.8.3  Adaptive Multimedia Document Models

Early work in the field of (adaptive) multimedia document models is the Amsterdam Hypermedia Model (AHM) [398] and its authoring system CMIFed [174, 950, 399] using events and timing-constrains. Constraints are also used in the adaptive document model of the multimedia authoring environment Madeus [482, 483]. The ZyX [136] multimedia document model provides, besides a temporal, spatial, and interaction model, an extensive support for the reuse (of parts) of multimedia presentations, and adaptability of the multimedia content. The ZyX model is used, e.g., for content authored with the domain-specific authoring wizard Cardio-OP [521] presented in Section 13.8.2.

With MM4U, we find a multimedia document model that allows for defining complex composition operators to encapsulate and abstract to higher level functionality [817]. It is based on ZyX and extends it by providing dynamic composition operators, where the structure of the resulting multimedia document is computed on-the-fly and depends on contextual parameters. The MM4U model is not yet another multimedia document model but has been derived by a backward analysis of the existing approaches [819].

The W3C standard SMIL [177] allows for the specification of adaptive multimedia documents by defining alternatives in the temporal course using the switch element. Some authoring tools for SMIL such as the GRiNS editor (not available any more) provided support for the switch element to define presentation alternatives. However, a comfortable interface for editing the different alternatives for many different contexts is not provided. SMIL State [466] is an external state engine for modeling adaptive time-based multimedia presentations on the Web. Unlike its name suggests, the state engine of the SMIL state can be used in XML-based multimedia document models such as SMIL and SVG. Different variants of petri nets are also used as formal models for multimedia documents [759, 995], with the focus on the temporal organization of the media objects.

A quasi-standard defined by the industry is the well-known and widespread proprietary Flash format [40] by Adobe. The recent W3C standard HTML 5 [970] challenges Flash with built-in support in web browsers through the use of Javascript and agreed-on application programming interfaces (APIs). Although the new standard is quickly gaining popularity, until today the proprietary Flash format is still predominant. For further discussions on multimedia document models please refer to the literature, such as [815, 136, 139].

### 13.8.4   Adaptive Hypermedia

Toward the creation of personalized multimedia documents, we find interesting work in the area of adaptive hypermedia systems (AHS) [527, 170, 169]. The adaptive hypermedia system AHA! [270, 268, 267] is a prominent example that also addresses the authoring aspect [873], e.g., in adaptive educational hypermedia applications [872]. Though these and further approaches integrate media assets in their adaptive hypermedia presentations, synchronized multimedia presentations are not in their focus. The main personalization techniques pursued are adaptive navigation support and adaptive presentation. With adaptive navigation [269] links are, e.g., enabled, disabled, annotated, sorted, and removed, according to the profile information about the user (also called link-adaptation [270]). The purpose of adaptive presentations [269] is to, e.g., show, hide, reorder, and highlight or dim specific fragments of the presented hypermedia content, according to the user profile information (also called content-adaptation [270]). Recently, the AHA! system has been extended to the Generic Adaptation Language and Engine (GALE) [157, 863]. Basically, GALE is a complete redesign of the AHA! system that allows for the use of distributed resources, and supports the distributed definition of adaptations.

Approaches for adaptive hypermedia have also been extended to make use of social media such as the Adaptive Retrieval and Composition of Heterogeneous INformation sources for personalized hypertext Generation (ARCHING) system. The ARCHING system allows for the authoring of adaptive hypermedia from different and heterogeneous data sources, including professional content and social media [875]. Another system making use of open resources on the Web is Slicepedia [580]. Further work on adaptive hypermedia systems include considering provenance modeling for adaptation [528]. A comprehensive study of adaptive hypermedia systems has been done by Knutov, De Bra, and Pechenizkiy [527].

### 13.8.5 Intelligent Tutoring Systems

Closely related to adaptive hypermedia systems are the so-called intelligent tutoring systems (ITS) [829]. ITS provide personalized content according to the learners' or students' knowledge. The aim of ITS is that the learners gain new knowledge and skills in a specific domain by independently solving problems in that domain. An ITS provides a model of the student, a model of the domain, and a model of educational strategies [829]. This means that it comprises explicit assumptions and information about the knowledge and the level of knowledge of the user in the considered domain (student model or diagnosis model), an expert's knowledge in the domain (domain model or expert model), and a didactic concept of how to convey and present the learning materials to the learners (tutor model or educational model). Such models are also defined with adaptive hypermedia systems, e.g., [267, 999, 1000], although they are named differently there. Consequently, AHS are sometimes considered integration of ITS and hypermedia systems [238].

### 13.8.6 Constraint-Based Multimedia Generation

A very early approach to the dynamic authoring of adapted multimedia content is the Coordinated Multimedia Explanation Testbed (COMET) [310]. It is based on an expert system and different knowledge bases and uses constraints and plans to generate the multimedia documents [328, 658, 310]. Another similar approach is the Multimedia Abstract Generation for Intensive Care (MAGIC) [253]. It is an expert system with static knowledge bases and a constraint-based content planner. Further approaches to automate the authoring of personalized multimedia documents are the Knowledge-Based Presentation of Information (WIP) and the Personalized Plan-Based Presenter (PPP). WIP is a knowledge-based presentation system that automatically generates instructions for the maintenance of technical devices by plan generation and constraint solving [68, 67, 66]. PPP enhances this system by providing a lifelike character to present the multimedia content, and by considering the temporal order in which a user processes a presentation [68, 67, 70, 69, 66].

Logics programming and constraints are used in the the Cuypers Multimedia Transformation Engine [606, 948, 947, 361, 949, 360] for the dynamic generation of multimedia presentations such as the example depicted in Figure 13.5. To this end, Cuypers makes use of its own internal representation model for multimedia content, called Hypermedia Formatting Objects (HFO) [948]. The HFOs are transformed to SMIL presentations using XSL style sheets [38]. The presentations Cuypers generates are adapted to user preferences, as well as limitations of the targeted presentation platform.

Little et al. [605] present an extension of Cuypers that generates personalized multimedia presentation through semantic inferencing. Subsequent to a keyword-based query, users start selecting media assets to be incorporated into the presentation. The media assets are automatically arranged in time and space using some *mapping rules*. The selected media assets are used to iteratively refine the query.

The Semi-Automatic Multimedia Presentation Generation Environment (SampLe) builds a narrative structure of the generated multimedia presentation based on genre-specific templates [319, 321]. The user can modify the structure of the presentation, adapt the temporal flow of the presentation, and modify the content selected for it, as well as determine the interaction possibilities of the users with the presentation. The final arrangement of the multimedia material is made available for consumption to the users in an HTML format.

### 13.8.7  Multimedia Calculi and Algebras

Multimedia query algebras and calculi can also be used to author multimedia documents. The multimedia presentation algebra (MPA) by Adali et al. [37, 36] considers a multimedia document as a tree. Each node represents a non-interactive presentation, e.g., a sequence of slides, a video element, or an HTML page. The branches reflect different possible playback variants of a set of presentations. The MPA provides extensions and generalizations of the `select` and `project` operations in the relational algebra. However, it also allows one to author new presentations based on the nodes and tree structure stored in the database using operators such as `merge`, `join`, `path-union`, `path-intersection`, and `path-difference`. These extend the relational algebraic join operators to the tree structures. Lee et al. [566] present a multimedia algebra where new multimedia presentations are created on the basis of a given query and a set of inclusion and exclusion constraints stored in the database. The Unified Multimedia Query Algebra (UMQA) aims at integrating different features for multimedia querying, such as traditional metadata like artist and author, as well as content-based features like image similarity

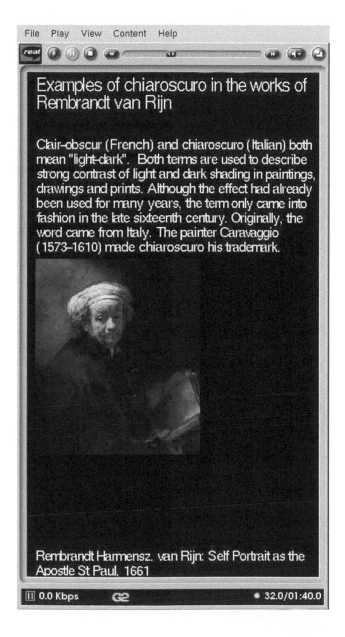

FIGURE 13.5: Screenshot of a multimedia presentation generated by the Cuypers Multimedia Transformation Engine (taken from [947]).

and spatio-temporal relations [189]. For the spatial relationships, Rectangle Algebra [729] is used. Regarding the temporal relations, only closed intervals, as defined in Allen's calculus [62] are supported. Open intervals, like those defined by Freksa [347], are not considered. Interaction relations are also out of scope. The Temporal Algebraic Operators (TAO) [758] allow for specifying multimedia documents along different sequential and parallel operators, equivalent to the closed intervals of Allen. In addition, TAO provides an alternative operator that is similar to the switch-tag of SMIL, and allows for an event-based synchronization of the media objects. Due to its focus on temporal relations, TAO does not support spatial relations or interactive relations. EMMA is a query algebra for Enhanced Multimedia Meta Objects (EMMOs) [1049]. It allows us to state queries about the media objects contained in EMMOs by following the typed edges of the EMMO graph. An edge type is similar to a property in the Resource Description Framework (RDF) [60] of the Semantic Web. For example, there can be an edge type saying that two movies are similar or that one movie is a remake of another, and the like. Fayzullin and Subrahmanian present the PowerPointAlgebra (pptA) for querying PowerPoint documents [327]. It is based on relational algebra and provides some new operators like APPLY. The APPLY operator conducts changes of attributes, which are defined by transformation functions. A transformation function can, e.g., change the color or font size. The APPLY operator can be used on different levels of granularity like slides, presentations, or the entire database. Like other query algebras, the semantics of PowerPoint presentations is not considered with pptA. The multimedia query model proposed by Meghini et al. [659] is based on fuzzy description logics, and aims at providing a unified approach to multimedia information retrieval for the two media types, text and image. Finally, SQL/MM is a standard for managing spatial information in a relational database [880]. It supports spatial queries on geometric shapes, as well as extended textual queries like stemming, and structural conditions such as finding an occurrence of query terms in the same paragraph [662].

### 13.8.8    Standard Reference Model for Intelligent Multimedia Presentation Systems

Finally, we find with the Standard Reference Model for Intelligent Multimedia Presentation Systems (SRM for IMMPS) [870, 142, 143, 318] a generalized architecture for the domain of so-called Intelligent MultiMedia Presentation Systems (IMMPS). The aim of IMMPS is to automate the authoring of multimedia presentations in order to enable on-the-fly personalization of presentations according to the individual needs of the user [870]. Hereby, IMMPS exploits techniques originating from the research area of artificial intelligence (AI) [796], such as knowledge bases, planning, user modeling, and the automated generation of media assets such as text, graphics, animation, and sounds

[870]. The goal of the SRM for IMMPS is to provide a common framework for the analysis, comparison, and benchmarking of IMMPS. An example of a system employing the SRM for IMMPS is the Berlage environment providing for a dynamic authoring of adaptive hypermedia content [798, 799].

## 13.9    Classification and Comparison of Authoring Support

Classifying the existing approaches and systems is a difficult and challenging task. Thus, it is not very surprising that there has been only little work so far. To the best of our knowledge, the only source is the work by Jourdan et al. [480, 481]. They provide a classification of existing systems and projects into different groups, and valuate these groups. In this work, we modify and extend the classification proposed by Jourdan et al. [480, 481] by suggesting the following six categories for classifying today's authoring support for personalized multimedia content:

- plain programming,

- templates and selection instructions,

- adaptive document models,

- transformations,

- constraints, rules, and plans, as well as

- calculi and algebras.

When considering the existing approaches and systems, it is not always easy to decide to which of the proposed categories a specific solution should be associated. In addition, some of the existing systems and projects explicitly combine or "synthesize" different approaches, and thus need to be associated to more than one category. This means that they employ more than one approach to actually implement their personalized multimedia functionality. Examples for such hybrid systems are found in [480] and [518].

The single approaches for multimedia personalization are described in the following Sections 13.9.1 to 13.9.6. For each personalization approach, we first introduce the characteristics of this approach. Subsequently, we refer to representative systems and projects for the considered approach. Finally, a valuation of the approach is given, i.e., the advantages and disadvantages of the approach are discussed. A summary of the different multimedia personalization approaches and their assessment is provided in Table 13.1.

**TABLE 13.1**

Evaluation of Different Approaches for Authoring Personalized
Multimedia Documents.

| Approach | Valuation |
|---|---|
| Programming | + Arbitrary personalization functionality can be programmed<br>− Providers develop their own (mostly) complex solution<br>− High effort necessary for extending or adapting the solution to a different domain or model |
| Templates & Selection Operators | + Well suited for applications where content selection can be split into a set of pre-defined database queries<br>− Static templates are limited in their expressiveness<br>− Difficult to handle global criteria spanning multiple queries for selecting and assembling media assets |
| Adaptive Document Models | + Provide extensive support for build-in adaptation and reuse of media assets<br>+ W3C standards exist<br>− Not practical to specify all presentation variants in advance |
| Transformations | + W3C standards exist<br>+ Transformation tools like XSLT are calculation complete<br>− Difficult to handle multiple transformations<br>− Often results in complex transformations |
| Constraints, Rules, & Plans | + Declarative description of the personalization functionality<br>− Expressiveness limited to declaratively describable personalization problems<br>− Additional programming required for complex and domain-specific personalization functionality |
| Algebras & Calculi | + Formal description of the multimedia authoring<br>− High effort to learn the algebraic operators and difficult to apply |

## 13.9.1   Authoring by Programming

In the authoring by programming approach, regular programming languages
are applied to develop the (multimedia) personalization functionality. Obviously, programming languages such as C++ and Java can be employed to
develop a system that generates personalized multimedia content [480]. However, logic programming, e. g., with Prolog, can also be used to implement the
personalization functionality such as in Cuypers. Programming is typically
also employed with the generalized multi-purpose authoring tools presented in

Section 13.8.1. Also domain specific authoring tools are typically programmed like the Cardio-OP Authoring Wizard for the medical domain.

With mere programming, every personalization functionality is feasible that can be implemented with a programming language. However, a well known disadvantage is the lack of independence between the programming code and the piece of information [480]. This makes it difficult to reuse one part of an application for another one. In addition, as providers today typically develop their own specific solution and data models, the personalization functionality is tailored and designed for their particular application domain. As a consequence, much effort is necessary when extending or adapting the solution to a different domain.

## 13.9.2   Authoring Using Templates and Selection Instructions

With the second category, personalized authoring takes place with templates and selection instructions. A template can be considered as the static part of a multimedia presentation. It is possibly designed in a concrete presentation language such as HTML or SMIL and is enriched with some selection instructions [480]. These selection instructions are executed when the user requests the template. Executing selection instructions means that information is extracted from external data sources [480] according to the users' profile information. The dynamically extracted information is then merged on-the-fly with the template. Merging the static part of the multimedia presentation with individually selected dynamic content on-demand allows the template approach to provide for a personalized composition and delivery of information to the users. The authoring approach by using templates and selection instructions is applied, e g., by the multimedia database METIS [517]. It uses XML-like data structures, where specific parts of the presentation are only filled in when the document is requested by the client. Rhetorical presentation patterns are used by Bocconi et al. [131] for the template-driven generation of video documentaries.

The template-based approach for multimedia personalization suits well for applications in which content selection can be split into several database requests [480]. However, static templates are limited in their expressiveness. In addition, it is possibly very difficult to handle global content selection and assembly criteria when considering multiple database requests, in order to choose those media assets that, e. g., best reflect the user's profile information or do not cross the maximum time limit of the presentation [480].

## 13.9.3   Authoring with Adaptive Document Models

Personalization by adaptive multimedia document models provides for specifying different presentation alternatives or variants of the multimedia content within the multimedia document. The presentation alternatives are statically defined within the adaptive multimedia document. During presentation, the document's alternative or variant is determined that best matches the user's profile information. Finally, the selected presentation alternative is presented on the end device. Consequently, with the personalization approach by adaptive document models, the multimedia player on the (mobile) end device decides within the range of the available presentation alternatives which variant of the multimedia document is presented. Examples of adaptive document models are the Amsterdam Hypermedia Model, SMIL, Z$_Y$X, and MM4U presented in Section 13.8.3.

Adaptive multimedia document models provide for an extensive support for the adaptation and reuse of multimedia presentations and parts of it. Another advantage of this approach is that there are W3C standards such as SMIL. However, adaptive document models are less practicable when a comprehensive support for personalization is needed because all of the different presentation alternatives have to be specified in advance within the same document.

### 13.9.4    Authoring by Transformations

With the authoring by transformations approach, two kinds of transformations can be distinguished [572]: the structural transformations and the media transformations. Structural transformations can be, e.g., the transformation of an XML-document into a (standardized) multimedia format such as SMIL or SVG. Structural transformations also include changing of the layout, and arrangement of the media assets to different presentation styles (cf. personalization by style sheets in [480]), e.g., changing the spatial layout of the visual media assets. With structural adaptation, an XML-document can be adapted from a desktop PC version to a mobile device, e.g., by dividing the content into different smaller screens or pages in the mobile situation. Consequently, structural transformations typically imply an adaptation of the temporal and/or spatial layout of the multimedia presentation as well as possibly changing the interaction design of the presentation with the user.

In contrast, media transformations change the media type, e.g., exchanging an image asset to a text asset describing the same content. It also includes adapting the media format, e.g., transcoding an image asset from PNG to JPG, or conducting other binary operations on the media assets, such as resizing a video asset or changing the color-depth of an image asset. Transformations are used by the Cuypers multimedia presentation engine to transform the multimedia content represented in their HFOs into the final multimedia format SMIL. XSL transformations (XSLT) are employed for generating SMIL documents within the Course Authoring and Management System (CAMS) [223] in the domain of e-learning. XSL-FO is used for providing different presentation styles [969]. Approaches that focus on media transformations are typically also found in the area of mobile multimedia presentation generation like the koMMa framework [465].

An advantage of the personalization by transformation approach is that the adaptation of the multimedia content can be described in the W3C standard XSL, employing XSLT, and XSL-FO. XSLT is supposed to be computationally complete [937, 945, 241, 807]. Thus, arbitrary transformations can be conducted with XSLT that can be described by an algorithm. However, due to recursive structures in XML, such transformations easily become very complex and difficult to handle.

### 13.9.5  Authoring by Constraints, Rules, and Plans

With authoring by constraints, rules, and plans, creating personalized multimedia content is considered an optimization problem [796]. The creation of the personalized multimedia content is explicitly described by using rules, constraints, and the like, which are, e. g., stored in different knowledge bases. The personalized multimedia presentation is generated on the basis of such a declarative description and by taking the profile information about the user into account. Such a presentation generation can also be regarded as a planning problem [796]. Here, the user's request is decomposed into some subgoals that are to be reached. The results of these subgoals are accumulatively assembled to the final multimedia presentation (cf. [480]). To respond to the requests, even those that were planned at design time, different knowledge bases are used [480]. A prominent example of a system that employs constraints and rules for the personalized multimedia content generation is again the Cuypers Multimedia Transformation Engine. Although it also employs transformation sheets, the main means for generating the personalized multimedia content are constraints and rules. With COMET, MAGIC, WIP, and PPP, we find several knowledge-based systems for personalization. The Standard Reference Model for Intelligent Multimedia Presentation Systems is a very generic approach for authoring personalized multimedia content by making use of multiple knowledge bases and constraints for layouting.

Systems applying personalization by constraints, rules, plans, or knowledge bases operate on a declarative level for describing the personalization functionality. However, due to their declarative description languages, only those multimedia personalization problems can be solved that can be covered by such a declarative specification. Consequently, these systems and projects find their limits when it comes to more complex or application-specific multimedia personalization functionality, and additional programming is required to solve that problem.

### 13.9.6  Personalization by Calculi and Algebras

With the last solution approach, calculi and algebras are applied to select media assets and merge them into a coherent multimedia presentation. This approach has emerged from the database community with the aim to store, process, and author multimedia presentations within databases. Consequently, work based on calculi and algebras are applied on the database level and provide for specifying queries that are sent to a database system. The database system executes the queries and determines the best match of the different media items and presentation alternatives stored in the database. The result is then sent back to the querying application. Examples of calculi and algebras for querying and automatically assembling multimedia content such as the multimedia presentation algebra are presented in Section 13.8.7.

The main advantage of the personalization by algebras approach is that the requested multimedia content is specified as a query in a formal language. However, typically, a lot of effort is necessary to learn the algebra and their operators. Consequently, it is very difficult to apply such a formal approach.

### 13.9.7 Summary of Multimedia Personalization Approaches

The classification of the existing systems and projects to the different categories of personalization approaches is not always easy and unambiguous. Nevertheless, we have presented a categorization of today's support for the authoring of personalized multimedia content. This provides for a more systematic management and examination of the tasks and challenges involved with the creation of such content.

## 13.10 Conclusion

In this chapter, we have defined the basic notions of media and multimedia. As a further prerequisite for our analysis, we have investigated and have defined the terms of multimedia document models and their instantiations, multimedia formats, as well as multimedia authoring and the personalization of multimedia content. Subsequently, we have conducted an extensive survey of existing authoring support for creating personalized multimedia content. This survey and classification will help in better understanding not only today's, but also future multimedia authoring approaches and support. Thus, it provides a more systematic introduction to the field of authoring personalized multimedia content.

# Part IV

# Human–Computer
# Affective Multimedia
# Interactions

# 14

## An Overview of Affective Computing from the Physiology and Biomedical Perspective

**Ilias Maglogiannis**

*University of Piraeus, imaglo@unipi.gr*

**Eirini Kalatha**

*University of Central Greece, ekalatha@ucg.gr*

**Efrosyni-Alkisti Paraskevopoulou-Kollia**

*University of Central Greece, ekolia@ucg.gr*

## CONTENTS

During the last few years, the field that deals with the identification, recording, interpreting, processing, and simulation of emotion and the affective state has won ground in the scientific community. It is a fact that many scientists are preoccupied with the domain called affective computing. In this chapter, we review the state-of-the-art methodologies, tools, and systems from the biomedical perspective, focusing on the collection and processing of biological signals. This chapter discusses the architecture of such systems, the acquisition and processing of biosignals, and the main methodologies used for emotion extraction. Finally, the chapter reports the statistics and results of the most important methodologies and systems found in literature.

## 14.1    Introduction

Affective computing is the science that deals with the research and development of special systems, techniques, and devices that can identify, record, interpret, process, and simulate various shades of human emotion. It is an interdisciplinary field that mostly pertains to the wider research field of human-computer interaction, and spans the fields of computer science, in particular the fields of artificial intelligence and recognition of patterns, psychology, philosophy (moral), and cognitive science.

While the origins of the field date back to early philosophical inquiries concerning emotion, the more modern branch of affective computing originates in Rosalind Picard's 1995 paper titled "Affective Computing"[744]. This paper sought the nature of emotions, researched various emotional states and the requirement for effective interaction, applied to a variety of conditions between humans and computers. The study of affective computing's "mother," Rosalind Picard, has contributed to the development of automation devices that can recognize human emotions, and the expansion of this field's research basis. According to Rosalind Picard, the exact definition of affective computing is:

"...computing that relates to, arises from, or deliberately influences emotion or other affective phenomena."

The purpose of a special system-device developed in the scope of affective computing is to wisely interact with the emotions of the user. What should primarily be done is for it to be able to survey the user, interpret the data from the surveillance, gather and conclude, and then interact in a manner that would fit each case (see Fig. 14.1).

In this way, the human-computer interaction can keep the interest of the user high and make this experience memorable, easily conceivable by human reason, more satisfactory, and more effective.

In summary, the goal of this chapter is to describe the concept of affective computing from the Physiology and Biomedical Perspective. In particular, an extensive reference to emotion, materials and methods of affective computing systems (i.e. system architecture, sensors and signal acquisition, processing and features extraction, classification modules, additional features, evaluation) and to some relative applications is attempted.

## 14.2    Background Information on Emotion

Affective computing was analyzed in the previous part; nevertheless, explaining the meaning of emotions has, for many years, not been addressed. For too long it has been known that the meaning of the term emotion is neither unique

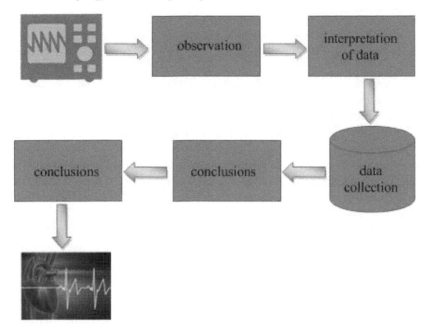

FIGURE 14.1: Affective computing related system/device interaction.

nor accurately defined [853]. Moreover, there are several, usually complex definitions of this term; their common characteristic is that they do not refer to a solid and clearly established definition [668]. In the literature relevant to philosophy and psychology, there is a wide range of definitions [853].

In particular, psychologists underline that emotion is the human's general response to facts, and not only an "emotional state", which is an inner feeling [668]. Another definition of the term is that it actually is a symptom of the characteristic vicissitudes of construction and not merely vague feelings or disturbances [508], [499]. Some researchers, apart from clarifying a definition, have dealt with the distinction of emotions in interpretable dimensions in order not to confuse the feelings between them. For example, in [732] an implicit model of emotions is proposed, according to which emotions are distinguished along dimensions of evaluation, direction, duration, and causation.

Given the absence of a clear definition, it would be appropriate to prevent any debate on whether computers are able to detect or have emotions [853]. In turn, new technically defined terms using architecture-based concepts can be introduced and utilized [846], [852], [109].

On this basis, we can distinguish the semantics of the following terms: "primary" (standard) and "secondary" (produced) emotions. These two concepts are the two main categories of emotions. The emotions which are experienced by the social mammals are named "Primary or standard" emotions [668].

They are related to particular types of expression, since the information from perceptual systems[1] when fed to a fast pattern recognition mechanism, are able to cause massive global changes (e.g., facial expressions, behavioral tendencies, physiological standards). Physiological standards are very useful since they combine a person's reaction when attacked or startled. By calculating physiological changes, sensors are able to detect such primary emotions [853].

It is a fact that neither the number nor the emotions in the primary emotions category are completely identified [668].

"Primary emotions" play an important role, due to the fact that they are represented as a direct consequence of encountering some event and they are easily recognizable through physiology. It is important to present some of their characteristics:

• They are unchangeable throughout all human societies, and for all social mammals.

• They occur both in humans and animals [806].

• They occurred during the evolutionary process [262].

• They are expressed in particular ways (e.g., facial expressions) [668].

• In humans, primary emotions appear within the first year of their life.

According to Damasio, "primary emotions" include six universal emotions namely happiness, sadness, fear, anger, surprise, and disgust [256], [687].

The variations or the combinations of the "primary emotions," which are perceived only by humans, are named "secondary or produced emotions" [668], [806]. Basically, they are "wealthy" emotional states, which emerge from cognitive procedures including the estimation of perceptible or imaginary conditions [853], [744], [257]. "Secondary emotions" include emotions such as pride, gratitude, sorrow, affection, irony, and even amazement [668].

It is worth mentioning some of the possible effects of these emotions:

• Response in the system of the primary emotion [853].

• Physiological changes such as, e.g., muscular stressing, crying, excitement.

• Fast, unintentional disorientation of thinking procedures [846], [854].

These emotions are divided into more categories. These are the "central secondary emotion" and the "peripheral secondary emotions":

**Central secondary emotions**: are emotions related to the unintentional disorientation of the evolving cognitive procedures as, for example, syllogism, repetitive thinking, etc. Moreover, diversions of attention completely occur cognitively without interference of emotional variations [853].

---

[1]Calculation system (biological or artificial) designed to come to conclusions regarding the properties of natural environment, according to the information arising from it.

**Peripheral secondary emotions**: they take place when cognitive procedures initiate situations, with no intention of redefining those related to thinking [853].

It is worth mentioning that there are emotions that combine characteristics of both the aforementioned subcategories; they are called "hybrid secondary emotions" [853]. An example of a "hybrid secondary emotion" is shyness, which is followed by obvious changes in posture or facial expressions.

But what happens with the detection of the emotions? As a matter of fact, emotions can be detected; nevertheless, opinions regarding what a sensor can detect differ. In [853] two assumptions are considered. The first is that a sensor can detect the general nature of an emotion (secondary) such as, for instance, anger or happiness, but not its semantic essence, that is, the reason someone is happy. The second is that both the nature and the semantic quintessence of emotions can be detected on basis of some standards.

In the same paper, the reasonable question of whether the above assumptions are true is answered. In reality, regarding the first assumption, the answer is that it depends on the person. Actually for most people "secondary emotions" set in motion mechanisms of "primary emotions," but that does not entail that they always happen or that they concern all people. According to Sloman, even civilization can actually make a difference; for instance, the British are characterized by the stiff upper lip [853]. Conclusively, the first assumption is not true, because it depends on the individual's personality, or even the given civilization.

On the other hand, the likelihood of the first assumption being true is the same as that the content of thoughts that can be deduced from observable standards, which seems rather impossible [853]. When someone is overwhelmed by mixed emotions, such as, e.g., a child on the last day of camping, it is a fact that his overall facial expressions, posture, etc., cannot describe his/her complex emotional state. In the next section, we will discuss all the technical details of affective and emotion recognition systems, along with the state-of-the-art, and prominent the methodologies in the domain, according to the recent literature.

## 14.3 Materials and Methods of Affective Computing Systems

### 14.3.1 System Architecture

It is widely known that a system receives inputs and produces outputs. Understandably, however, someone might wonder what the inputs and outputs are in the case of a system that is developed so that an affective computer would wisely interact with the emotions of the user.

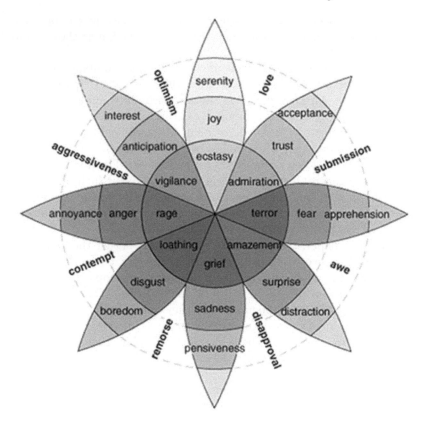

FIGURE 14.2: Three-dimensional model by Robert Plutchik [752].

An emotion-recognizing device has as its primary purpose the classification and identification of emotional states [668]. This implies that the inputs will be various signals which directly or indirectly provide information about the user's emotional state. Such information could be either biosignals, sound, speech signals, or even continuous images (video).

For example, [363] and [811] mention a unique method of assessing human balance based on electromyography (EMG) signals by using a wrist-worn biosensor, that enables comfortable measurement of skin conductance, skin temperature and motion, sleep quality, and examining the performance of learning and memory tasks.

In Fig. 14.2 the description of the relations among emotion concepts, which are analogous to the colors on a color wheel, is illustrated. The cone's vertical dimension represents intensity, and the circle represents degrees of similarity among emotions. Eight sectors are designed, to indicate that there are eight primary emotion dimensions defined by theory, arranged as four pairs of oppo-

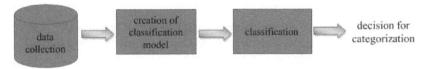

FIGURE 14.3: The architecture of an emotion recognition system including the procedure of training and the operation of classification.

sites. In the expanded model, feelings in the blank spaces are the main dyads, i.e., emotions that consist of two basic emotions [668]. After the signal has been recorded, the system faces the challenge of pattern recognition. For this reason, this stage is analyzed using two separate procedures:

- the training process, and

- the classification process.

Initially, the identification system should be "trained" to serve the intended purpose. During the training procedure, the input of the system is a number of representative examples from an appropriate dataset or database, and the class of each example. It is worth emphasizing that the choice of an appropriate database depends on the goal for which the identification system will be used. Then, a generalized model of classification is created by the system, which processes the training examples according to the chosen classification algorithm and the desired output. Finally, after the system's training is completed and the classification model has been created, the system is ready to function normally and sort the entries that interest the user [668] (see Fig. 14.3).

## 14.3.2 Sensors and Signal Acquisition

One of the key components that gives substance to the concept of affective computing is the use of sensors. Sensors are devices whose primary function is to detect a signal or stimulus and produce a measurable output. Typically, modern sensors respond to the detected stimulus through an electrical signal. They convert a physical quantity input into a signal compatible with the use of electronic circuits.

The technology is used in two distinct areas: information collection and control systems. Gathering information provides data aimed at incessant presentation and the best comprehension of the present situation, or the evolution of system parameters. However, while controlling the system, the sensor's signal feeds a controller, which creates an output that adjusts the value of the measured parameter. Combining their use with the suitable processing algorithms and the information given, adds to mechanical devices the essential "intelligence" so as to be qualified as robots.

It is worth mentioning that some of the main applications record biosignals, such as blood pressure, temperature, sweating, blood sugar level, and muscle tension, or record speech, facial expressions or gestures, using video and other multimedia. Specifically, in [998] a closed-loop system is able to provide real-time evaluation and manipulation of a user's affective and cognitive state, and identified and classified physiological responses such as skin conductance level, respiration, electrocardiography (ECG), and electroencephalography (EEG). Another case of recording biosignals using sensors is found in [848]. In this paper, functional near-infrared spectroscopy (NIRS) and functional magnetic resonance imaging (fMRI) with the implementation of a brain-computer interface (BCI), was used for the acquisition, decoding, and regulation of hemodynamic signals in the brain.

The sensors are divided into five main categories:

- Optical,

- Acoustic wave,

- Temperature,

- Piezoelectric accelerometers,

- Electrochemical.

Optical sensors are divided into proximity sensors and ambient light sensors. Proximity sensors detect the presence of an object or motion in various industrial, mobile, electronic appliances, and retail automations [22].

On the other hand, ambient light sensors provide precise light detection for a wide range of ambient brightness [22] in a manner similar to that of the human eye. In order to be accurate, they are used wherever the settings of a system have to be adjusted to the ambient light conditions as perceived by humans.

Acoustic wave sensors measure the frequency or phase characteristics of an acoustic wave and associate these measurements with the respective physical quantity being measured. [28]

The thermal sensor operates on the principal conversion of the thermal energy into electric energy which, in turn, can be processed further. In essence, a non-thermal signal is converted into a heat flow. The heat flow causes a change in temperature, and is finally transformed back into an electrical signal. [32], [496]

The piezoelectric accelerometers convert one form of energy into another and provide an electrical signal in response to a quantity or condition that is being measured. Using the general sensing method upon acceleration, a seismic mass that is restrained by a spring or suspended on a cantilever beam converts a physical force into an electrical signal. [23]

Electrochemical sensors consist of a working or sensing electrode, a counter electrode, and, usually, a reference electrode. These electrodes are enclosed in the sensor, which is in contact with a liquid electrolyte. The working electrode lies on the inner face of a membrane that is porous to the measured substance, but can not be penetrated by the electrolyte. The substance is diffused into the sensor and via the membrane to the working electrode. An electrochemical reaction takes place the moment the measured substance reaches the working electrode. [8]. There is a number of electrochemical sensors based on variations of the above basic concept, such as electrochemical immunosensors, modern glucose biosensors for diabetes management, biosensors based on nanomaterials (e.g., nanotubes or nanocrystals), biosensors for nitric oxide and superoxide, or biosensors for pesticides. More information can be found in [1041].

Depending on the recorded signal, the appropriate sensor is selected. For example, in [206] an optical sensor is used due to the fact that the researchers want to record the patient's emotional expression in real time. In addition, in [337] several different types of sensors are used, and they provide information on postural transitions and walk periods. The sensors used are: Infrared Presence Sensors, door contacts, temperature and hygrometry sensors, microphones, and wearable kinematic sensors [953].

All types of sensors display some main measuring features such as "precision," "accuracy," "repeatability," "reproducibility," "fault tolerance," "range," "linear response," and "sensitivity" [76]. The term "accuracy" refers to whether the sensor is approaching the physical reality within a reasonable range, and it is usually given as a percentage of the sensor's operating range. The term "precision" reflects the sensor's degree of freedom from accidental errors. In a large sample, the variance between measurements of an accurate sensor will be small. It is important to stress that sometimes precision is misinterpreted as accuracy, and therefore it should be made clear that a precise sensor is not necessarily highly accurate.

The term "range" describes the minimum and maximum values of the natural size that a sensor can measure. Furthermore, it is desirable that the response of a sensor fluctuates in a linear mode vs. the measured size. "Nonlinearity" is usually expressed as the deviation of the sensor's range.

Finally, "repeatability" indicates whether the results of a sensor are similar when this sensor measures the same fixed size in stable conditions. Another term which is identical to the previous one is "reproducibility," with the sole difference that the conditions are not stable, but changing. "Fault tolerance" is another feature that is closely related to "accuracy" and it is defined as the maximum expected error of a value. In Fig. 14.4 the reader can see the variety of sensors used in the surveyed works, along with their frequency. The brain based sensors such as EEG and fMRI are the most frequently used.

### 14.3.3 Processing and Feature Extraction

In order to draw conclusions related to the affective state of the user, receiving biosignals, video, audio, etc., the use of sensors solely does not suffice

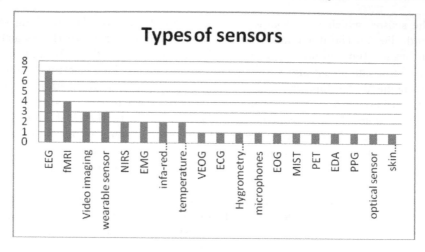

FIGURE 14.4: Types of sensors.

[692]. Instead, further processing of these signals must take place so that precise conclusions regarding emotional computing may be drawn. One of the usual ways for signal processing to be used in several applications, as for example in [495] and [718], is the Fourier transformation. This is a useful tool for examining signals in three dimensions: Width, Time, and Frequency. The contribution of this method is that it helps by developing effective techniques that deal with the noise introduced by the channel to the broadcast signal [493].

The Fourier transformation $X(f)$ of a function $x(t)$ is defined as:

$$F\{x(t)\} = X(f) = \int\limits_{t=-\infty}^{\infty} x(t)e^{-j2\pi ft}dt, \qquad (14.1)$$

and the inverse Fourier transformation is given by:

$$x(t) = F^{-1}\{X(f)\} = \int\limits_{-\infty}^{\infty} X(f)e^{-j2\pi ft}df. \qquad (14.2)$$

The Hilbert transformation constitutes a different way of processing signals which has been extensively used for the recovery of the analytic signal $f_{+}(t)$ associated with the real signal $f(t)$ and it leads directly into the match of $f(t)$ and $f_{+}(t)$ without containing information on the heuristic methodology [258].

The Hilbert Transformation of a function is defined by:

$$f_H(t) = -\frac{1}{\pi} \int_{+\infty}^{-\infty} \frac{f(x)}{t-x} dx, \tag{14.3}$$

where the notation indicates the Cauchy principal value of the integral.

If $F(\omega)$ and $F_H(\omega)$ are the corresponding Fourier transformations of $f(t)$ and $f_H(t)$, the above relation is expressed in the frequency domain as:

$$F_H = jF(\omega)\,\text{sgn}\,\omega \tag{14.4}$$

where $\text{sgn}\,\omega$ is the sign function [258].

The Matching Pursuit technique produces a sum of functions, which approaches the optimal (optimal adjustment) by repetitively selecting waveforms from a general dictionary. The choice of functions is performed by progressively improving the signal in order to approach the original signal through a repeated process [258]. The main objective is to demodulate a given signal $S$ into a linear combination of $N$ functions $(g_i)$, which have been selected by a predefined dictionary (i.e., a redundant set of functions) so that:

$$S = \sum_{i=1}^{N} a_i g_i + R_N, \tag{14.5}$$

where $R_N$ is the rest of the demodulation.

Great advantages of the matching pursuit when analyzing biomedical signals emerge from the meticulous parameterization of transients, the robust estimate of the time-frequency energy density and the combination of those two [302]. An extensive summary of the matching pursuit-based frameworks successfully applied to the EEG analysis is given in [642].

Another heuristic technique for deducing the analytic signal is complex demodulation. The accuracy of this technique is based on choosing the optimal demodulation and cutting-off frequency of the lowpass filter [258]. It includes the following steps:

1. The frequency spectrum of a signal $f_t$ is shifted to the principle of the axes by a quantity also called frequency demodulation $(f_0)$.

2. The complex signal is convoluted with a lowpass filter, with cutoff frequency $f_c$.

3. The frequency spectrum of the filtered signal is shifted back to its original position.

The result is to eliminate the negative spectrum of the frequency and remain with a signal $f_+(t)$, which is defined only in the positive half-axe of its frequency spectrum [258].

The Wavelet Transformation $W(\alpha, \tau)$ of a signal $f(t)$ is defined as the convolution between $f(t)$ and the dilated versions of a complex wavelet function $\psi(t)$ [258]:

$$W(\alpha, \tau) = \frac{1}{\sqrt{a}} \int_{-\pi}^{+\pi} \psi^* \left( \frac{t - \tau}{\alpha} \right) f(t) dt \qquad (14.6)$$

where $\alpha \in \mathbb{R}^+$ is the parameter of the scale of expansion and $\tau \in \mathbb{R}$ is the parameter of the relocation. In [754] the wavelet transformation is used due to the fact that the wavelet coherence produces an estimate of the coupling between non-stationary neural signals as a function of time [550]. Furthermore, in [346] the DWT is used as described in [365] instead of traditional filtering techniques.

Finally, video processing is a special case of signal processing, where input and output signals are video files or video streams. It is worth mentioning that in [387] the existence of a number of early efforts to detect nonbasic affective states, such as attentiveness [309], fatigue [385], and pain [104] from face video, is described.

### 14.3.4   Classification Modules

The aim of affective computing applications is mainly sorting out the user's emotions, expressions or physiological responses. Accordingly, "learning" is an extremely important process. Specifically, according to Simon (1983):

> "Learning signals adaptive changes in a system in the sense that they allow to do the same work or work in the same category, more efficiently and effectively the next time."

Classification methods use various machine learning methodologies, e.g., neural networks, Support Vector Machines' classification, or decision trees. First and foremost, artificial neural networks or, simply, neural networks, which are mechanical learning methods, are a special approach to create systems with intelligence. They are based on biological standards, structures, and processes that imitate those of the human brain. Their four basic properties are:

- ability to learn by example,

- high tolerance for noisy learning data, i.e., data with occasionally incorrect values,

- capability of pattern recognition,

- being considered as distributed memory and memory correlation.

These properties have contributed to their great popularity in the field of emotion recognition and thus emotional computing [958].

In [387] it is mentioned that the multilayer perceptron (MLP) can be seen as the simplest kind of feedforward neural network that uses back propagation to train. An MLP is comprised of various layers of neurons, each layer being completely connected to the next one. Also, in [389] it has been shown that an MLP and a large amount of data when combined will produce high classification accuracy for both the arousal and valence dimensions.

The decision tree is one of the most popular learning algorithms used to predict a model's output values, based on input values that are presumed to be independent [958]. The most notable advantages of this method are:

- it is easy to interpret, at least when the tree size is quite small

- it is inexpensive to construct,

- it provides an extremely fast method for classifying unknown records, and

- for many simple datasets, its accuracy is comparable to that of other classification techniques.

There are many applications in which this method was used as, for example, in [345]. In this paper it is mentioned that the emotional biosignals are primarily distinguished according to their valence dimension through a data mining approach, which is the C4.5 decision tree algorithm. The C4.5 classifier is a generator of supervised symbolic classifiers based on the notion of entropy, since its output consists of nodes created by minimizing an entropy cost function [387]. Furthermore, in [514], an online touch pattern algorithm based on a Temporal Decision Tree (TDT) is suggested in order to ensure reactive responses to various touch patterns, and in [434] an efficient single-camera multidirectional wheelchair detector based on a cascaded decision tree (CDT) is proposed to detect a wheelchair and its moving direction in tandem, from video frames, for a healthcare system.

The Support Vector Machines classification constitutes a relatively new method of supervised machine/mechanical learning, which can be applied to classification problems. In their simplest form, the SVMs learn to distinguish between two categories. With small algorithmic differences, however, they can also separate examples from several categories. The most notable advantages of this method are:

- it has a capacity for quality of generalization and ease of training that exceeds that of more traditional methods such as neural networks,

- it can model complex, real-world problems, and

- it tends to perform well even on multivariate datasets.

It is well mentioned that this method is quite common in applications relevant to affective computing. For example, in [337] the researchers have created a Health Smart Home which includes, in a real apartment, infrared presence sensors, door contacts, a temperature and hygrometry sensor in the bathroom, microphones, and a wearable kinematic sensor. The data collected from the various sensors are used to categorize each temporal frame into one of the international activities of daily living (ADL) that were previously acquired. This classification is carried out using SVM. Furthermore, in [695] the use of twin SVM for gesture classification based on sEMG, which is exceptionally suited to such applications, is reported. Also, in [998] a VRST was used from the Virtual Reality Cognitive Performance Assessment Test (VRCPAT) in order to identify the optimal arousal level that can act as the affective/ cognitive state goal. The arousal classification is carried out using the SVM.

Hidden Markov Models (HMMs) and Gaussian Mixture Models (GMMs) constitute generative models[2], which are probability density estimators [742]. In particular, for the recognition of affective states, HMM [766] and its variations [114], [723] are the most commonly used techniques. Such models can also be used for the detection of the temporal segments, provided that one can enforce some correspondence between the state values of HMMs and the temporal segments of the affective state.

A number of studies, such as [234] and [719], have detected the temporal parts of facial expressions by using HMMs [387]. In addition, in [867] a tripled HMM is introduced to perform recognition which allows the state asynchrony of the audio and visual observation sequences, while preserving their natural correlation over time. The experimental findings indicate that this approach outperforms the conventional approaches of emotion recognition [867].

On the other hand, the most common approach for modeling emission probabilities is by making use of mixtures of GMMs [747], [387]. One of the applications in which this method of classification is used, is described in [718], where the resulting information theoretic effect size maps are supplemented by a statistical evaluation based on Gaussian null model simulations using a false-discovery rate procedure. As noticed for emotional state modeling, a variety of pattern recognition methods are utilized to construct a classifier, such as the GMMs, SVMs and decision trees. However, a base-level classifier may not perform well on all emotional states [997]. There are studies, which have proven that hybrid-based approaches can achieve higher recognition performance than individual classifiers. For example, [616] refers to a technique to enhance emotion classification in spoken dialog systems by means of two fusion modules. The first combines emotion predictions generated by a set of classifiers and produces other predictions on the emotional state of the user which are the input to the second fusion module, where they are combined to deduce the user's emotional state. The results of this research show that the inclusion of the second fusion module enhances the classification rate by 2.25% absolute.

---

[2] A generative model is a descriptive model for parameter estimation. It is a popular tool for machine learning because of its powerful classication capacity, which is based on the utilization of Bayesian rules.

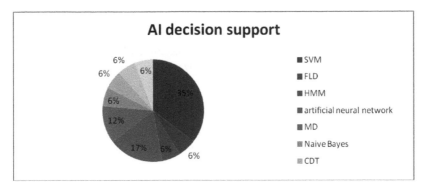

FIGURE 14.5: AI decision support.

In Fig. 14.5 we summarize the most common AI decision support techniques. According to this figure SVM is the most usual AI decision support technique, taking into consideration that 35% of the applications which are studied in this chapter are using it. Furthermore, artificial neural networks and MD are also common techniques, since they come up with an average of 17% and 12% respectively. The FLD, HMM, Nave Bayes, CDT, C4.5 decision tree, and the boosted cascade classifier are the least usual classification methods, as each of them has an average of 6%.

## 14.3.5 Additional Features

The findings from psychology, neuroscience and sociology on emotions and its effects call for innovation in various sectors [953]. Some of them are information and communications technology, privacy security and trust, novel patient modeling, and multimodal intelligent affective user interfaces. Many of the applications related to affective computing either already fulfill some of these requirements — requirements vary depending on the application — or researchers refer to future approaches toward this direction.

For example, in [894], a Pervasive Environment for Affective Healthcare (PEACH) is proposed. PEACH is a innovative/original framework for quickly formulating and managing ad hoc rescue groups in the same locality as the one of the victims. Using this system, rescue teams have the ability to perform rescue operations and provide life-saving assistance to a victim. Researchers indicate that they are working on extending PEACH services to embody modules that will ensure data integrity, privacy, and also non repudiation.

Furthermore, some systems have already been characterized as pervasive. In [595] a pervasive and ubiquitous computing application is proposed. Specifically, a probabilistic framework based on the dynamic Bayesian networks

(DBNs) is introduced, which first formulates an initial hypothesis about the user's current affective state and then actively selects the most informative sensory strategy in order to quickly and economically confirm or refute the hypothesized user state. In [827] a middleware is referred to, which is intended to leverage existing insights into reconfiguration of component-based software and adaptive applications [802] for the development of human-centered pervasive adaptive applications. In addition, a new chance for affective and pervasive computing by detecting early warning signs of poor heart health is presented in [78]. Particularly, this opportunity is related to a continuous monitoring of stress levels during office work.

It is worth mentioning that the characteristics of pervasion are extremely important for some applications. For instance, in [741] an innovative emotion evocation and EEG-based feature extraction method using Higher Order Crossings (HOC) is presented. The researchers of this system note that the integration of the HOC-emotion classifier (HOC-EC) in a wide healthcare system could contribute to ameliorated human-machine interactions (HMIs).

Additionally, some systems are based in supervised techniques and others in unsupervised techniques. For example in [1033] a Support Vector Machine is used to conduct the supervised classification of affective states between "stressed" and "relaxed".

Finally, the modeling multimodality is another important feature for some systems. In [1032], many examples of affect-sensitive multimodal Human-Computer Interaction (HCI) systems are mentioned. Some of them are those in [604], [301] and [491]. In particular, the system described in [604] associates facial expressions and physiological signals to identify the users emotions and, consequently, to adjust an animated interface agent to mirror. [301] refers to a multimodal system which applies a model of embodied cognition that can be perceived as a detailed mapping between the user's affective states and the types of interface adaptations. Finally, the automated Learning Companion [491] combines information from cameras, a sensing chair, and a mouse, a wireless skin sensor and a task state, so as to detect frustration in order to predict or anticipate when the user needs help [1032].

---

## 14.4   Evaluation

In literature associated with affective computing, it is observed that during the assessment of an application we should observe the accuracy and reliability of the results, and also the usability and acceptance known to the public. For example in [833] an ambulatory system for clinical gait analysis via ultrawideband radios (UWB) is proposed. The accuracy of this system is far better than the accuracy reported in literature for the intermarker-distance measurement. This system characteristic makes it appropriate for clinical gait analysis, characterization, and estimation of unstable mobility diseases.

In [435] it is mentioned the fact that the usability of the Virtual reality (VR) interface design is very significant, since poor usability limits the effectiveness in delivering instructions [889]. Moreover, in [769] the researchers identified that there are three artifact aspects that have an impact on emotions. One of them is instrumentality, i.e., the usability in the HCI (Human-Computer Interaction) context [917].

The characteristics of acceptance are extremely important for an application. The complexity of the interfaces of many existing computer games has an impact on their acceptance by creating a threshold for their widespread acceptance [35].

In order to verify those characteristics (accuracy, reliability, usability, acceptance etc.) numerical results are cited in several cases. These results verify the existence of those characteristics in the best way.

Furthermore, quantifying provides a more comprehensive picture of the application. It is well mentioned that the chapter contains numerical results since the estimators for their research are several. Below some of them are mentioned.

In [666] a behavioral approach to deception detection in the context of national security is presented. Its recognition accuracy approaches the rate of 71% for the two-class problem [387]. In particular, this automated system records a set of features extracted from head and hand movements in a video, and using them it can infer deception or truthfulness.

Another work where numerical results are presented is in [98]. In this paper in order to extract expression related features, and infer an emotional condition, a systematic approach is used. The recorded features are a combination of facial expressions and hand gestures. The accuracy rate which was achieved for emotion recognition from facial features was 85%; using hand gestures only, it was 94.3%.

In [497] a dynamic approach to detect emotion in video sequences that are taken from nearly real-life situations is described. The recognition accuracy for a four-class problem from the data used are 67% for visual, 73% for prosody, and 82% with all modalities combined. Also, in [65] where a metanalysis is presented, it is reported that human judgment of behaviors based solely on a face is less accurate than the one based on both the face and the body (35% more accurate) [387].

Furthermore, in [492] a multi-sensory recognition system which classifies three affective states (high interest, low interest and refreshing) with 85% accuracy has been reported [387]. During the user's activity on the computer, the system records facial expressions and postural shifts, and thereafter combines them with information on his activity thus performing the evaluation.

As we mentioned above, in [741], a novel emotion evocation and an EEG-based feature extraction technique using HOC is presented. In the same paper, the HOC technique is compared to other feature extraction methods. This comparison suggests that HOC predominates over the other methods during differentiating among the six basic emotions. As a matter of fact, the rate of classification accuracy for single channel cases reaches 62.3%, while for combined-channel cases is 83.33%.

In [997], an approach related to emotion recognition of affective speech based on multiple classifiers using acoustic-prosodic information (AP) and semantic labels (SLs) is presented. The recognition accuracy of this approach, considering the individual personality trait for personalized application, is 85.79%. It is notable that the combination of AP and SLs can reach 83.5% accuracy. As a matter of fact, this rate is superior to AP-based or SL-Based approaches.

In [78] a new approach to detecting early warning signs is proposed. For instance, this approach associates affective states with features derived from the pressure distribution on a chair. The recognition accuracy of this system reaches the rate of 73.75%. Finally, another research with numerical results is that of [35]. This paper reports on creating and evaluating a context-aware yet situation-adaptive Bayesian inference framework that predicts human mental states [35]. The recognition accuracy of this network is more than 85%.

## 14.5 Applications

Undoubtedly the applications of affective computing vary. There are applications with a learning and instructive character, others with a diagnostic character, and many others such as learning, games, HCI, virtual reality and virtual environments, and diagnostic, treatment, and support/assistance coaching. They dispose their particular characteristics, use, structure, and effects. For example, in paper [848] studies with NIRS and fMRI for the implementation of BCI, in order to acquire, decode and regulate hemodynamic signals in the brain as well as investigate their behavioral consequences are mentioned. To be more precise, it consists of a review which considers fundamental principles, recent developments, applications, and future directions and challenges of NIRS-based and fMRI-based BCIs. In papers [542], [547], [434] and [894] support/assistance coaching applications are described. Specifically, in [542] a Web2OHS that provides omni-bearing homecare and patient care services for medical staff and caregivers is mentioned. This application is capable of not only assisting families, physicians, and nurses in obtaining physiological information on patients using healthcare sensors, but is also capable of monitoring their behaviors using monitoring-based services.

In [547] a visual context-aware-based sleeping-respiration measurement system that measures the respiration information of elderly sleepers is proposed. This support application takes into account all possible contexts for the sleeping person, in order to take an accurate respiration measurement.

Precisely, a body-motion-context-detection subsystem, a respiration-context-detection subsystem, and a fast motion-vector-estimation-based respiration measurement subsystem constitute this rare sleeping care system system. In [434] an efficient single-camera multi-directional wheelchair detector based on a cascaded decision tree is developed. A wheelchair and its moving direction are detected in tandem from video frames for a healthcare system. In particular, the cascaded decision tree, which is developed, is able to materialize early confidence decisions to rapidly discard non-wheelchairs and decide the moving directions, reducing detection time and increasing detection's accuracy. In [894], as mentioned above, a novel framework for quickly processing and managing ad hoc rescue groups in the same locality as the victims is proposed. This pervasive framework provides a set of services for integrating wearable biosensors, which are able to detect and monitor changes of the patients' psychophysical conditions, to aggregate sensed information, to detect potentially emergency cases for the patient, and to provide assistance to the victim.

Another category, which is studied as part of this chapter, is diagnostic applications. Forty percent of the applications studied belong to this category (see Fig. 14.5). Some of them are described in [550], [652], [206], [365], [833] and [894]. To be more specific in [550] the method of wavelet coherence and its statistical properties are completely developed. The fact that the wavelet coherence which can detect short, significant episodes of coherence between non-stationary neural signals is also described.

In [652] the contributions of cerebro-cerebellar circuitry to executive verbal working memory using event-related fMRI is studied. Many important findings related to the cerebro-cerebellum's relevance to executive verbal working memory arose from this study. Moreover, in paper [365] WE, relative WE and WE changes were evaluated from the recorded EEG and ERP signals, through a working memory task, from dyslexic children. The findings of this study show that there are revealed differentiations primarily in relative WE and WE change that take into account the variability of EEG between dyslexics and non-dyslexics. According to the research the WE may provide a useful tool in analyzing electrophysiological signals related to dyslexia. There is one more application which has been analyzed that belongs both to diagnostic and assistance coaching applications. This application is described in [894] and we referred to it above.

Moreover, activity recognition applications are analyzed in [754], [998], [206], [337], [495], [642], [78], [336], [654] and [345].

In [754] a single-trial analytic framework for EEG analysis and practice, in order to target detection and classification, is mentioned. To be more precise, the present single-trial framework which is based on effective noise mitigation and transformation of scalp data to source space, extends the current state-of-the-art in single-trial analysis of neural activity. In [998] approaches for assessing both the optimal affective/cognitive state and the user's observed state are presented. Specifically, the VRST is a virtual environment for assessment of neurocognitive and affective functioning that is used in order to identify the optimal arousal level in mental tasks.

Furthermore, the researchers of this paper claim that their investigation reflects progress toward the implementation of a closed-loop affective computing system. Also, the application belongs to the category of diagnostic, HCI, and virtual reality applications.

In [206] an overview of the approaches to building a reliable affective model, based on affective trust theory for high accurate health monitoring systems, is proposed. In general terms, in the proposed affective model, a specific patient's facial expressions and biomedical signals were captured by a sensor for a swift trusted comparison with emotions. Another diagnostic application is described in [337]. More specifically, researchers have designed a smart home filled with sensors that are recording data related to daily living activities — as mentioned earlier. The data which are collected from the varied sensors are used to categorize each temporal frame into one of the international activities of daily living that were earlier acquired. These activities are: hygiene, toilet use, eating, resting, sleeping, communication, and dressing/undressing. In [495] the capability of gait to show a person's affective state is analyzed. To be more specific, this study has proved that from gait, different levels of arousal and dominance can be recognized, rendering the gait as an additional modality for the recognition of affect. In [642] the evaluation of changes of power, and frequency of sleep spindles and delta waves related to sleep depth, is presented. The researches exploit the fact that adaptive time-frequency approximations, implemented via the matching pursuit algorithm, construct explicit filters for finding EEG waveforms, so as to quantitatively evaluate the changes of power and frequencies of sleep spindles and delta waves with the depth of the sleep.

Moreover, in [78] researchers concluded that the posture channel contains affective information related to stress, and that such information can be detected. Precisely, on the basis of a laboratory stress test which is similar to an office scenario, they investigated whether affective information related to stress can be found in the posture channel during office work. Through the test they observed that most of the subjects became nervous during stress, a fact which consequently led to an increased chair movement variation. In another paper, [336], the design of compact wearable sensors for long-term measurement of electrodermal activity, temperature, motor activity, and blood volume pulse is described. In particular, the developed sensors provide an important contribution in presenting systems for collecting data in long-term naturalistic settings. In research, this new technology for continuous long-term monitoring of the autonomic nervous system, and motion data obtained from active infants, children, and adults, has been widely used.

In [654] techniques to improve communication in BSN that collects data on the affective states of the patient are proposed. The data which relate to the patients' lifestyle can be associated with her/his physiological conditions in order to identify how various stimuli can trigger symptoms. According to researchers, this study constitutes a step toward ubiquitous systemic monitoring that will enable the analysis of dynamic human physiology, in order to understand, diagnose, and treat mood disorders. Finally, in [345] a novel architecture for the evident differentiation of emotional physiological signals caused/triggered by viewing pictures chosen from the IAPS is presented. It is worth mentioning that researchers discuss how future developments may be steered toward serving affective health care applications, such as the monitoring of the elderly or chronically ill people.

Another category of applications is that of applications related to HCI. For example, in [998], [346] and [710] applications of this type are cited. In [346] a methodology for evident differentiation of neurophysiological data is proposed. More specifically, this novel methodology initially elicits neurophysiological emotional responses which are classified, using data mining approaches, into four emotional states. Responses are recorded during the passive viewing of emotionally evocative pictures selected from the IAPS. n [710] BCI for HCI and potential game applications. This is a study of the state-of-the-art brain-computer interfaces (BCI) in the context of games and entertainment. Game design research indicates that BCI provides possible means for a gamer to modify input variables in order to control game situations.

Finally, in [710] (which is described above) and in [190], game applications are mentioned. In particular, in [190] a novel device with the ability to guarantee not only a greater gaming experience for everyone, but also a better quality of life for sight-impaired people is proposed. The present study discusses a multipurpose system that allows blind people to play chess or other board games over a network, reducing their disability obstruction. The information provided in this section is summarized in Table 14.1 below.

TABLE 14.1: This Table Summarizes the Applications That Are Studied in This Chapter. The Conclusions Arising from Tt Are Described in the Form of Diagrams (Figs. 14.4–14.6) and a Detailed Report of Them is Done Above.

| Ref | Sensor | Type of processing | Features used | AI decision support | Application | Evaluation |
|-----|--------|-------------------|---------------|---------------------|-------------|------------|
| [848] | NIRS, fMRI | hemo-dynamic signals analysis | oxyHB, dy-oxyHB, con-centration | SVM, MNN, FLD, HMM | BCI, clin-ical treat-ment | over 80% accuracy in NRIS signal capturing |
| | | | | | | Continued on next page |

**TABLE 14.1 – continued from previous page**

| Ref | Sensor | Type of processing | Features used | AI decision support | Application | Evaluation |
|-----|--------|--------------------|---------------|---------------------|-------------|------------|
| [754] | EEG | Wavelet transform, DCA, source space transformation | electrical signal (resulting from eye blinks, eye and head movements), heart beat | | activity recognition (tracking neural activity captured by EEG) | 87% accuracy in classification of neural signatures, reliability in classification |
| [998] | VEOG, ECG, EEG | VRST, SFS | physiological signals (SCL, RSP, VEOG, electrocardiographic activity, encephalographic activity) | SVM | HCI, virtual reality, recognition activity, clinical treatment | useful, 84% accuracy in distinguishing using physiological signals |
| [550] | | wavelet transform | statistical properties wavelet feautures | Neural network | diagnostic in episodes of coherence between non-stationary neural signals | The reported accuracy is sufficient, reliability |
| [346] | EEG | discrete wavelet transform | EEG parameters, MD | SVM | HCI | 79.5% accuracy in classification for MD, 81.3% accuracy in classification for SVM |
| | | | | | | Continued on next page |

**TABLE 14.1 – continued from previous page**

| Ref | Sensor | Type of processing | Features used | AI decision support | Application | Evaluation |
|-----|--------|-------------------|---------------|---------------------|-------------|------------|
| [652] | fMRI | statistical computational ANOVA | cerebro-cerebellar activity parameters, verbal working memory | | diagnostic in the contributions of cerebro-cerebellar circuitry to executive verbal working memory | |
| [718] | EEG, fMRI | Fourier transform | EEG parameters, fMRI features | | providing evidence for the topographically of brain | |
| [710] | EEG, fNIRS, EMG | | Brain activity, electrical activity (face/ body muscle movements) | | HCI, BCI, games | User acceptance, sufficient |
| [206] | video imaging | video/image processing | biomedical signals, facial extracted expressions features | | activity recognition | high accuracy information report on patients health situation |

Continued on next page

**TABLE 14.1** – continued from previous page

| Ref | Sensor | Type of processing | Features used | AI decision support | Application | Evaluation |
|---|---|---|---|---|---|---|
| [337] | smart home, infrared presence sensors, temperature and hygrometry sensor, microphones, wearable kinematic sensor | wavelet-based pattern recognition | biosignals, sound, speech, postural transitions, walk periods | SVM | emergency event detection, activity recognition | 86% accuracy of well-classified frames |
| [495] | video | Fourier transform | kinematic parameters | naive Bayes, NN, SVM | recognition of emotional activity | 95% accuracy in recognition of motion capture data |
| [642] | EEG | matching pursuit algorithm | sleep spindles, delta waves | | activity recognition | |
| [365] | EEG, EOG | wavelet transform | EEG parameters, ERP signals, eye movements | | diagnostic in dyslexia | useful, accuracy of results |

Continued on next page

**TABLE 14.1 – continued from previous page**

| Ref | Sensor | Type of processing | Features used | AI decision support | Application | Evaluation |
|-----|--------|--------------------|--------------|--------------------|-----------|-----------|
| [78] | fMRI, PET | MIST, ANOVA | Physiological signals, acceleration, sitting pressure | SOM, neural networks | recognition of emotional activity | 73.75% accuracy in supervised variant of self-organizing map |
| [336] | EDA, PPG | | electrodermal activity, temperature, motor activity, BVP, HRV | | activity recognition | privacy and data security |
| [542] | healthcare sensors, monitoring-based services | | physiological information | | assistance coaching | compatible with related medical information systems |
| [654] | BSN-based sensors | linear programming techniques | physiological information | | recognition of mood disorders | |
| [190] | board navigator, move selector, optical sensor, CMOS signals | DSP algorithm | | | games | practical and low-cost system architecture |
| | | | | | | Continued on next page |

**TABLE 14.1 – continued from previous page**

| Ref | Sensor | Type of processing | Features used | AI decision support | Application | Evaluation |
|-----|--------|--------------------|--------------|---------------------|-------------|------------|
| [547] | CBSATCS, near-infrared camera and lighting | | respiration information, respiratory and nonrespiratory body movement parameters | | assistance coaching | |
| [833] | wearable UWB transceivers | clinical gait analysis | movement parameters and patterns | | diagnostic in unstable mobility diseases (Parkinson's disease) | low-cost, low-complexity, high accuracy in assessment of clinical gait analysis |
| [434] | single-camera multidirectional wheelchair detector | video processing | moving direction of a wheelchair | CDT, boosted cascade classifiers | assistance coaching | over 92% detection rate |
| [695] | sEMG | | muscle activity, strength of muscle contraction, body gestures | SVM | activity recognition | |
| | | | | | | Continued on next page |

**TABLE 14.1 – continued from previous page**

| Ref | Sensor | Type of processing | Features used | AI decision support | Application | Evaluation |
|---|---|---|---|---|---|---|
| [345] | | image processing and analysis | emotional physiological signals (multi-channel recordings from both the central and the autonomic nervous systems), MD | C4.5 decision tree algorithm | recognition of emotional activity | 77.68% average recognition (success) rate |
| [894] | SP, blood pressure sensors, respiration sensors, skin conductivity sensors | | blood pressure, body temperature, hormone levels, heartbeats, breathing, sweat in the palms, moving information | | diagnostic in psychophysical alteration, assistance coaching | |

TABLE 14.2: Explanation of Acronyms.

| Acronym | Explanation |
|---|---|
| BCI | Brain—Computer Interface |
| BSN | Body Sensor Network |
| BVP | Blood Vol. Pulse |
| CDT | Cascaded Decision Tree |
| DCA | Directed Components Analysis |
| DSP | Digital Signal Processing |
| DWT | Discrete Wavelet Transformation |
| ECG | Electrocardiography |
| EDA | Electrodermal Activity |
| EEG | Electroencephalography |

Continued on next page

**TABLE 14.2 – continued from previous page**

| Acronym | Explanation |
|---------|-------------|
| EOG | Electro—Oculogram |
| ERP | Event—Related Potentials |
| FLD | Fisher Linear Discriminant |
| fMRI | Functional Magnetic Resonance Imaging |
| fNIRS | Functional Near—InfraRed Spectroscopy |
| HCI | Human- Computer Interaction |
| HMM | Hidden Markov Models |
| HRV | Heart Rate Variability |
| IAPS | International Affective Picture System |
| MD | Mahalanobis Distance |
| MIST | Montreal Imaging Stress Task |
| MNN | Nevral Network |
| NIRS | Near—Infrared Spectroscopy |
| NN | Nearest Neighbor |
| PET | Positron Emission Tomography |
| PPG | Photoplethysmography |
| RCP | Respiration |
| RSVP | Rapid Serial Visual Presentation |
| SCL | Skin Conductance Level |
| sEMG | Surface Electromyogram |
| SFS | Sequential Forward Selection |
| SOM | Self—Organizing Map |
| SVM | Support Vector Machines |
| UWB | Ultra Wideband |
| VEOG | Vertical Electrooculograph |
| VRST | Virtual Reality Stroop Task |
| WE | Wavelet Entropy |
| Web2OHS | Web2.0—Based Omni-Bearing Homecare System |
| WSNs | Wireless Sensors Networks |

## 14.6 Conclusions

This chapter focused on research and development of the most important methodologies on affective and emotional computing from the biomedical perspective in the context of health monitoring. The processing type that is used in each application and the applications' characteristics are also described here. We tried to concisely refer to the general context of the applications

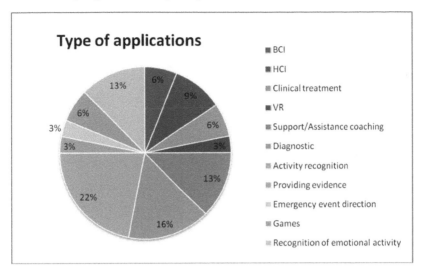

FIGURE 14.6: Type of applications: Brain–Computer Interface (BCI), Human–Computer Interaction (HCI).

overviewed. Additionally, we sought to analyze the AI classification modules used.

In total, it was observed that researchers are mainly dealing with recognition activity applications which come up with an average of 22% accuracy, while diagnostic applications reach 16%, support/assistance coaching applications reach 13%, and HCI reaches 9%. To a lesser degree they examine and materialize applications related with BCI (6%), virtual reality (3%) and games (6%) (see Fig. 14.6). Furthermore, the most common types of sensors are EEG, fMRI, video imaging, EMG, ERP, and temperature sensors. The most upcoming AI decision support methodology is SVM, which comes up with an average of 35% accuracy. To conclude, the current study suggests that affective computing is gaining ground within the scientific community. As a result of many scientists working on that domain, rapid progress in its theory and applications has been forthcoming of late. There are distinct challenges that still need to be addressed (such as the design and production of more comfortable wearable sensors, or, indeed, the recognition and validation of the user's state), and this is the subject of work in the immediate and foreseeable future. The further aim is to produce acceptable, reliable, user-friendly real-life systems that can increase their user's quality of living [727].

# 15

## Affective Natural Interaction Using EEG: Technologies, Applications and Future Directions

**Charline Hondrou**

*National Technical University of Athens, charline@image.ece.ntua.gr*

**George Caridakis**

*National Technical University of Athens, gcari@image.ece.ntua.gr*

**Kostas Karpouzis**

*National Technical University of Athens, kkarpou@cs.ntua.gr*

**Stefanos Kollias**

*National Technical University of Athens, stefanos@cs.ntua.gr*

## CONTENTS

Electroencephalography signals have been studied in relation to emotion even prior to the establishment of Affective Computing as a research area. Technological advancements in the sensor and network communication technology have allowed EEG collection during interaction with low obtrusiveness levels, as opposed to earlier work which classified physiological signals as the most obtrusive modality in affective analysis. This chapter provides a critical survey of research work dealing with broadly affective analysis of EEG signals collected during natural or naturalistic interaction. It focuses on sensors that allow such natural interaction (namely NeuroSky and Emotiv), related technological features, and affective aspects of applications in several application domains. These aspects include an emotion representation approach, the induction method, and stimuli and annotation chosen for the application. Additionally, machine learning issues related to affective analysis (such as incorporation of multiple modalities and related issues, feature selection for dimensionality reduction and classification architectures) are revised. Finally, future directions of EEG incorporation in affective and natural interaction are discussed.

---

## 15.1    Introduction

The use of Electroencephalography (EEG) to study electrical activity in the human brain was demonstrated for the first time approximately 90 years ago by the German physiologist and psychiatrist Hans Berger (1873–1941)[390]. Since then, a constant progress in the field has led to the everyday use of the EEG for diagnostic and research purposes. While hardware technology advances, we are now at the point that EEG sensors are found in user-friendly applications used by everyone in their own home without any prior knowledge needed on the subject. So EEG sensors, having started as complicated to use, expensive equipment found in laboratories, are now embedded in commercial packages that anyone can use. This development has had far-reaching implications for the study of the human brain's activity changes in response to changes in affect and emotion. Numerous studies have taken place in recent years, aiming to understand the correlation between brain signals and emotional states.

In EEG, the electrical activity of the brain is observed through scalp electrodes positioned according to the 10–20 system [708] for most clinical and research applications. The 10–20 system is an internationally recognized method to describe the location of the 19 electrodes (plus ground and system reference) used originally. The "10" and "20" refer to the fact that the actual distances between adjacent electrodes are either 10 % or 20 % respectively of the total front-back or right-left distance of the skull. Each scalp location is assigned a letter to identify the lobe and a number to identify the hemisphere location. Letters F, T, C, P and O stand for frontal, temporal, central, parietal, and occipital lobes, respectively. Note that there exists no central lobe, the "C" letter is only used for identification purposes. A "Z" letter (zero) refers to an electrode placed on the midline. Even numbers (2,4,6,8) refer to electrode positions on the right hemisphere, whereas odd numbers (1,3,5,7) refer to those on the left hemisphere. The electrodes placed on the ear lobe or mastoid are usually considered as reference channels (Fig. 15.1 [33]). A smaller number of electrodes are typically used when recording EEG from neonates. Additional electrodes can be added to the standard setup when a clinical or research application demands increased spatial resolution for a particular area of the brain. High-density arrays (typically via cap or net) can contain up to 256 electrodes more-or-less evenly spaced around the scalp. After being detected by the sensors, the signal needs to be amplified 1,000–100,000 times. Finally, it is important to mention that a typical adult human EEG signal is about 10V to 100V in amplitude when measured from the scalp and is about 10–20 mV when measured from subdural electrodes.

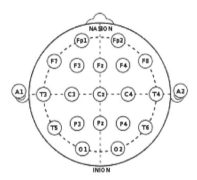

FIGURE 15.1: The International 10-20 system.

Brain signals are classified in 5 frequency bands, associated with different mental states. The most common of them are:

- Delta waves are the slowest waves (0–3.5 Hz) and occur in deep sleep [394]. They are the slowest [557], but highest amplitude brainwaves and can arise either in the thalamus or in the cortex (frontally in adults, posteriorly in children).

- Theta waves (3.5–7.5 Hz) can be separated into two categories: The "hippocampal theta rhythm" which is a strong oscillation that can be observed in the hippocampus and other brain structures in numerous species of mammals including rodents, rabbits, dogs, cats, bats, and marsupials, and the "Cortical theta rhythms" that are low-frequency components, usually recorded in EEG studies. Cortical theta waves are associated with drowsiness, daydreaming, creative inspiration, and meditation. The function of the hippocampal theta rhythm is not clearly understood. Among others, there have been studies that propose that it is related to arousal [378], sensorimotor processing [985], and mechanisms of learning and memory [402].

- Alpha waves (7.5–12 Hz) predominantly originate from the occipital lobe during wakeful relaxation with closed eyes [598]. They are reduced with open eyes, drowsiness, and sleep. In a similar frequency band (8–13 Hz) mu waves can be found. This wave activity appears to be associated with the motor cortex (central scalp), and is diminished with movement or an intent to move, or when others are observed performing actions.

- Beta waves that range from 12 to 30 Hz are split into 3 or 4 bands (there are different categorizations in bibliography) and are of relatively low amplitude. They can be found in both sides of the brain (they are most evident frontally) and have symmetrical distribution. The lowest frequencies of beta rythm are associated with focus and concentration, the middle frequencies are associated with high alertness, and the higher ones with agitation and anxiety. Beta activity is increased when movement has to be resisted or voluntarily suppressed [1042].

- Gamma waves begin at 25 Hz and can go up to 100 Hz [438] (frequencies up to 200 Hz have also been measured — called hyper-gamma). These waves are found in the somatosensory cortex, and are associated with very high states of consciousness, focus, and intellectual acuity. This is why this rhythm is very present during meditation.

Furthermore, there are two important characteristics measured in EEG studies. The first one is the Event-Related Potential (ERP)[164]. An ERP is any measured brain response that is a direct result of a thought or perception. More formally, it is any stereotyped electrophysiological response to an internal or external stimulus. Increased interest has been shown in P300 which is one of the components of an ERP elicited by task-relevant stimuli. It is considered to be an endogenous potential, as its occurrence links to a person's reaction to the stimulus. It is a positive deflection in voltage (2-5 V) with a latency of about 300-600 ms from the stimulus onset. The second one is the Steady-State Visual Evoked Potential (SSVEP). The SSVEP is a periodic response elicited in the brain by visual spatial attention on a flickering stimulus at frequency of 6 Hz and above. SSVEPs have the same fundamental frequency as the stimulating frequency, but they also include higher and/or subharmonic frequencies in some situations. SSVEPs are usually recorded from the occipital region of the scalp. Compared to other types of EEG features, SSVEPs have a better Signal-to-Noise-Ratio (SNR) [794].

Finally, another angle in examining the correlation of EEG activity and affective/cognitive parameters is that of Frontal Asymmetry. It expresses the asymmetry in electrical activity between the left and right side of the brain in the frontal area, and is measured with the Asymmetry Relation Ratio (ARR) [121], which is calculated as ARR=(PL-PR)/(PL+PR) where PL=Left band power and PR=Right band power. The positive value of ARR represents the Left Frontal Activity (LFA), and the negative value represents the Right Frontal Activity (RFA). Greater left frontal EEG activity measured during resting conditions has been associated with a disposition for a behavioral approach and the expression of approach-related emotions such as happiness and anger, whereas relatively greater resting right frontal EEG activity has been associated with a disposition for a behavioral withdrawal and the expression of avoidance-related emotions such as sadness, fear, and anxiety [278].

## 15.2 Sensors Technology Overview

The EEG sensors started off containing many wires with the electrodes placed directly on the scalp, went to caps with built-in electrodes still connected through wires to the amplifier, and got to wireless sensors that become more and more user-friendly with time.

### 15.2.1 Wired Sensors

In conventional scalp EEG, the recording is obtained by placing electrodes, each one attached to an individual wire, on the scalp with a conductive gel or paste, usually after preparing the scalp area by light abrasion to reduce impedance due to dead skin cells. An example of this type of EEG can be seen in Fig. 15.2 [869]. In the effort to make the recording of EEG easier, less time-consuming, and more comfortable for the user, many companies created caps with built-in sensors. Also, in many cases, all the electrodes' cables are merged inside the cap to one cable that connects the cap to the processing unit. Different types of caps are shown in Fig. 15.3 [3], [20], [7], [14]. Some of the companies producing these types of EEG caps are: BioSemi, Psychlab, NeuroScan, Electro-Cap International Inc., Mind Media and MindPeak.

Mind Media's NeXus cap has 21 EEG channels placed according to the 10–20 system. This cap does not require the purchase of separate electrodes, since the electrodes are already built into the cap. The rubber suction cups on the inside of the cap keep it in place and prevent unwanted leakage of the conductive gel. All the cables are carbon coated and use active noise cancellation technology to reduce artifacts to a minimum. The NeXus Microcap is an alternative for the NeXus Minicap if the chinstrap needs to be avoided [14]. Neuroscan's Quik-Caps contain up to 256 electrodes. They are manufactured of elastic breathable Lycra material with neoprene electrode gel reservoirs, and all the electrodes are placed according to the International 10–20 elec-

FIGURE 15.2: Conventional EEG.

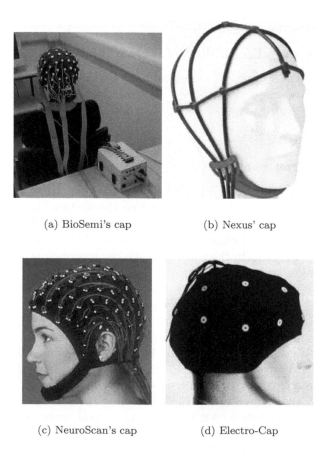

(a) BioSemi's cap        (b) Nexus' cap

(c) NeuroScan's cap       (d) Electro-Cap

FIGURE 15.3: Examples of wired EEG caps.

trode placement standard. Quik-Caps are available in a variety of electrode configurations from 12 to 256 channels and with a variety of electrode materials. The use of Ag/Ag/Cl-sintered electrodes is strongly recommended because of their durability and the ease of cleaning and re-use [20]. Electro-Caps are made of an elastic spandex-type fabric with recessed, pure tin electrodes attached to the fabric. The electrodes on the standard caps are positioned to the International 10–20 method of electrode placement. More specifically, the ECI Surgical Cap is a strapless Electro-Cap designed to be used for intraoperative EEG monitoring. The cap fits the head securely, thereby eliminating the need for chin straps or body harness. The Surgical Caps are available with any electrode configuration and are compatible with all EEG equipment [7]. The BioSemi headcap consists of an elastic cap with plastic electrode holders. The cap itself does not contain electrodes. It is placed on the subject's head, and the electrode holders are filled with electrode gel with a syringe. Because of the active electrode setup, high electrode impedances can be tolerated, so no skin preparation is required. After the buttons are filled with paste, the active electrodes are plugged into the cap one by one. Different electrode layouts are possible. It is also possible to assemble your own cap layout [3]. MindPeak's WaveRider cap connects to the head with sterling silver electrodes, and can be used with any standard protected pin electrode [15]. Finally, PsychLab's electrode cap is also compliant to the 10–20 system and comes with a full kit of accessories (gel, ear electrodes) [25].

## 15.2.2 Wireless Sensors

Recently, the wireless sensors' technology has evolved, making it possible for various systems to be developed. These sensors are inexpensive, easy to set up, and are accompanied with out-of-the-box applications or easy to use SDK's, and provide freedom of movement for the user. The industry has shown strong interest in this field. Some examples are Neurofocus; OCZ Technology; QUASAR; Cortech Solutions; Imec, Panasonic, and Holst centre (Fig. 15.4 [6], [19], [26], [10]); as well as Neurosky and Emotiv. In this section we will briefly present the characteristics of the first sensors but will devote the next section to analyzing the latter ones more, because they are the most popular ones, in terms of integration in NI/EEG studies: Emotiv and Neurosky.

Cortech Solutions' Mind-Fi detects frequencies from 0.8 to 2kHz, has 63 channels and comes in a wide range of standard and customizable versions. It uses the standard EDF data file format, and communicates with the Windows, MAC and Linux drivers through USB 2.0 [5]. NeuroFocus' Mynd has dry "smart" easily-replaceable electrodes, eliminating the use of gels and is capable of interfacing with any Bluetooth-enabled mobile communications device [18]. OCZ Technology's Neural Impulse Actuator (NIA) consists of 3 electrodes. Its name implies that the signals originate from some neuronal activity; however, what is actually captured is a mixture of muscle, skin, and nerve activity including sympathetic and parasympathetic components that have to be sum-

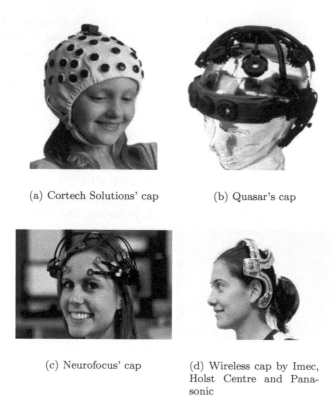

(a) Cortech Solutions' cap (b) Quasar's cap

(c) Neurofocus' cap (d) Wireless cap by Imec, Holst Centre and Panasonic

FIGURE 15.4: Examples of wireless EEG caps.

marized as biopotentials rather than as pure neural signals. As of 2011, the OCZ website says that the NIA is no longer being manufactured and has been end-of-lifed [31]. QUASAR's Dry Sensor Interface (DSI) 10/20 is designed to position 21 dry EEG sensors according to the 10–20 International System. Independent testing of the technology showed wet/dry signal correlations above 80%. Its patented shielding and circuit design reduce environmental noise, and its proprietary artifact reduction technologies reduce motion artifacts [26]. Finally, Imec, Holst Centre and Panasonic developed an 8-channel, wireless EEG headset with low power electronics and noise ($< 6\mu Vpp, 0.5 - 100Hz$) [10].

### 15.2.2.1 Popular Wireless EEG Sensors in Natural Interaction Studies

Depending on the version of these sensors and their SDK there are some variations in the characteristics and properties provided. Here we are presenting the ones that are relative to our theme without separating them according to the edition.

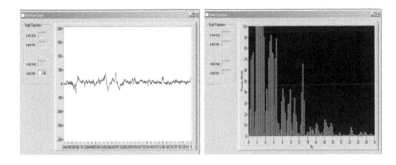

(a) Raw EEG                    (b) Power spectrum of EEG

(c) The user's emotions displayed in a
scale from 0-100

FIGURE 15.5: Instances of Neurosky's interface.

*Neurosky*

The Neurosky sensor is able to measure frequencies in the range of 0.5–50
Hz. Through a simple interface, the raw EEG as well as its power spectrum
can be displayed (Fig. 15.5(a) and 15.5(b) [21]). There is also the possibility
to observe the variations of the waves of different frequencies (alpha, beta,
gamma, delta, theta), as well as four mental states (attention, meditation,
anxiety and drowsiness) on a scale from 0 (lowest) to 100 (highest) (Fig.
15.5(c) [16]). Finally, the user's blinking is tracked in real time presenting
an "explosion" animation for each blink. The 10–20 system doesn't apply to
Neurosky due to the fact that there is only one sensor on the forehead and
one "ground" sensor on the earlobe.

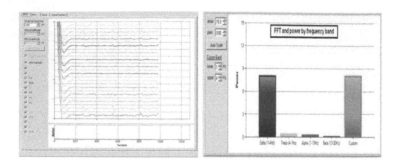

(a) Raw EEG signal recorded by all 14 electrodes

(b) Power of the most important brainwaves

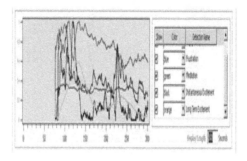

(c) Each one of the six emotions is represented by a graph

FIGURE 15.6: Instances of Emotiv's interface: Signal processing.

*Emotiv*

The Emotiv sensor measures frequencies ranging from 0.2 to 43 Hz. Through its interface we have access to the raw EEG (Fig. 15.6(a)) and its Fast Fourier Transform (FFT) (Fig. 15.6(b)), as well as the graph of the different mental states (long-term excitement, instantanious excitement, engagement, frustration, meditation, boredom) (Fig. 15.6(c)). There is also the possibility to monitor the facial expressions (look left/right, blink, left/right wink, raise brow, clench teeth, smile) (Fig 15.7(a)) and the movements of the head, calculated from the headset's gyroscope (Fig. 15.7(b)). The images of the Emotiv sensor presented here have been retrieved from [29]. Emotiv is more compliant to the 10–20 system than Neurosky (Electrodes AF3, AF4, FC5 and FC6 are regarded as intermediate 10% locations of the 10–20 system).

In the following sections various (conventional and wireless) EEG studies on mental/emotional state are presented and analyzed according to the in-

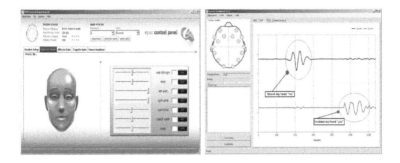

(a) The user's facial movements are detected and displayed

(b) The user's head movement in the x-axis and y-axis

FIGURE 15.7: Instances of Emotiv's interface: Facial movements and gyroscope.

duction, annotation, pre-processing, classification and application domains of their data, as well as the combination in with other signals.

## 15.3 Affective Aspects of EEG Application

### 15.3.1 Detection and Output

*Obtrusive EEG*

In [392] the SSVEP is examined and used in a computer game in order to improve the user experience. Here the stimulus that triggers the SSVEP gives the coordinates of the point where the mouse would otherwise manually go. In [822] the frontal brain activity is studied in order to see if it distinguishes valence and intensity of emotions induced by music. The emotions induced are characterized as intense-unpleasant, intense-pleasant, calm-pleasant, and calm-unpleasant. The NeuroScan system detects in [841] 4 levels of vigilance: wakefulness, middle state 1, middle state 2, and sleepiness. Finally [594] presents a study on frequency and amplitude analysis of alpha waves while the subject browses a study web site.

*Natural Interaction Approach in EEG*

Because of the user-friendly interface these sensors provide, the emotion representation is often quite straightforward. Emotiv provides as direct output frustration and excitement in [446], long-term excitement, instantaneous excitement and engagement in [538], and [370] excitement, engagement, boredom, meditation, and frustration in [374] through its simple interface (Fig. 15.6(c)). In some cases, information from the Emotiv sensor about the raw EEG and wave variations are combined with the information provided by other biological (or not) sensors, resulting in different types of emotion representation: anger, disgust, fear, happiness, sadness, joy, despise, stress, concentration, and excitement in [393], positive, negative, and neutral states of mind in [314] and [740]. In cases where the cognitive load of the subject during a task is being studied [388], changes in power of lower frequency (alpha, theta) brain waves detected by Emotiv are analyzed in order to determine whether the cognitive load is low, medium, or high. Finally, a different approach is adopted when attention or the wink movement is detected. In this case, the raw EEG provided by the sensor is processed in order to detect the P300 signal [186], [772] or the signal caused by the winking [186]. Neurosky sensor outputs, through its simple interface as well, attention and meditation in [245], attention level in [776], and relaxation and irritation in [546]. In a fatigue detection system in [485] delta and theta waves are monitored. When either of these signals maintains the highest value of all frequency bands, attention decreases below a certain threshold, and meditation increases above a threshold, the user is considered to be fatigued.

## 15.3.2 Ground Truth

*Obtrusive EEG*

The most common way of finding the ground truth of experimental corpora is through self assessment. This includes questionnaires or tasks prior to and/or after the experiment that vary. A commonly used tool is the Self-Assessment Manikin (SAM) which is a non-verbal pictorial assessment technique that directly measures the pleasure, arousal, and dominance [660] associated with a person's affective reaction to a wide variety of stimuli [554]. An example can be seen in [822]. Here the goal is to see if the EEG distinguishes the valence and intensity of musical emotions, so the stimuli used (four orchestral musical excerpts) were pre-rated by a different group of subjects using the SAM. In [392], where the intention is to augment the feeling of immersion while playing a computer game, the subject is requested to fill out a SAM to measure their affective reaction, as well as a questionnaire on immersion by Jennett et al. [467]. In the case of [841], in order to measure their vigilance, subjects that are lying down in a relaxing environment are asked to open their eyes every time they hear music, while also being recorded by a camera. For this experiment, the data from the subject's self assessment is taken into consideration only when it is compatible with what the video shows.

*Natural Interaction Approach in EEG*

Questionnaires or tasks concerning the emotion or the cognitive feature examined are often used in EEG studies [245], [546], [643], [388]. But more specifically, certain well-known tasks are given to the subject in order to evaluate the results. As seen in [393] and [740] the International Affective Picture System (IAPS) is incorporated in order to provide a set of normative emotional stimuli for the experimental investigation of emotion and attention. The goal is to develop a large set of standardized, emotionally-evocative, internationally accessible, color images that includes contents across a wide range of semantic categories [554]. In order to measure engagement, in [370] the Independent Television Comission-Sense of Presence Inventory (ITC-SOPI) questionnaire [574] was used. This questionnaire offers a valid cross-media presence measure that allows results from different laboratories to be compared. It studies four factors: Sense of Physical Space, Engagement, Ecological Validity, and Negative Effects. In the same study the SAM was used as well.

Sometimes annotation is performed through the experiment itself. An example of that is [314] where positive, negative, and neutral state of mind are measured. During this experiment the participants are asked to control a robot that is in a maze by thinking about the direction it must take (in reality they can't). The direction in which the robot should move is shown with an arrow. The robot follows a predefined path, sometimes the right one, sometimes the wrong one. When it follows the right path the participant is assumed satisfied (positive), when it follows the wrong path the participant is assumed not satisfied (negative), and while the participant is looking at the arrow (before the robot moves), a neutral state is assumed. Another example is [772] where the P300 signal responding to a visual stimuli is measured. In this case the subject is asked to press a button when the stimulus appears. Another article [776] where attention is measured builds on the results of [273] that suggest that attention is correlated with the errors made and the speed used during the activity, as well as whether the participant gave up or not.

## 15.3.3 Stimuli/Induction

*Obtrusive EEG*

The induction of the different mental states detected in EEG studies can be obtained by audio stimuli, such as orchestral excerpts in [822] or soft music in [841]. Visual stimuli are used in [392] where various white circles flicker with different frequencies on different parts of the screen while the subject is playing a game called "Mind the Sheep!. When the user focuses on one of the stimuli, the periodic signal of that stimulus can be traced back in the user's brain signals. Finally, another commonly used method is web browsing, as can be seen in [594] where the subject is using a study web site.

*Natural Interaction Approach in EEG*

Studies with wireless sensors also use audio stimuli such as the sounds of a game in CAVE environment [538] or wind, sea waves, and a composition of slow violins, tinkling bells, and oboes to induce a positive feeling, and musical pop tracks that the subject strongly disliked to induce a negative feeling, as in the case of [546]. Visual stimuli can include displaying pleasant, neutral, and unpleasant pictures from the International Affective Picture System [740], visuals in game-playing [538], or a robot's movement [314]. If the P300 signal is to be detected, examples of stimuli can be found in [772] where the subject is looking at a monitor where a ship appears at times, or in [186] where the photo of each person of a mobile phone's adress book flashes and when the photo of the person we want to call flashes the signal is detected. Furthermore, examples of interactive Brain Computer Interfaces (BCI) used to induce different emotions are encountered in [393], [446] and [370]. Finally, learning activities and tasks such as the Stroop Test, Hanoi Towers, Berg's Card Sorting Task, and information seeking on the Web can also be used as stimuli [245], [776], [643], and [388].

## 15.4    Machine Learning Aspects

### 15.4.1    Multimodal Input Streams

Studies within the bioinformatics research area, in their effort to be more accurate, combine multiple biological signals as well as information coming from the subject's face, voice, body movements, and actions on the computer.

*Obtrusive EEG*

The electrooculograph (EOG) is used to measure the resting potential of the retina, and its data is often combined with the EEG data. Such studies [822] are examining the valence and intensity of emotions, and [841] examining the level of vigilance.

*Natural Interaction Approach in EEG*

Often the EEG is studied in combination with Galvanic Skin Response (GSR), which is a method of measuring the electrical conductance of the skin, which varies with its moisture level. This is of interest because the sweat glands are controlled by the sympathetic nervous system, so skin conductance is used as an indication of psychological or physiological arousal. Examples can be found in [538], [546], [388] and [374].

Heart rate is another measurement that can be an index of phychological state of mind, and fusing it with EEG has been studied in [538], [546] and [393].

EEG information is combined with information taken from the subject's face using visual input or observations in [446], [245], [393], [643] and [374]. For the latter, "MindReader," a system developed at MIT Media Lab, infers emotions from facial expressions and head movements in real-time. The eyes' movement is another cue that provides information about the user's attention or focus point. In [374] the Tobii Eye Tracking System was incorporated in the system, providing data about the user's attention and focus time while the user was performing a task on the computer. This information is combined with the information provided by the EEG sensor. In the case of [388], the eyes' movements and the EEG were used to predict mental states of a person engaged in interactive information seeking.

Other inputs used to determine mental states can be mouse clicks, keyboard strokes, screen cam recordings [388], body posture, finger pressure (usually on the mouse) [374], and acoustic features [393].

## 15.4.2   Feature Processing and Selection

Due to the huge variability of features collected in biometric studies, in order to reduce the dimensionality of the problem, eliminate the noise, or extract particular characteristics from the signal, the need to preprocess them is common.

*Obtrusive EEG*

In the case of SSVEP detection, a new approach called Canonical Correlation Analysis (CCA) is being used in [392] because of its improved SNR and lower inter-subject variability. Sets of reference signals are constructed for each one of the SSVEP stimulation frequencies. Each set contains the sine and cosine for the first, second, and third harmonic of the stimulation frequency. The re-referenced EEG data and each set with reference signals are used as input for the CCA. The CCA tries to find pairs of linear transformations for the two sets so that when the transformations are applied the resulting sets have a maximal correlation. The stimulation frequency with the highest maximum correlation is classified as the frequency the participant was focusing on.

In [841], Principal Component Analysis (PCA) is used. PCA is a mathematical procedure that uses an orthogonal transformation to convert a set of observations of possibly correlated variables into a set of values of linearly uncorrelated variables called principal components. The number of principal components is less than or equal to the number of original variables. This transformation is defined in such a way that the first principal component has the largest possible variance (that is, accounts for as much of the variability in the data as possible), and each succeeding component in turn has the highest variance possible under the constraint that is orthogonal to (i.e., uncorrelated with) the preceding components [24].

In [841], Independent Component Analysis (ICA) is implemented. It is a computational method for separating a multivariate signal into additive subcomponents supposing the mutual statistical independence of the non-Gaussian source signals [678].

Noise reduction is also a very important issue in biosignal processing. While measuring the EEG, all sorts of artifacts make the isolation of the signal harder, such as the ones caused by eye blinks, eye movements, and other motor movements [822]. Notch filters (50 and 60 Hz) are widely used, as well as all sorts of other types of filters, in order to isolate the signal we want.

Finally, depending on the characteristics of the signal that need to be studied, DFT and natural log transformation are implemented [822].

*Natural Interaction Approach in EEG*

Several wireless EEG studies have focused on the P300 signal, but because the strength of an ERP signal is very low, it is usually covered by the noise and is not visible in a typical EEG recording. Therefore, in order to extract the actual ERP waveform, one has to bandpass filter the EEG signals (typically 1–20 Hz) and average them over multiple trials.

The ICA and PCA that are explained above are also present in [740], [772] and [740], [546] respectively.

### 15.4.3    AI: Tools and Methods

*Obtrusive EEG*

The K-means algorithm, which is used in [841], outlines a method used to cluster a particular set of instances into K different clusters, where K is a positive integer. It should be noted here that the K-means clustering algorithm requires knowing the number of clusters from the user. It cannot identify the number of clusters by itself. The algorithm starts by placing K centroids as far away from each other as possible within the available space. Then each of the available data instances is assigned a particular centroid, depending on a metric like Euclidian distance, Manhattan distance, Minkowski distance, etc. The position of the centroid is recalculated every time an instance is added to the cluster, and this continues until all the instances are grouped into the final required number of clusters. Since recalculating the cluster centroids may alter the cluster membership, the cluster memberships are also verified once the position of the centroid changes. This process continues until there is no further change in the cluster membership, and there is as little change in the positions of the centroids as possible [4].

Another classification method, which is reminiscent of the Support Vector Machines that will be explained in the next section, and that is implemented in [841], is the Normalized Cuts clustering algorithm. This algorithm views the dataset as a graph, where nodes represent data points, and the edges are weighted according to the similarity, or affinity, between data points [770] so that the problem becomes one of graph partitioning. In other words, the normalized cut criterion measures both the total dissimilarity between the different groups as well as the total similarity within the groups [840].

In [841], a dynamic clustering algorithm is proposed in order to take into consideration the temporal aspect (non-stationary features) of the EEG signal during the phases of the experiment.

### Natural Interaction Approach in EEG

Regression Analysis, which is a statistical technique for estimating the relationships among variables, is often used as a classification method. A common type of Regression Analysis is Linear Regression, which models the relationship of two variables with a linear function. If the goal is prediction, linear regression can be used to fit a predictive model to an observed dataset of X and Y. After developing such a model, if an additional value of X is then given without its accompanying value of Y, the fitted model can be used to make a prediction of the value of Y. An example can be found in [446] where Linear Regression was incorporated in order to create predictive models of frustration and excitement.

Whereas in Linear Regression the dependent variable is a numerical quantity, Linear Discriminant Analysis (LDA) can be used, as in [314], to determine the satisfaction level as a categorical representation (satisfied, neutral, unsatisfied).

A very simple classifier can be based on a nearest-neighbor approach. In this method, one simply finds in the N-dimensional feature space the closest object from the training set to an object being classified [30]. Since the neighbor is nearby, it is likely to be similar to the object being classified, and so is likely to be in the same class as that object. Closeness is usually defined in terms of Euclidean distance [743].

A variation of this method is the k-nearest neighbor algorithm where the unknown sample is assigned the most common class among its k nearest neighbors. This algorithm has been used in [446] to create predictive models of frustration and excitement, in [546] to distinguish between positive and negative states of mind, and in [643] to classify numerous features of raw EEG in order to create a model of human academic emotion (boredom, confusion, engagement, and frustration).

Another supervised learning method used for classification is the Support Vector Machine (SVM). The advantage of SVM's is that they can make use of certain kernels in order to transform the problem, such that we can apply linear classification techniques to non-linear data [4]. Applying the kernel equations arranges the data instances in such a way within the multidimensional space, that there is a hyperplane that separates data instances of one kind from those of another. The kernel equations may be any function that transforms the linearly non-separable data in one domain into another domain where the instances become linearly separable. Kernel equations may be linear, quadratic, Gaussian, or anything else that achieves this particular purpose. Once we manage to divide the data into two distinct categories, our aim is to get the best hyper-plane to separate the two types of instances. This hyperplane is important because it decides the target variable value for future predictions. We should decide upon a hyperplane that maximizes the margin between the support vectors on either side of the plane. Support vectors are those instances that are either on the separating planes on each side, or a little bit on the wrong side. This was one of the classification methods used in [446] to create an intelligent tutoring system that would predict the students' appraisal of the given feedback. This method can also be found in [546] for the creation of the Advanced Multimodal Biometric Emotion Recognition System, as well as in [643].

In [546] we can also find Decision Trees. Decision trees classify instances by sorting them based on feature values. Each node in a decision tree represents a feature in an instance to be classified, and each branch represents a value that the node can assume. Instances are classified starting at the root node and sorted based on their feature values [743].

In the same publication [546], the Naive Bayes classifier has also been used. This classifier is based on the Bayes rule of conditional probability. It makes use of all the attributes contained in the data, and analyzes them individually as though they are equally important and independent of each other [4].

For the case of a categorical output $y$, Logistic Regression can be used as a classifier. The fitting of the training data can be done by using the logistic function which is a monotonic, continuous function between 0 and 1. This technique was also used in [546].

In [546] and [643], they also made use of a multilayer perceptron (MLP). An MLP is a feedforward artificial neural network model that maps sets of input data onto a set of appropriate output. It consists of multiple layers of nodes in a directed graph, with each layer fully connected to the next one. Except for the input nodes, each node is a neuron with a nonlinear activation function. MLP utilizes a supervised learning technique called backpropagation for training the network, and is a modification of the standard linear perceptron, which can distinguish data that is not linearly separable [13].

In [546] the idea of ensemble learning has also been applied. It consists of implementing multiple learners and combining their predictions [832].

Linear discriminant analysis (LDA) is used in statistics, pattern recognition, and machine learning to find a linear combination of features that characterizes or separates two or more classes of objects or events. The resulting combination may be used as a linear classifier, or for dimensionality reduction before later classification. It attempts to express one dependent variable as a linear combination of other features or measurements [27]. For this method, the independent variables have to be normally distributed, and the dependent variable has to be a categorical variable (i.e., the class label). An example of this technique can be seen in [314].

In [772] the Adaptive Neuro Fuzzy Inference System (ANFIS) is used to detect the P300-rhythm. This is a kind of neural network that is based on the Takagi–Sugeno fuzzy inference system. Since it integrates both neural networks and fuzzy logic principles, it has the potential to capture the benefits of both in a single framework.

Finally, the K-means algorithm that has been explained above is used in [740].

At this point, it is worth mentioning that different types of open-source data-mining software with embedded algorithms have been a very useful tool for EEG studies. Important examples are "RapidMiner" in [446] and [643], and "Weka" in [546].

## 15.5 Application Domains

Various systems have been created for research purposes in order to study emotion recognition in BCI's [245], [546], [374], affective valence and intensity [822], vigilance [841], fatigue [485], and attention detection [772]. The main fields in which the EEG sensors find application are games (subsection 15.5.1), E-Learning (subsection 15.5.2) and health (subsection 15.5.3). They are also used in mobile phones [740], [186], Human-Robot Interaction (HRI) [314], and interactive information seeking [388].

### 15.5.1 Games

*Obtrusive EEG*

The detection of SSVEPs and their use to make decisions in the game "Mind the Sheep!" in order to make the game more immersive is analyzed in [392]. As the following section shows, the introduction of EEG in games has had much greater success since the wireless sensors became widely commercialized.

*Natural Interaction Approach in EEG*

Often the EEG signal is used as feedback to the game so that the game scenario adjusts to the player's needs. For example, detection of boredom will cause changes in the game to make it more challenging, whereas detection of anxiety will cause the game to slow down or decrease the levels of difficulty. The portability of wireless sensors is even more useful in cases where they are combined with games played in virtual reality environments. In these cases the player's immersion is augmented even more. A good example can be seen in [538] where the environment used is CAVE (projectors are directed to three, four, five or six of the walls of a room-sized cube).

FIGURE 15.8: Screenshot of the learning environment presented in [446].

## 15.5.2   E-Learning

*Obtrusive EEG*

In order to create learning experiences with better outcomes for the students, the EEG has been a very helpful tool. [121] examines the subject's learning style and personality traits, and [594] proposes an interactive learning system that recognizes the student's emotions. [594] proposes the merging of two ontologies – the first one consists of the user's actions while browsing a study web site, and the second one is created by the user's alpha waves' data. Through this merging, the scientists are hoping to create an efficient system for distance learning that will take into consideration the user's interests.

*Natural Interaction Approach in EEG*

In [643] and [370] a virtual learning environment is created which simulates and enhances conventional learning environments. In [446] a tutoring system predicts the student's appraisal of the feedback given (Fig. 15.8), and in [776] the system adapts and modifies the learning activity according to the student's level of attention. [393] is proposing a virtual environment for job interview training that senses the applicant's emotions. A snapshot of the interview simulation can be seen in (Fig. 15.9).

## 15.5.3   Health and Universal Access

*Obtrusive EEG*

One of the most popular tools that has been developed in this area is the P300 Speller. It is a 6x6 matrix of alphanumeric characters where one of its rows or columns flashes randomly. When the character the subject wants to see flashes, the P300 signal is measured [541] (Fig. 15.10 [11]). The Mindwalker project, which is funded by the European Commission, has presented a robotic exoskeleton controlled by EEG signals [17] (Fig. 15.11).

FIGURE 15.9: Job interview simulation.

FIGURE 15.10: The P300 Speller.

FIGURE 15.11: The exoskeleton created for the Mindwalker project.

(a) Cart controlled by the Neu-     (b) Wheelchair controlled by
rosky sensor                             the Emotiv sensor

FIGURE 15.12: Ways of transportation via BCI.

*Natural Interaction Approach in EEG*

Wireless sensors are also being introduced to the P300 Speller, contributing to
its portability [308]. People with mobility problems can also benefit from the
new BCI's used for self transportation and object movement. Such examples
are: fully controlling a Lego Mindstroms NXT Robot via Emotiv and Neurosky
sensors [964], a cart on rails via Neurosky (Fig. 15.12(a)) [9], and a wheelchair
via an Emotiv sensor as an open-source project (Fig. 15.12(b))[12].

## 15.6 Conclusions and Future Directions

The current article provides a critical survey of research work dealing with
broadly affective analysis of EEG signals collected during natural or naturalistic interaction. It presents a great variety of them, but focuses on sensors that
allow such natural interaction (namely NeuroSky and Emotiv) more easily,
related technological features, and affective aspects of applications in several
application domains. These aspects include the emotion, affect, and cognition
representation approach, the induction method, and stimuli and annotation
chosen for the application. Additionally, machine learning issues related to affective analyses are revised, such as incorporation of multiple modalities and
related issues, feature selection for dimensionality reduction, and classification
architectures. Finally, future directions of EEG incorporation in affective and
natural interactions are discussed.

    Biometric systems for emotion recognition and attention detection is a field
that is evolving with great speed. The arrival of inexpensive wireless sensors

participated to a great degree in this evolution and has still more to offer. The hardware can still improve in order to maximize freedom of the user and eliminate unwanted noise. Furthermore, systems that can deal with numerous biological signals will offer more detailed and analytical results. As these systems improve, the real-time aspect will also. Delays will be minimized so the communication will get more direct and accurate. As for the software, the SDK's can add more features so that more detail is added to the comprehension of the user's mental state, and more emotions are recognized. This will bring the machine-human interaction closer to the human-human communication. EEG analysis in relation to emotion, while having a long history, has always been considered an extremely obtrusive emotional cue capture technique, establishing it as unsuitable for natural interaction. Recent releases of sensors that enable less obtrusive and/or wireless collection of EEG signals enable affect-aware applications using natural interaction. It is expected that application domains such as gaming and e-learning will dominate the field. A related milestone could be the release of a gaming platform with EEG sensors included, or at least available as an add-on. As Microsoft's Kinect sensor and OpenNI framework boosted research and applications on Natural Interaction, a similar explosion could follow if a major industry released a gaming platform (similar to the Xbox 360) which would collect EEG data. On the other hand, enabling low-cost and widely available collection of EEG activity during learning activities through e-learning applications would boost research on correlation of brain and cognitive functions as well as adaptive, educational interfaces.

## 15.7    Acknowledgments

This research is supported by the European Union FP7 ICT TeL project SIREN (project No: 258453).

# Bibliography

[1] ISO/IEC 15938-3:2001 Information Technology - Multimedia Content Description Interface - Part 3: Visual, Ver. 1.

[2] Amazon Mechanical Turk – Artificial Artificial Intelligence, http://www.mturk.com.

[3] Biosemi – http://www.biosemi.com/index.htm.

[4] Classification methods – http://www.d.umn.edu/ padhy005/chapter5.html.

[5] Cortech solutions – http://www.cortechsolutions.com/.

[6] E provider medical – http://www.eprovider54.com/medical.html.

[7] Electro-Cap International, Inc. – http://www.electro-cap.com/.

[8] Electrochemical sensors. Industrial Scientific. available from `http://www.indsci.com/electrochemicalsensor/`.

[9] Iee spectrum – http://spectrum.ieee.org/geek-life/tools-toys/mind-over-matter.

[10] Imec – http://www2.imec.be.

[11] Institut national de la santé et de la recherche médicale – http://www.inserm.fr/thematiques/technologies-pour-la-sante/dossiers-d-information/interface-cerveau-machine.

[12] Instructables – http://www.instructables.com/id/brain-controlled-wheelchair/.

[13] Java neural network framework – http://tinyurl.com/a2btd9y.

[14] Mind media – http://www.mindmedia.nl/cms/index.php.

[15] Mindpeak – http://www.mindpeak.com/.

[16] mindupdate – http://www.mindupdate.com/category/neuro-games/.

[17] mindwalker-project – https://mindwalker-project.eu/.

[18] Neurofocus inc. – http://www.neurofocus.com/.

[19] Neurogadget    –    http://neurogadget.com/2011/03/21/neurofocus-reveals-mynd

[20] Neuroscan – http://www.neuroscan.com/.

[21] nnm – http://tinyurl.com/b9pawpr.

[22] Optical sensors. Avago Technologies. available from `http://www.avagotech.com/pages/en/optical_sensors/`.

[23] Piezoelectric accelerometer. available from `http://en.wikipedia.org/wiki/Piezoelectric_accelerometer`.

[24] Princeton University – http://tinyurl.com/auovb5s.

[25] Psychlab – http://www.psychlab.com/.

[26] Quasar – http://www.quasarusa.com/.

[27] rapid-i – http://tinyurl.com/aajgzt5.

[28] Sensing and sensors: Acoustic sensors. MediaRobotics Lab. available from `http://www.realtechsupport.org/UB/MRII/docs/sensing/Acoustic%20Sensors.pdf`.

[29] Smart device central – http://tinyurl.com/a6ncl7s.

[30] Space telescope science institute – http://sundog.stsci.edu/rick/scma/node2.html.

[31] The techreport, neural impulse actuator – http://techreport.com/review/14957/ocz-neural-impulse-actuator.

[32] Temperature sensor tutorial! Adafruit Industries. available from `http://www.instructables.com/id/Temperature-Sensor-Tutorial/`.

[33] Uom brain computer interaction project – https://sites.google.com/site/uombci/about-bci/10—20-system.

[34] Exchangeable image file format for digital still cameras: Exif version 2.2 (2002). 2002.

[35] A. R. Abbasi, A. Hussain, and N. V Afzulpurkar. Towards context-adaptive affective computing. In *Electrical Engineering/Electronics Computer Telecommunications and Information Technology (ECTI-CON), 2010 International Conference on*, pages 122–126. IEEE, 2010.

[36] S. Adali, M. L. Sapino, and V. S. Subrahmanian. A multimedia presentation algebra. In *Management of data*, pages 121–132. ACM, 1999.

[37] S. Adali, M. L. Sapino, and V. S. Subrahmanian. An algebra for creating and querying multimedia presentations. *Multimedia Syst.*, 8(3):212–230, 2000.

[38] S. Adler, A. Berglund, J. Caruso, S. Deach, T. Graham, P. Grosso, E. Gutentag, A. Milowski, S. Parnell, J. Richman, and S. Zilles. Extensible Stylesheet Language (XSL) Version 1.0: W3C Recommendation, October 2001. http://www.w3.org/TR/2001/REC-xsl-20011015/xslspecRX.pdf, last accessed: 22/1/2013.

[39] Adobe, Inc. Edge tools and services, 2013. http://html.adobe.com/edge/, last accessed: 20/1/2013.

[40] Adobe, Inc. SWF file format specification (version 10), 2013. http://www.adobe.com/devnet/swf.html, last accessed: 22/1/2013.

[41] Adobe, Inc., USA. Authorware 7, January 2013. http://www.adobe.com/products/authorware/, last accessed: 22/1/2013.

[42] Adobe, Inc., USA. Director 12, 2013. http://www.adobe.com/products/director.html, last accessed: 22/1/2013.

[43] Adobe Systems, Inc., USA. Flash Professional, 2013. http://www.adobe.com/products/flash.html, last accessed: 22/1/2013.

[44] G. Adomavicius and A. Tuzhilin. Toward the next generation of recommender systems: A survey of the state-of-the-art and possible extensions. *IEEE Transactions on Knowledge and Data Engineering*, 17(6):734–749, 2005.

[45] G. Adomavicius and A. Tuzhilin. Context-aware recommender systems. In *RecSys*, pages 335–336, 2008.

[46] Adomavicius G. and Tuzhilin A. Towards the next generation of recommender systems: a survey of the state-of-the-art and possible extensions. *IEEE Transactions on Knowledge and Data Engineering*, 17(6):739–749, June 2005.

[47] V. Adzic, H. Kalva, and B. Furht. A survey of multimedia content adaptation for mobile devices. *Multimedia Tools and Applications*, 51(1):379–396, January 2011.

[48] S. Agarwal and P. Agarwal. A fuzzy logic approach to search results' personalization by tracking user's web navigation pattern and psychology. In *Tools with Artificial Intelligence, 2005. ICTAI 05. 17th IEEE International Conference on*, pages 318–325, 2005.

[49] S. Agarwal and S. Lamparter. A semantic matchmaking portal for electronic markets. In *Proceedings of the Seventh IEEE international Conference on E-Commerce Technology (July 19 – 22)*, 2005.

[50] H. Ahmadi and J. Kong. Efficient web browsing on small screens. In *Proc. of the ACM working conference on Advanced visual interfaces (AVI'08), Napoli, Italy,* pages 23–30, May 2008.

[51] J. Ajmera and C. Wooters. A robust speaker clustering algorithm. In *Proceedings of the IEEE Workshop on Automatic Speech Recognition and Understanding, ASRU'03,* pages 411–416, St. Thomas, Virgin Islands, IEEE Press, November 2003.

[52] S. Aksoy, K. Koperski, C. Tusk, G. Marchisio, and J. C. Tilton. Learning Bayesian classifiers for scene classification with a visual grammar. *Geoscience and Remote Sensing, IEEE Transactions on,* 43(3):581–589, 2005.

[53] M. Y. H. Al-Shamri and K. K. Bharadwaj. Fuzzy-genetic approach to recommender systems based on a novel hybrid user model. *Expert Syst. Appl.,* 35(3):1386–1399, October 2008.

[54] R. Albatal, P. Mulhem, and Y. Chiaramella. Visual phrases for automatic images annotation. In *International Workshop on Content-Based Multimedia Indexing (CBMI'10), June 23–25, Grenoble, France,* 2010.

[55] S. Alcic and S. Conrad. A clustering-based approach to web image context extraction. In *Proceedings of 3rd International Conferences on Advances in Multimedia,* pages 74–79, Budapest, Hungary, April 17–22 2011.

[56] P. Alexopoulos, J. Pavlopoulos, and Ph. Mylonas. Learning vague knowledge from socially generated content in an enterprise framework. In Lazaros S. Iliadis, Ilias Maglogiannis, Harris Papadopoulos, Kostas Karatzas, and Spyros Sioutas, editors, *AIAI (2),* vol. 382 of *IFIP Advances in Information and Communication Technology,* pages 510–519. Springer, 2012.

[57] P. Alexopoulos, M. Wallace, K. Kafentzis, and D. Askounis. Utilizing Imprecise Knowledge in Ontology-Based CBR Systems through Fuzzy Algebra. *International Journal of Fuzzy Systems,* 12(1), 2010.

[58] P. Alexopoulos, M. Wallace, K. Kafentzis, and D. Askounis. Ikarusonto: a methodology to develop fuzzy ontologies from crisp ones. *Knowledge and Information Systems,* 32(3):667–695, September 2012.

[59] P. Alexopoulos, M. Wallace, K. Kafentzis, and A. Thomopoulos. A fuzzy knowledge-based decision support system for tender call evaluation. In *Proceedings of the 5th IFIP Conference on Artificial Intelligence Applications & Innovations (AIAI 2009),* 2009.

[60] D. Allemang and J. Hendler. *Semantic Web for the Working Ontologist.* Morgan Kaufmann, 2008.

[61] C. Allen. Personalization vs. Customization, July 1999. http://www.clickz.com/experts/archives/mkt/precis_mkt/ article.php/814811, last accessed: 22/1/2013.

[62] J. F. Allen. Maintaining knowledge about temporal intervals. *Commun. ACM*, 26(11):832–843, November 1983.

[63] J. F. Allen. Towards a general theory of action and time. *Artificial Intelligence*, 23:225–255, July 1984.

[64] X. Amatriain, J. M. Pujol, and N. Oliver. I like it... i like it not: Evaluating user ratings noise in recommender systems. In *Proceedings of the 17th International Conference on User Modeling, Adaptation, and Personalization: formerly UM and AH*, UMAP '09, pages 247–258, Berlin, Heidelberg, Springer–Verlag, 2009.

[65] N. Ambady and R. Rosenthal. Thin slices of expressive behavior as predictors of interpersonal consequences: A meta-analysis. *Psychological Bulletin*, 111(2):256, 1992.

[66] E. André, W. Finkler, W. Graf, T. Rist, A. Shauder, and W. Wahlster. WIP: the automatic synthesis of multimodal presentations. In Maybury [655], Chapter 3, pages 75–93.

[67] E. André, J. Müller, and T. Rist. WIP/PPP: automatic generation of personalized multimedia presentations. In *Proc. of the 4th ACM Int. Conf. on Multimedia; Boston, MA, USA*, pages 407–408. ACM, 1996.

[68] E. André, J. Müller, and T. Rist. WIP/PPP: Knowledge-Based Methods for Fully Automated Multimedia Authoring. In *Proc. of EURO-MEDIA; London, UK*, pages 95–102, 1996.

[69] E. André and T. Rist. Generating coherent presentations employing textual and visual material. *Artif. Intell. Rev.*, 9(2-3):147–165, 1995.

[70] E. André and T. Rist. Coping with Temporal Constraints in Multimedia Presentation Planning. In *Proc. of the 13th National Conf. on Artificial Intelligence; Portland, OR, USA*, pages 142–147, August 1996.

[71] N. Andrienko, G. Andrienko, and P. Gatalsky. Exploratory spatio-temporal visualization: an analytical review. *Journal of Visual Languages and Computing*, 14(6):503–541, December 2003.

[72] D. Androutsos, K. N. Plataniotis, and A. N. Venetsanopoulos. Directional detail histogram for image retrieval. *Electronics Letters*, 33(23):1935–1936, 1997.

[73] M. C. Angelides. Guest Editor's Introduction: Multimedia Content Modeling and Personalization. *IEEE Multimedia*, 10(4):12–15, October/December 2003.

[74] P. M. G. Apers, H. M. Blanken, and M. A. Houtsma. *Multimedia Databases in Perspective*. Springer–Verlag, 1997.

[75] A. Arampatzis, K. Zagoris, and S. A. Chatzichristofis. Dynamic two-stage image retrieval from large multimedia databases. *Information Processing & Management*, 49(1):274–285, January 2013.

[76] A. A. Argyriou. Sensors semiconductor, sensors, thermal, mechanical, magnetic, radiation sensors and chemical sensors, 2004.

[77] D. Arijon. *Grammar of the Film Language*. Silman-James Press, Los Angeles, CA, US, September 1991.

[78] B. Arnrich, C. Setz, R. La Marca, G. Troster, and U. Ehlert. What does your chair know about your stress level? *Information Technology in Biomedicine, IEEE Transactions on*, 14(2):207–214, 2010.

[79] B. Arons. SpeechSkimmer: Interactively skimming recorded speech. In *Proceedings of UIST'93: ACM Symposium on User Interface Software Technology*, pages 187–196, Atlanta, GA, November 1993. ACM Press.

[80] L. Aroyo, P. Bellekens, M. Björkman, J. Broekstra, and G.-J. Houben. Ontology-based personalization in user-adaptive systems. In *Proc. of the 2nd International Workshop on Web Personalization, Recommender Systems and Intelligent User Interfaces (WPRSIUI'06)*, pages 229–238, 2006.

[81] D. Arthur and S. Vassilvitskii. k-means++: the advantages of careful seeding. In *Proceedings of the Eighteenth Annual ACM-SIAM Symposium on Discrete Algorithms, January 7–9, New Orleans, Louisiana, USA*, pages 1027–1035, 2007.

[82] K. M. Asadi and J.-C. Dufour. Context-aware semantic adaptation of multimedia presentations. In *Proc. of the IEEE International Conference on Multimedia and Expo (ICME 2005), Amsterdam, The Netherlands*, pages 362–365, July 2005.

[83] K. Athanasakos, V. Stathopoulos, and J. Jose. A framework for evaluating automatic image annotation algorithms. *Lecture Notes in Computer Science*, 5993:217–228, 2010.

[84] T. Athanasiadis, P. Mylonas, Y. Avrithis, and S. Kollias. Semantic image segmentation and object labeling. *Circuits and Systems for Video Technology, IEEE Transactions on*, 17(3):298–312, March 2007.

[85] V. Athitsos, J. Alon, S. Sclaroff, and G. Kollios. Boostmap: A method for efficient approximate similarity rankings. In *Computer Vision and Pattern Recognition, 2004. CVPR 2004. Proceedings of the 2004 IEEE Computer Society Conference on*, pages II–268–II–275 Vol.2, 2004.

[86] S. Auer, C. Bizer, G. Kobilarov, J. Lehmann, R. Cyganiak, and Z. Ives. Dbpedia: A nucleus for a web of open data. *The Semantic Web*, pages 722–735, Berlin, Heidelberg, Springer–Verlag, 2007.

[87] M. Avlonitis, K. Chorianopoulos, and D. A. Shamma. Crowdsourcing user interactions within web video through pulse modeling. In *Proceedings of the ACM multimedia 2012 workshop on Crowdsourcing for multimedia*, pages 19–20, 2012.

[88] J. Ayars, D. Bulterman, A. Cohen, K. Day, E. Hodge, P. Hoschka, et al. Synchronized Multimedia Integration Language (SMIL 2.0) Specification – W3C Recommendation, August 2001. http://www.w3.org/TR/2001/REC-smil20-20010807/, last accessed: 22/1/2013.

[89] B. Ayers and M. Boutell. Home interior classification using sift keypoint histograms. In *Proceedings of IEEE International Conference on Computer Vision and Pattern Recognition*, pages 1–6, Minneapolis, Minnesota, USA, June 17-22 2007.

[90] M. Z. Aziz and B. Mertsching. Fast and robust generation of feature maps for region-based visual attention. *IEEE Transactions on Image Processing*, 17(5):633–644, 2008.

[91] I. A. Azzam, A. G. Charlapally, C. H. C. Leung, and J. F. Horwood. Content-based image indexing and retrieval with xml representations. In *Intelligent Multimedia, Video and Speech Processing, 2004. Proceedings of 2004 International Symposium on*, pages 181–185, 2004.

[92] I. A. Azzam, C. H C Leung, and J. F. Horwood. Implicit concept-based image indexing and retrieval. In *Multimedia Modelling Conference, 2004. Proceedings. 10th International*, pages 354–359, 2004.

[93] J. R. Bach, C. Fuller, A. Gupta, A. Hampapur, B. Horowitz, R. Humphrey, R. C. Jain, and C. F. Shu. Virage image search engine: an open framework for image management. vol. 2670, pages 76–87. SPIE, 1996.

[94] B. Bailey, J. A. Konstan, R. Cooley, and M. Dejong. Nsync – A toolkit for building interactive multimedia presentations. In *Proc. of the 6th ACM Int. Conf. on Multimedia; Bristol, UK*, pages 257–266. ACM, 1998.

[95] M. Balabanović and Y. Shoham. Fab: content-based, collaborative recommendation. *Commununications of the ACM*, 40(3):66–72, March 1997.

[96] N. Ballas, B. Labbé, A. Shabou, H. Le Borgne, P. Gosselin, M. Redi, B. Merialdo, H. Jégou, J. Delhumeau, R. Vieux, B. Mansencal, J. Benois-Pineau, S. Ayache, A. Hamadi, B. Safadi, F. Thollard, N. Derbas, G. Quenot, H. Bredin, M. Cord, B. Gao, C. Zhu, Y. Tang, E. Dellandrea, C.-E. Bichot, L. Chen, A. Benoit, P. Lambert, T. Strat, J. Razik, S. Paris, H. Glotin, T. N. Trung, D. Petrovska-Delacrétaz, G. Chollet, A. Stoian, and M. Crucianu. IRIM at TRECVID 2012: Semantic Indexing and Instance Search. In *Proceedings of the workshop on TREC Video Retrieval Evaluation (TRECVID'12)*, 2012.

[97] T. Ballendat, N. Marquardt, and S. Greenberg. Proxemic interaction: designing for a proximity and orientation-aware environment. In *ACM International Conference on Interactive Tabletops and Surfaces*, ITS '10, pages 121–130, New York, NY, USA, 2010. ACM.

[98] T. Balomenos, A. Raouzaiou, S. Ioannou, A. Drosopoulos, K. Karpouzis, and S. Kollias. Emotion analysis in man-machine interaction systems. In *Machine Learning for Multimodal Interaction*, pages 318–328. Springer, 2005.

[99] M. Banerjee, M. K. Kundu, and P. K. Das. Image retrieval with visually prominent features using fuzzy set theoretic evaluation. In *Visual Information Engineering (VIE). IET International Conference on*, pages 298–303. IET, 2006.

[100] S. Banerjee, C. Rose, and A. I. Rudnicky. The necessity of a meeting recording and playback system, and the benefit of topic-level annotations to meeting browsing. In *Proceedings of the 10th International Conference on Human-Computer Interaction (INTERACT'05)*, pages 643–656, Rome, Italy, September 2005. Springer.

[101] Z. Bangzuo, G. Yu, S. Haichao, L. Qingchao, and K. Jun. Survey of user behaviors as implicit feedback. In *Computer, Mechatronics, Control and Electronic Engineering (CMCE), 2010 International Conference on*, volume 6, pages 345–348, August 2010.

[102] K. Barnard, P. Duygulu, and D. Forsyth. Exploiting text and image feature co-occurrence statistics in large datasets. *Trends and Advances in Content-Based Image and Video Retrieval, Lecture Notes in Computer Science*, Springer, 2005.

[103] K. Barnard, P. Duygulu, D. Forsyth, N. de Freitas, D. M. Blei, and M. I. Jordan. Matching words and pictures. *J. Mach. Learn. Res.*, 3:1107–1135, March 2003.

[104] M. S. Bartlett, G. Littlewort, M. Frank, C. Lainscsek, I. Fasel, and J. Movellan. Fully automatic facial action recognition in spontaneous behavior. In *Automatic Face and Gesture Recognition, 2006. FGR 2006. 7th International Conference on*, pages 223–230. IEEE, 2006.

[105] C. Basu, H. Hirsh, and W. Cohen. Recommendation as classification: using social and content-based information in recommendation. In *Proceedings of 15th National Conference on Artificial Intelligence*, pages 714–720, Wisconsin, USA, July 1998.

[106] G. E. A. P. A. Batista, X. Wang, and E. J. Keogh. A complexity-invariant distance measure for time series. In *Proc. SIAM Conference on Data Mining*, pages 699–710, 2011.

[107] H. Bay, A. Ess, T. Tuytelaars, and L. V. Gool. Surf: Speeded up robust features. *Computer Vision and Image Understanding*, 110(3):346–359, June 2008.

[108] H. Bay, A. Ess, T. Tuytelaars, and L. Van Gool. Surf: Speeded up robust features. *Computer Vision and Image Understanding*, 110(3):346–359, 2008.

[109] L. J. Beaudoin, A. Sloman, and I. Wright. Towards a design-based analysis of emotional episodes. *Philosophy, Psychiatry, & Psychology*, 3(2):101–126, 1996.

[110] S. Bechhofer, F. van Harmelen, J. Hendler, I. Horrocks, D. McGuinness, P. Patel-Schneider, and L.A. Stein. *OWL Web Ontology Language Reference*. W3C Recommendation 10 February 2004. Latest version: http://www.w3.org/TR/owl-ref/, 2004.

[111] T. Becker. Visualizing time series data using web map service time dimension and SVG interactive animation. master thesis, International Institute for Geo-Information Science and Earth Observation (ITC), Twente, Netherlands, February 2009.

[112] D. Beckett and B. McBride. RDF/XML Syntax Specification (Revised). Recommendation, W3C, February 2004.

[113] M. Belk. Adapting generic web structures with semantic web technologies: A cognitive approach. In *Proceedings of the 4th International Workshop on Personalized Access, Profile Management, and Context Awareness in Databases, in conjunction with VLDB*, pages 35–40, 2010.

[114] S. Bengio. Multimodal speech processing using asynchronous hidden Markov models. *Information Fusion*, 5(2):81–89, 2004.

[115] T. L. Berg, A. Sorokin, G. Wang, D. A. Forsyth, D. Hoiem, I. Endres, and A. Farhadi. It's all about the data. *Proceedings of the IEEE*, 98(8):1434–1452, 2010.

[116] M. Berman. Modeling and Visualizing Historical GIS Data. In *Spatio-Temporal Workshop*, Cambridge, MA, US, April 2009.

[117] T. Berners-Lee. SWAP Pim Contacts Vocabulary. http://www.w3. org/2000/10/swap/pim/contact.

[118] M. Bertini, A. Del Bimbo, C. Torniai, C. Grana, and R. Cucchiara. Dynamic pictorial ontologies for video digital libraries annotation. In *Workshop on multimedia information retrieval on The many faces of multimedia semantics*, MS '07, pages 47–56, New York, NY, USA, ACM, 2007.

[119] J. Bidner and R. Burden. *Amphoto's Complete Book Of Photography: How to Improve Your Pictures With A Film Or Digital Camera.* Amphoto Books, 2004.

[120] I. Biederman. Recognition-by-components: A theory of human image understanding. *Psychological Review*, 94:115–147, 1987.

[121] N. bin Abdul Rashid, M. N. Taib, S. Lias, and N. Sulaiman. EEG analysis of frontal hemispheric asymmetry for learning styles. In *Control and System Graduate Research Colloquium (ICSGRC), 2011 IEEE*, pages 181–184, Shah Alam, Malaysia, June 2011. IEEE.

[122] Y. Blanco, J. J. Pazos, A. Gil, M. Ramos, A. Fernández, R. P. Díaz, M. ín López, and B. Barragáns. Avatar: an approach based on semantic reasoning to recommend personalized tv programs. In *Special interest tracks and posters of the 14th international conference on World Wide Web*, pages 1078–1079. ACM, 2005.

[123] Y. Blanco-Fernández, J. J. Pazos-Arias, A. Gil-Solla, M. Ramos-Cabrer, M. López-Nores, J. García-Duque, A. Fernández-Vilas, and R. P. Díaz-Redondo. Exploiting synergies between semantic reasoning and personalization strategies in intelligent recommender systems: A case study. *Journal of Systems and Software*, 81(12):2371–2385, 2008.

[124] D. M. Blei and M. I. Jordan. Modeling annotated data. In *Proceedings of the 26th annual international ACM SIGIR conference on Research and development in informaion retrieval*, SIGIR '03, pages 127–134, New York, NY, USA, ACM, 2003.

[125] D. M. Blei and P. J. Moreno. Topic segmentation with an aspect hidden Markov model. In *SIGIR '01: Proceedings of the 24th annual international ACM SIGIR conference on Research and development in information retrieval*, pages 343–348, New York, NY, USA, ACM, September 2001.

[126] A. Blum and T. Mitchell. Combining labeled and unlabeled data with co-training. In *Proceedings of the 11th Annual Conference on Computational Learning Theory*, pages 92–100, July 24–26 1998.

[127] A. F. Bobick and J. W. Davis. The recognition of human movement using temporal templates. *IEEE Transactions on Pattern Analysis Machine Intelligence*, 23(3):257–267, 2001.

[128] F. Bobillo, M. Delgado, and J. Gomez-Romero. DeLorean: A Reasoner for Fuzzy OWL 1.1. In *14th International Workshop on Uncertainty Reasoning for the Semantic Web (URSW 2008)*, 2008.

[129] F. Bobillo and U. Straccia. fuzzydl: An expressive fuzzy description logic reasoner. In *Fuzzy Systems, 2008. FUZZ-IEEE 2008.(IEEE World Congress on Computational Intelligence). IEEE International Conference on*, pages 923–930. IEEE, 2008.

[130] F. Bobillo and U. Straccia. Fuzzy ontology representation using OWL 2. *International Journal of Approximate Reasoning*, 52(7):1073–1094, October 2011.

[131] S. Bocconi, F. Nack, and L. Hardman. Automatic generation of matter-of-opinion video documentaries. *J. Web Sem.*, 6(2):139–150, 2008.

[132] O. Boiman, E. Shechtman, and M. Irani. In defense of nearest-neighbour based image classification. In *IEEE Conference on Computer Vision and Pattern Recognition (CVPR'08), June 24–26, Anchorage, Alaska, USA*, 2008.

[133] C. Bolchini, C. A. Curino, E. Quintarelli, F. A. Schreiber, and L. Tanca. A data-oriented survey of context models. *SIGMOD Rec.*, 36(4):19–26, December 2007.

[134] C. Bolchini, G. Orsi, E. Quintarelli, F. A. Schreiber, and L. Tanca. Context modeling and context awareness: steps forward in the context-addict project. *IEEE Data Engineering Bulletin*, 34(2):47–54, June 2011.

[135] S. Boll. *ZYX – Towards flexible multimedia document models for reuse and adaptation*. PhD thesis, Vienna University, Austria, 2001.

[136] S. Boll and W. Klas. ZYX – A Multimedia Document Model for Reuse and Adaptation. *IEEE Trans. on Knowledge and Data Engineering*, 13(3):361–382, 2001.

[137] S. Boll, W. Klas, C. Heinlein, and U. Westermann. Cardio-OP: Anatomy of a Multimedia Repository for Cardiac Surgery. Technical Report TR-2001301, University of Vienna, Austria, August 2001.

[138] S. Boll, W. Klas, and M. Löhr. Integrated database services for multimedia presentations. In S. M. Chung, editor, *Multimedia Information Storage and Management*, Chapter 16. Kluwer Academic Publishers, August 1996.

[139] S. Boll, W. Klas, and U. Westermann. Multimedia Document Formats – Sealed Fate or Setting Out for New Shores? *Proc. of the IEEE Int. Conf. on Multimedia Computing and Systems; Florence, Italy*, pages 604–610, June 1999. Republished in *MULTIMEDIA–TOOLS AND APPLICATIONS*, 11(3):267-279, 2000.

[140] K. Bollacker, C. Evans, P. Paritosh, T. Sturge, and J. Taylor. Freebase: a collaboratively created graph database for structuring human knowledge. In *Proceedings of the 2008 ACM SIGMOD international conference on Management of data*, pages 1247–1250. ACM, 2008.

[141] G. Boone. Concept features in RE:Agent, an intelligent email agent. In *Proceedings of 2nd International Conference on Autonomous Agents)*, pages 141–148, Minneapolis, USA, May 1998.

[142] M. Bordegoni, G. Faconti, S. Feiner, M. T. Maybury, T. Rist, S. Ruggieri, P. Trahanias, and M. Wilson. A standard reference model for intelligent multimedia presentation systems. *Computer Standards & Interfaces*, 18(6-7):477–496, December 1997.

[143] M. Bordegoni, G. Faconti, T. Rist, S. Ruggieri, P. Trahanias, and M. Wilson. Intelligent Multimedia Presentation Systems: A Proposal for a Reference Model. In *The Int. Conf. on Multi-Media Modeling; Toulouse, France*, pages 3–20, November 1996.

[144] A. Borji and L. Itti. State-of-the-art in visual attention modeling. *IEEE Transactions on Pattern Analysis Machine Intelligence*, 35(1), 2013.

[145] A. Bothin and P. Clough. Participants personal note-taking in meetings and its value for automatic meeting summarisation. *Information Technology and Management*, 13(1):39–57, March 2012.

[146] M. Bouamrane, A. Rector, and M. Hurrell. Using OWL ontologies for adaptive patient information modelling and preoperative clinical decision support. *Knowledge and Information Systems*, 29(2):405–418, November 2011.

[147] M.-M. Bouamrane, D. King, S. Luz, and M. Masoodian. A framework for collaborative writing with recording and post-meeting retrieval capabilities. *IEEE Distributed Systems Online*, page 6pp, November 2004. Special issue on the 6th International Workshop on Collaborative Editing Systems.

[148] M.-M. Bouamrane and S. Luz. Navigating multimodal meeting recordings with the meeting miner. In Gabriella Pasi Henrik Legind Larsen, Daniel Ortiz-Arroyo, Troels Andreasen, and Henning Christiansen, editors, *Flexible Query Answering Systems: 7th International Conference, FQAS 2006*, volume 4027 of *Lecture Notes in Artificial Intelligence*, pages 356–367, Milan. Italy, Springer, June 2006.

[149] M.-M. Bouamrane and S. Luz. Meeting browsing. *Multimedia Systems*, 12(4–5):439–457, March 2007.

[150] J.-Y. Bouguet. Pyramidal implementation of the Lucas–Kanade feature tracker. *Intel Corporation, Microprocessor Research Labs*, 2000.

[151] N. Boujemaa, J. Fauqueur, and V. Gouet. *Trends and Advances in Content-Based Image and Video Retrieval*, Chapter What's beyond query by example. 2005.

[152] H. Boujut, J. Benois-Pineau, T. Ahmed, O. Hadar, and P. Bonnet. A metric for no-reference video quality assessment for HD TV delivery based on saliency maps. In *IEEE International Conference on Multimedia and Expo (ICME'11), 11–15 July, Barcelona, Catalonia, Spain*, 2011.

[153] H. Boujut, J. Benois-Pineau, and R. Megret. Fusion of multiple visual cues for visual saliency extraction from wearable camera settings with strong motion. In *European Conference on Computer Vision - Workshops and Demonstrations (ECCV'12), October 7–13, Florence, Italy*, 2012.

[154] K. Boukadi, C. Ghedira, S. Chaari, L. Vincent, and E. Bataineh. CWSC4EC: How to employ context, web service, and community in enterprise collaboration. In *Proc. of the 8th International Conference on New Technologies in Distributed Systems (NOTERE'08)*, Lyon, France, pages 22–33, June 2008.

[155] Y. L. Boureau, F. Bach, Y. LeCun, and J. Ponce. Learning mid-level features for recognition. In *IEEE Conference on Computer Vision and Pattern Recognition (CVPR'10), June 13–18, San Francisco, CA, USA*, 2010.

[156] S. Bouyakoub and A. Belkhir. Smil builder: An incremental authoring tool for smil documents. *ACM Trans. Multimedia Comput. Commun. Appl.*, 7(1):2:1–2:30, February 2011.

[157] P. De Bra and D. Smits. A fully generic approach for realizing the adaptive web. In M. Bielikov, G. Friedrich, G. Gottlob, S. Katzenbeisser, and G. Turn, editors, *SOFSEM 2012: Theory and Practice of Computer Science*, vol. 7147 of *Lecture Notes in Computer Science*, pages 64–76. Springer Berlin Heidelberg, 2012.

[158] P. De Bra, J. Kay, and S. Weibelzahl. Introduction to the special issue on personalization. *IEEE Transactions on Learning Technologies*, 2(1):1–2, 2009.

[159] B. Braverman. *Video Shooter: Storytelling With DV, HD, And HDV Cameras*. DV Expert Series. Taylor & Francis Group, London, UK, October 2005.

[160] T. Bray, J. Paoli, C. M. Sperberg-McQueen, E. Maler, and F. Yergeau. Extensible Markup Language (XML) 1.0 (Fifth Edition). http://www.w3.org/TR/2008/REC-xml-20081126/, last accessed: 23/1/2013.

[161] J. S. Breese, D. Heckerman, and C. Kadie. Empirical analysis of predictive algorithms for collaborative filtering. In *Proceedings of the Fourteenth conference on Uncertainty in Artificial Intelligence*, UAI'98, pages 43–52, San Francisco, CA, USA, Morgan Kaufmann Publishers Inc, 1998.

[162] L. Breiman. Random forests. *Machine Learning*, 45:5–32, October 2001.

[163] L. Breiman, J. Friedman, C. J. Stone, and R. A. Olshen. *Classification and regression trees*. Chapman and Hall/CRC, 1984.

[164] S. L. Bressler and M. Ding. Event-related potentials. In Metin Akay, editor, *Wiley encyclopedia of biomedical engineering*. Wiley Online Library, Hoboken, New Jersey, USA, June 2006.

[165] P. Brezillon. Context-based modelling of operators practices by contextual graphs. In *Proc. of the 14th Mini Euro Conference in Human Centered Processes, Luxembourg*, pages 129–137, May 2003.

[166] D. Brickley. Basic Geo (WGS84 lat/long) Vocabulary. http://www.w3.org/2003/01/geo/.

[167] D. Brickley, L. Miller, T. Inkster, Y. Zeng, Y. Wang, D. Damljanovic, Z. Huang, S. Kinsella, J. Breslin, and B. Ferris. The Weighted Interests Vocabulary. http://smiy.sourceforge.net/wi/spec/weightedinterests.html.

[168] O. Brouard, V. Ricordel, and D. Barba. Cartes de Saillance Spatio-Temporelle basées Contrastes de Couleur et Mouvement Relatif. In *Compression et representation des signaux audiovisuels (CORESA)*, *Toulouse, France*, 2009.

[169] P. Brusilovsky. Adaptive Hypermedia: An Attempt to Analyze and Generalize. In P. A. M. Brusilovsky, P. Kommers and N. A. Streitz, editors, *First Int. Conf. Multimedia, Hypermedia, and Virtual Reality: Models, Systems, and Applications; Moscow, Russia*, vol. 1077 of *Lecture Notes in Computer Science*, pages 288–304. Springer–Verlag, September 1994.

[170] P. Brusilovsky. Methods and techniques of adaptive hypermedia. *User Modeling and User-Adapted Interaction*, 6(2-3):87–129, 1996.

[171] S. Buchholz, T. Hamann, and G. Hübsch. Comprehensive structured context profiles (CSCP): Design and experiences. In *Proc. of the Second IEEE Annual Conference on Pervasive Computing and Communications Workshops (PERCOMW), Washington, USA*, pages 43–47, 2004.

[172] S. Bugaj, D. Bulterman, B. Butterfield, W. Chang, G. Fouquet, C. Gran, et al. Synchronized Multimedia Integration Language (SMIL 1.0) Specification: W3C Recommendation, June 1998. http://www.w3.org/TR/REC-smil/, last accessed: 22/1/2013.

[173] D. Bulterman, J. Jansen, P. Cesar, S. Mullender, E. Hyche, M. De-Meglio, J. Quint, H. Kawamura, D. Weck, X. G. Paeda, D. Melendi, S. Cruz-Lara, M. Hanclik, D. F. Zucker, and T. Michel. Synchronized Multimedia Integration Language (SMIL 3.0): W3C Recommendation, December 2008. http://www.w3.org/TR/2008/REC-SMIL3-20081201/, last accessed: 22/1/2013.

[174] D. C. A. Bulterman. Embedded video in hypermedia documents: supporting integration and adaptive control. *ACM Trans. Inf. Syst.*, 13(4):440–470, 1995.

[175] D. C. A. Bulterman and L. Hardman. Structured multimedia authoring. *ACM Trans. on Multimedia Computing, Communications, and Applications*, 1(1):89–109, 2005.

[176] D. C. A. Bulterman, L. Rutledge, L. Hardman, and J. van Ossenbruggen. Supporting Adaptive and Adaptable Hypermedia Presentation Semantics. In *The 8th IFIP 2.6 Working Conf. on Database Semantics: Semantic Issues in Multimedia Systems; Rotorua, New Zealand*, January 1999.

[177] D. C. A. Bulterman and L. W. Rutledge. *SMIL 3.0: Flexible Multimedia for Web, Mobile Devices and Daisy Talking Books*. Springer Publishing Company, Incorporated, 2nd edition, 2008.

[178] D. C. A. Bulterman, G. van Rossum, and R. van Liere. A Structure of Transportable, Dynamic Multimedia Documents. In *Proc. of the Summer 1991 Usenix Conf.; Nashville, TN, USA*, 1991.

[179] A. Bur and H. Hügli. Optimal cue combination for saliency computation: A comparison with human vision. In *International Work-Conference on Nature Inspired Problem-Solving Methods in Knowledge Engineering, June 18–21, La Manga Del Mar Menor, Spain*, 2007.

[180] S. Burger, V. MacLaren, and H. Yu. The ISL meeting corpus: The impact of meeting type on speech style. In *Seventh International Conference on Spoken Language Processing (ICSLP)*, pages 301–304, Denver, CO, September 2002.

[181] L. Buriano, M. Marchetti, F. Carmagnola, F. Cena, C. Gena, and I. Torre. The role of ontologies in context-aware recommender systems. In *Proc. of the 7th International Conference on Mobile Data Management (MDM 2006)*, pages 80–80, Nara, Japan, May 2006.

[182] R. Burke. Hybrid recommender systems: Survey and experiments. *User Modeling and User-Adapted Interaction*, 12:331–370, 2002.

[183] R. Burke. *The Adaptive Web*. Chapter Hybrid web recommender systems, pages 377–408. Springer–Verlag, Berlin, Heidelberg, 2007.

[184] D. Cai, X. He, Z. Li, W.-Y. Ma, and J.-R. Wen. Hierarchical clustering of www image search results using visual, textual and link information. In *Proceedings of the 12th Annual ACM International Conference on Multimedia*, MULTIMEDIA '04, pages 952–959, New York, NY, USA, ACM, 2004.

[185] S. Calegari and E. Sanchez. A Fuzzy Ontology-Approach to improve Semantic Information Retrieval. In *Third ISWC Workshop on Uncertainty Reasoning for the Semantic Web – URSW'07*, 2007.

[186] A. Campbell, T. Choudhury, S. Hu, H. Lu, M. K. Mukerjee, M. Rabbi, and R. D. S. Raizada. Neurophone: brain-mobile phone interface using a wireless eeg headset. In *Proceedings of the Second ACM SIGCOMM Workshop on Networking, Systems, and Applications on Mobile Handhelds*, pages 3–8, New Delhi, India, Aug–Sep, ACM, 2010.

[187] N. W. Campbell, W. P. J. Mackeown, B. T. Thomas, and T. Troscianko. Interpreting image databases by region classification, 1997.

[188] I. Cantador, M. Fernández, D. Vallet, P. Castells, J. Picault, and M. Ribière. A multi-purpose ontology-based approach for personalised content filtering and retrieval. In *Advances in Semantic Media Adaptation and Personalization*, number 93 in *Studies in Computational Intelligence*, pages 25–51. Springer, 2008.

[189] Z. Cao, Z. Wu, and Y. Wang. UMQL: A unified multimedia query language. In *Signal-Image Technologies and Internet-Based System*, pages 109–115. IEEE, 2007.

[190] N. Caporusso, L. Mkrtchyan, and L. Badia. A multimodal interface device for online board games designed for sight-impaired people. *Information Technology in Biomedicine, IEEE Transactions on*, 14(2):248–254, 2010.

[191] F. Carbone, J. Contreras, and J. Z. Hernández. Enterprise 2.0 and semantic technologies for open innovation support. In *Proceedings of the 23rd International Conference on Industrial Engineering and Other Applications of Applied Intelligent Systems - Volume Part II*, IEA/AIE'10, pages 18–27, Berlin, Heidelberg, Springer–Verlag, 2010.

[192] G. Carneiro, A. B. Chan, P. J. Moreno, and N. Vasconcelos. Supervised learning of semantic classes for image annotation and retrieval. *IEEE Transactions on Pattern Analysis and Machine Intelligence*, 29(3):394–410, March 2007.

[193] G. Carneiro, A. B. Chan, P. J. Moreno, and N. Vasconcelos. Supervised learning of semantic classes for image annotation and retrieval. *Pattern Analysis and Machine Intelligence, IEEE Transactions on*, 29(3):394–410, 2007.

[194] J. J. Carroll, C. Bizer, P. J. Hayes, and P. Stickler. Semantic web publishing using named graphs. In *ISWC Workshop on Trust, Security, and Reputation on the Semantic Web*, 2004.

[195] C. Carson, S. Belongie, H. Greenspan, and J. Malik. Region-based image querying. In *Proceedings of the 1997 Workshop on Content-Based Access of Image and Video Libraries (CBAIVL '97)*, CAIVL '97, pages 42–, Washington, DC, USA, IEEE Computer Society, 1997.

[196] C. Carson, S. Belongie, H. Greenspan, and J. Malik. Blobworld: Image segmentation using expectation-maximization and its application to image querying. *IEEE Transactions on Pattern Analysis and Machine Intelligence*, 24:1026–1038, 1999.

[197] C. Carson, S. Belongie, H. Greenspan, and J. Malik. Blobworld: Image segmentation using expectation-maximization and its application to image querying. *IEEE Transactions on Pattern Analysis and Machine Intelligence*, 24:1026–1038, 1999.

[198] A. Celentano and O. Gaggi. Context-aware design of adaptable multimodal documents. *Multimedia Tools and Applications*, 29:7–28, 2006.

[199] M. Cha, H. Kwak, P. Rodriguez, Y.-Y. Ahn, and S. Moon. I tube, you tube, everybody tubes: analyzing the world's largest user generated content video system. In *Proc. ACM SIGCOMM Conference on Internet Measurement*, pages 1–14, 2007.

[200] B. Chandrasekaran, J. R. Josephson, and V. R. Benjamins. What are ontologies, and why do we need them? *IEEE Intelligent Systems*, pages 20–26, 1999.

[201] E. Chang, Kingshy G., G. Sychay, and Gang W. CBSA: content-based soft annotation for multimodal image retrieval using Bayes point machines. *Circuits and Systems for Video Technology, IEEE Transactions on*, 13(1):26–38, January 2003.

[202] S. F. Chang, T. Sikora, and A. Purl. Overview of the MPEG-7 standard. *Circuits and Systems for Video Technology, IEEE Transactions on*, 11(6):688–695, 2001.

[203] S.-K. Chang and A. Hsu. Image information systems: where do we go from here? *Knowledge and Data Engineering, IEEE Transactions on*, 4(5):431–442, October 1992.

[204] S. K. Chang and E. Jungert. Pictorial data management based upon the theory of symbolic projections. *Journal of Visual Languages and Computing*, 2(2):195–215, 1991.

[205] S. K. Chang, C. W. Yan, D. C. Dimitroff, and T. Arndt. An intelligent image database system. *Software Engineering, IEEE Transactions on*, 14(5):681–688, 1988.

[206] X. Chao and F. Zhiyong. A trusted affective model approach to proactive health monitoring system. In *Future BioMedical Information Engineering, 2008. FBIE'08. International Seminar on*, pages 429–432. IEEE, 2008.

[207] S. A. Chatzichristofis, A. Arampatzis, and Y. S. Boutalis. Investigating the behavior of compact composite descriptors in early fusion, late fusion and distributed image retrieval. *Radioengineering*, 19(4):725–733, December 2010.

[208] S. A. Chatzichristofis and Y. S. Boutalis. CEDD: Color and edge directivity descriptor: A compact descriptor for image indexing and retrieval. In A. Gasteratos, M. Vincze, and J. K. Tsotsos, editors, *Computer Vision Systems*, vol. 5008 of *Lecture Notes in Computer Science*, pages 312–322. Berlin Heidelberg, Springer–Verlag, May 2008.

[209] S. A. Chatzichristofis and Y. S. Boutalis. FCTH: fuzzy color and texture histogram - a low level feature for accurate image retrieval. In *Proceedings of 9th International Workshop on the Image Analysis for Multimedia Interactive Services*, pages 191–196, Klagenfurt, Austria, May 7–9 2008.

[210] S. A. Chatzichristofis and Y. S. Boutalis. Content based radiology image retrieval using a fuzzy rule based scalable composite descriptor. *Multimedia Tools and Applications*, 46:493–519, January 2010.

[211] S. A. Chatzichristofis and Y. S. Boutalis. *Compact Composite Descriptors for Content Based Image Retrieval: Basics, Concepts, Tools*. VDM Verlag, 2011.

[212] S. A. Chatzichristofis, Y. S. Boutalis, and M. Lux. SPCD - spatial color distribution descriptor – a fuzzy rule based compact composite descriptor appropriate for hand drawn color sketches retrieval. In *Proceedings of the 2nd International Conference on Agents and Artificial Intelligence*, pages 58–63, Valencia, Spain, January 22–24 2010.

[213] C. L. Chen, F. S. C. Tseng, and T. Liang. An integration of fuzzy association rules and wordnet for document clustering. *Knowledge and Information Systems*, 28(3):687–708, 2011.

[214] H. Chen, B. Schatz, T. Ng, J. Martinez, A. Kirchhoff, and C. Lin. A parallel computing approach to creating engineering concept spaces for semantic retrieval: The Illinois digital library initiative project. *IEEE Transactions on Pattern Analysis and Machine Intelligence*, 18:771–782, 1996.

[215] H. L. Chen. An analysis of image queries in the field of art history. *Journal of the American Society for Information Science and Technology*, 52(3):260–273, 2001.

[216] J. L. Chen and G. C. Stockman. Indexing to 3d model aspects using 2d contour features. In *Computer Vision and Pattern Recognition, 1996. Proceedings CVPR '96, 1996 IEEE Computer Society Conference on*, pages 913–920, 1996.

[217] L. Chen, G.-C. Chen, C.-Z. Xu, J. March, and S. Benford. Emoplayer: A media player for video clips with affective annotations. *Interact. Comput.*, 20(1):17–28, January 2008.

[218] L. Chen, R. Rose, Y. Qiao, I. Kimbara, F. Parrill, H. Welji, T. Han, J. Tu, Z. Huang, M. Harper, F. Quek, Y. Xiong, D. McNeill, R. Tuttle, and T. Huang. VACE multimodal meeting corpus. In Steve Renals and Samy Bengio, editors, *Proceedings of Machine Learning for Multimodal Interaction (MLMI)*, vol. 3869 of *Lecture Notes in Computer Science*, pages 40–51, Bethesda, MD, Springer, May 2006.

[219] S. Chen and P. Gopalakrishnan. Speaker, environment and channel change detection and clustering via the Bayesian information criterion. In *Proc. of DARPA Broadcast News Transcription and Understanding Workshop*, pages 127–132, Herndon, VA, February 1998.

[220] S. M. Chen and C. H. Chang. A new method to construct membership functions and generate weighted fuzzy rules from training instances. *Cybernetics and Systems*, 36(4):397–414, 2005.

[221] Y. Chen and J. Z. Wang. A region-based fuzzy feature matching approach to content-based image retrieval. *Pattern Analysis and Machine Intelligence, IEEE Transactions on*, 24(9):1252–1267, 2002.

[222] E. Cheng, F. Jing, M. Li, W. Ma, and H. Jin. Using implicit relevane feedback to advance web image search. In *Multimedia and Expo, 2006 IEEE International Conference on*, pages 1773–1776. IEEE, 2006.

[223] S. N. Cheong, H. S. Kam, K. M. Azhar, and M. Hanmandlu. Course Authoring and Management System for Interactive and Personalized

Multimedia Online Notes through XSL, XSLT, and SMIL. In *Interactive Computer Aided Learning; Villach, Austria*, September 2002.

[224] M. Chevalier, C. Soul-Dupuy, and P. Tchienehom. Semantics-based profiles modeling and matching for resources access. *Journal des Sciences pour l'Ingnieur*, 1(7):54–63, June 2006.

[225] J. S. Cho and J. Choi. Contour-based partial object recognition using symmetry in image databases. In *Proceedings of the 2005 ACM symposium on Applied computing*, SAC '05, pages 1190–1194, New York, NY, USA, ACM, 2005.

[226] M. Chock, A. F. Cardenas, and A. Klinger. Database structure and manipulation capabilities of a picture database management system (picdms). *Pattern Analysis and Machine Intelligence, IEEE Transactions on*, PAMI-6(4):484–492, 1984.

[227] Y. Choi and E. M. Rasmussen. Searching for images: the analysis of users' queries for image retrieval in American history. *J. Am. Soc. Inf. Sci. Technol.*, 54(6):498–511, April 2003.

[228] S. Christodoulakis, M. Foukarakis, L. Ragia, H. Uchiyama, and T. Imai. Semantic maps and mobile context capturing for picture content visualization and management of picture databases. In *Proceedings of the $7^{th}$ International Conference on Mobile and Ubiquitous Multimedia*, MUM '08, pages 130–136, New York, NY, USA, ACM, 2008.

[229] S. Christodoulakis, M. Foukarakis, L. Ragia, H. Uchiyama, and T. Imai. Picture Context Capturing for Mobile Databases. *MultiMedia, IEEE*, 17(2):34–41, April–June 2010.

[230] K. K. W. Chu and M. H. Wong. Fast time-series searching with scaling and shifting. In *PODS*, pages 237–248, 1999.

[231] E. F. Churchill, J. Trevor, S. Bly, L. Nelson, and D. Cubranic. Anchored Conversations: chatting in the context of a document. In *Proceedings of the CHI 2000 Conference on Human Factors in Computing Systems*, pages 454–461, The Hague, The Netherlands, ACM Press, April 2000.

[232] P. Cingolani and J. Alcala-Fdez. jfuzzylogic: a robust and flexible fuzzy-logic inference system language implementation. In *2012 IEEE International Conference on Fuzzy Systems*, pages 1–8, June 2012.

[233] H. H. Clark and S. E. Brennan. Grounding in communication. *Perspectives on Socially Shared Cognition*, 13(1991):127–149, 1991.

[234] I. Cohen, N. Sebe, A. Garg, L. S. Chen, and T. S. Huang. Facial expression recognition from video sequences: temporal and static modeling. *Computer Vision and Image Understanding*, 91(1):160–187, 2003.

[235] W. W. Cohen. Fast effective rule induction. In *12th International Conference on Machine Learning*, pages 115–123, Tahoe City, CA, USA, July 9–12 1995.

[236] C. Colombo, A. Del Bimbo, and P. Pala. Semantics in Visual Information Retrieval. *IEEE MultiMedia*, 6(3):38–53, 1999.

[237] P. Compieta, S. Di Martino, M. Bertolotto, F. Ferrucci, and T. Kechadi. Exploratory spatio-temporal data mining and visualization. *Journal of Visual Languages & Computing*, 18(3):255–279, 2007.

[238] O. Conlan. Novel components for supporting adaptivity in education systems - model-based integration approach. In *Proc. of the 8th ACM Int. Conf. on Multimedia; Marina del Rey, CA, USA*, pages 519–520. ACM, 2000.

[239] G. F. Cooper and E. Herskovits. A Bayesian method for the induction of probabilistic networks from data. *Machine Learning*, 9(4):309–347, 1992.

[240] C. Cornelis, J. Lu, X. Guo, and G. Zhang. One-and-only item recommendation with fuzzy logic techniques. *Information Sciences*, 177(22):4906–4921, 2007.

[241] P. Cousot. Methods and logics for proving programs. In *Handbook of Theoretical Computer Science, Volume B: Formal Models and Sematics (B)*, pages 841–994. Elsevier, 1990.

[242] I. J. Cox, M. L. Miller, T. P. Minka, T. V. Papathomas, and P. N. Yianilos. The Bayesian image retrieval system, pichunter: theory, implementation, and psychophysical experiments. *Image Processing, IEEE Transactions on*, 9(1):20–37, January 2000.

[243] D. Cremers, M. Rousson, and R. Deriche. A review of statistical approaches to level set segmentation: Integrating color, texture, motion and shape. *International Journal of Computer Vision*, 72:215, 2007.

[244] C. Crockford and H. Agius. An empirical investigation into user navigation of digital video using the VCR-like control set. *Int. J. Hum.-Comput. Stud.*, 64(4):340–355, 2006.

[245] K. Crowley, A. Sliney, I. Pitt, and D. Murphy. Evaluating a brain-computer interface to categorise human emotional response. In *Advanced Learning Technologies (ICALT), 2010 IEEE 10th International Conference on*, pages 276–278, Sousse, Tunisia, IEEE, July 2010.

[246] B. Cuenca-Grau, I. Horrocks, B. Motik, B. Parsia, P. F. Patel-Schneider, and U. Sattler. Owl2: the next step for OWL. *Journal of Web Semantics*, 6(4):309–322, 2008.

[247] C. Cusano, G. Ciocca, and R. Schettini. Image annotation using svm. In *Proceedings of Internet Imaging V*, vol. SPIE 5304, pages 330–338, December 22, 2003.

[248] Y. Zeng, Y. Wang, D. Damljanovic, Z. Huang, S. Kinsella, J. Breslin, B. Ferris, D. Brickley, L. Miller. The cognitive characteristics ontology 0.2., 2010. http://smiy.sourceforge.net/cco/spec/cognitivecharacteristics.html.

[249] D. Lin. An information-theoretic definition of similarity. In *Proceedings of 15th International Conference on Machine Learning*, pages 296–304, Madison, WI, USA, July 1998.

[250] E. Gonçalves da Silva, L. Ferreira Pires, and M. van Sinderen. Towards runtime discovery, selection and composition of semantic services. *Elsevier Computer Communications Journal, Special Issue: Open network service technologies and applications*, 34(2):159–168, February 2011.

[251] C. Dagli and T. S. Huang. A framework for grid-based image retrieval. In *Pattern Recognition, Proceedings of the 17th International Conference on*, vol. 2, pages 1021–1024. IEEE, 2004.

[252] Y. Dai. Semantic tolerance-based image representation for large image/video retrieval. In *Signal-Image Technologies and Internet-Based System, 2007. SITIS '07. Third International IEEE Conference on*, pages 1005–1012, 2007.

[253] M. Dalal, S. Feiner, K. McKeown, S. Pan, M. Zhou, T. Hllerer, J. Shaw, Y. Feng, and J. Fromer. Negotiation for automated generation of temporal multimedia presentations. In *Proc. of the 4th ACM Int. Conf. on Multimedia; Boston, MA, USA*, pages 55–64. ACM, 1996.

[254] N. Dalal and B. Triggs. Histograms of oriented gradients for human detection. In *Proceedings of International Conference on Computer Vision & Pattern Recognition*, pages 886–893, San Diego, CA, USA, June 20–25, 2005.

[255] S. J. Daly. Engineering observations from spatiovelocity and spatiotemporal visual models. In *IS&T/SPIE Conference on Human Vision and Electronic Imaging (HVEI'98), January 26–29, San Jose, CA, USA*, 1998.

[256] A. Damasio. The feeling of what happens: Body and emotion in the making of consciousness. 1999.

[257] A. R. Damasio. Descartes' error: Emotion, reason, and the human brain, 1994.

[258] G. Damaskos. Computer System for Analysis of Time-varying Micro-architecture of EEG sleep spindles. Diploma thesis, University of Central Greece, Greece, 2008.

[259] M. Damova, A. Kiryakov, K. Simov, and S. Petrov. Mapping the central LOD ontologies to proton upper-level ontology. In *Proceedings of the Fifth International Workshop on Ontology Matching*, pages 61–72, 2010.

[260] P. Y. Danet, editor. *Future Internet Strategic Research Agenda*. Future Internet X-ETP Group, Version 1.1 Edition, January 2010.

[261] M. d'Aquin, S. Elahi, and E. Motta. Semantic Monitoring of Personal Web Activity to Support the Management of Trust and Privacy. 2010.

[262] C. Darwin. *The expression of the emotions in man and animals*, volume 526. University of Chicago Press, 1965.

[263] R. Datta, D. Joshi, J. Li, and J. Z. Wang. Image retrieval: Ideas, influences, and trends of the new age. *ACM Computing Surveys*, 40(2):5:1–5:60, April 2008.

[264] R. Datta, D. Joshi, J. Li, and J. Z. Wang. Image retrieval: Ideas, influences, and trends of the new age. *ACM Computing Surveys*, 40(2):5, 2008.

[265] R. Datta, D. Joshi, J. Li, and J. Z. Wang. Image retrieval: Ideas, influences, and trends of the new age. *ACM Comput. Surv.*, 40(2):5:1–5:60, May 2008.

[266] I. Davis and E. Vitiello Jr. RELATIONSHIP: A vocabulary for describing relationships between people. RDF Vocabulary Specification (2005) http://vocab.org/relationship/.

[267] P. De Bra, A. Aerts, B. Berden, B. De Lange, B. Rousseau, T. Santic, D. Smits, and N. Stash. AHA! The adaptive hypermedia architecture. In *Proc. of the 14th ACM Conf. on Hypertext and hypermedia; Nottingham, UK*, pages 81–84, August 2003.

[268] P. De Bra, A. Aerts, D. Smits, and N. Stash. AHA! Version 2.0: More Adaptation Flexibility for Authors. In *Proc. of the AACE ELearn'2002 Conf.*, pages 240–246. Association for the Advancement of Computing in Education, VA, USA, October 2002.

[269] P. De Bra, P. Brusilovsky, and G.-J. Houben. Adaptive hypermedia: from systems to framework. *ACM Computing Surveys*, 31(4):12, December 1999.

[270] P. De Bra, G.-J. Houben, and H. Wu. AHAM: A Dexter-based Reference Model for Adaptive Hypermedia. In *Proc. of the 10th ACM Conf. on Hypertext and hypermedia: returning to our diverse roots; Darmstadt, Germany*, pages 147–156. ACM, 1999.

[271] M. De Gemmis, P. Lops, G. Semeraro, and P. Basile. Integrating tags in a semantic content-based recommender. In *Proceedings of the 2008 ACM conference on Recommender systems*, pages 163–170. ACM, 2008.

[272] F. M. G. De Jong, T. Westerveld, and A. P. De Vries. Multimedia search without visual analysis: the value of linguistic and contextual information. *Circuits and Systems for Video Technology, IEEE Transactions on*, 17(3):365–371, 2007.

[273] A. de Vicente and H. Pain. Informing the detection of the students motivational state: an empirical study. In *Intelligent tutoring systems*, pages 933–943, Biarritz, France and San Sebastian, Spain, Springer, June 2002.

[274] M. Dean. OWL-S: Semantic markup for web services. Submission, W3C, 2004.

[275] A. Del Bimbo, P. Pala, and S. Santini. Image retrieval by elastic matching of shapes and image patterns. In *Multimedia Computing and Systems, 1996., Proceedings of the Third IEEE International Conference on*, pages 215–218, 1996.

[276] B. Delezoide, F. Precioso, P. Gosselin, M. Redi, B. Merialdo, L. Granjon, D. Pellerin, M. Rombaut, H. Jégou, R. Vieux, B. Mansencal, J. Benois-Pineau, S. Ayache, B. Safadi, F. Thollard, G. Quénot, H. Bredin, M. Cord, A. Benoit, P. Lambert, T. Strat, J. Razik, S. Paris, and H. Glotin. IRIM at TRECVID 2011: Semantic Indexing and Instance Search. In *Proceedings of the workshop on TREC Video Retrieval Evaluation (TRECVID'11)*, 2011.

[277] Y. Deng, D. Mukherjee, and B. S. Manjunath. Netra-v: Towards an object-based video representation. *IEEE Transactions on Circuits and Systems for Video Technology*, 8:616–627, 1998.

[278] T. A. Dennis and B. Solomon. Frontal EEG and emotion regulation: Electrocortical activity in response to emotional film clips is associated with reduced mood induction and attention interference effects. *Biological Psychology*, 85(3):456, July 2010.

[279] M. Derdour, P. Roose, M. Dalmau, and N. Ghoualmi Zine. An adaptation platform for multimedia applications – CSC (component, service, connector). *Journal of Systems and Information Technology*, 14(1):4–22, June 2012.

[280] M. Derdour, N. Ghoualmi Zine, P. Roose, M. Dalmau, and A. Alti. UML-profile for multimedia software architectures. *International Journal of Multimedia Intelligence and Security*, 1(3):209–231, 2010.

[281] M. Deshpande and G. Karypis. Item-based top-$n$ recommendation algorithms. *ACM Trans. Inf. Syst.*, 22(1):143–177, 2004.

[282] A. K. Dey and G. D. Abowd. Towards a Better Understanding of Context and Context-Awareness. In *Proc. of the CHI 2000 Workshop on the What, Who, Where, When, Why and How of Context-Awareness)*. ACM Press, 2000.

[283] S. Dickinson, A. Pentland, and S. Stevenson. Viewpoint-invariant indexing for content-based image retrieval. In *Proc. IEEE Int. Workshop on Content-Based Access of Image and Video Database*, pages 20–30, 1998.

[284] A. Dielmann and S. Renals. Automatic meeting segmentation using dynamic Bayesian networks. *IEEE Transactions on Multimedia*, 9(1):25–36, 2007.

[285] A. Dielmann and S. Renals. Recognition of dialogue acts in multiparty meetings using a switching dbn. *IEEE Transactions on Audio, Speech, and Language Processing*, 16(7):1303–1314, 2008.

[286] H. Ding, G. Trajcevski, P. Scheuermann, X. Wang, and E. Keogh. Querying and mining of time series data: experimental comparison of representations and distance measures. *Proc. VLDB Endowment*, 1(2):1542–1552, 2008.

[287] Y. Ding and X. Li. Time weight collaborative filtering. In *Proceedings of 14th ACM International Conference on Information and Knowledge Management*, pages 485–492, Bremen, Germany, October 2005.

[288] Y. Djaghloul, V. Groués, and Y. Naudet. Combining Situation and Content similarities in Fuzzy based Interest Matchmaking Mechanism. In *Proc. of the 7th International Workshop on Semantic and Social Media Adaptation and Personalization (SMAP2012)*, Luxembourg, December 2012.

[289] C. Djeraba. Association and content-based retrieval. *Knowledge and Data Engineering, IEEE Transactions on*, 15(1):118–135, January–February 2003.

[290] C. Djeraba. Association and content-based retrieval. *Knowledge and Data Engineering, IEEE Transactions on*, 15(1):118–135, 2003.

[291] D. Djordjevic and E. Izquierdo. An object- and user-driven system for semantic-based image annotation and retrieval. *Circuits and Systems*

*for Video Technology, IEEE Transactions on*, 17(3):313–323, March 2007.

[292] A. H. Doan, J. Madhavan, R. Dhamankar, P. Domingos, and A. Halevy. Learning to match ontologies on the semantic web. *The VLDB Journal*, 12(4):303–319, 2003.

[293] M. Doller, A. Yakou, R. Tous, J. Delgado, M. Gruhne, Miran Choi, and Tae-Beom Lim. Semantic MPEG query format validation and processing. *MultiMedia, IEEE*, 16(4):22–33, 2009.

[294] M. Dorr, T. Martinetz, K. R. Gegenfurtner, and E. Barth. Variability of eye movements when viewing dynamic natural scenes. *Journal of Vision*, 10(10), 2010.

[295] C. Dromzee, S. Laborie, and P. Roose. A semantic generic profile for multimedia documents adaptation. *Intelligent Multimedia Technologies for Networking Applications: Techniques and Tools*, pages 225–246, January 2013.

[296] S. Drucker, T. Galyean, and D. Zeltzer. Cinema: a system for procedural camera movements. In *Proceedings of the 1992 Symposium on Interactive 3D graphics*, I3D '92, pages 67–70, New York, NY, USA, ACM, 1992.

[297] S. M. Drucker, A. Glatzer, S. D. Mar, and C. Wong. Smartskip: consumer level browsing and skipping of digital video content. In D. R. Wixon, editor, *CHI*, pages 219–226. ACM, 2002.

[298] A. Duda and C. Keramane. Structured Temporal Composition of Multimedia Data; Blue Mountain Lake, NY, USA. In *Proc. of the IEEE Int. Workshop Multimedia-Database-Management Systems*, pages 136–142. IEEE, 1995.

[299] L. Duen-Ren and S. Ya-Yueh. Hybrid approaches to product recommendation based on customer lifetime and purchase preferences. *The Journal of Systems and Software*, 77(1):181–191, August 2005.

[300] S. Dumais, E. Cutrell, J. J. Cadiz, G. Jancke, R. Sarin, and D. C. Robbins. Stuff I've seen: a system for personal information retrieval and re-use. In *Proceedings of the 26th annual international ACM SIGIR conference on Research and development in information retrieval*, SIGIR '03, pages 72–79, New York, NY, USA, ACM, 2003.

[301] Z. Duric, W. D Gray, R. Heishman, F. Li, A. Rosenfeld, M. J Schoelles, C. Schunn, and H. Wechsler. Integrating perceptual and cognitive modeling for adaptive and intelligent human-computer interaction. *Proceedings of the IEEE*, 90(7):1272–1289, 2002.

[302] P. J. Durka. Matching pursuit. *Scholarpedia*, 2(11):2288, 2007.

[303] P. Duygulu, K. Barnard, J. F. G. de Freitas, and A. D. Forsyth. Object recognition as machine translation: learning a lexicon for a fixed image vocabulary. In *Proceedings of the 7th European Conference on Computer Vision-Part IV*, pages 97–112, Copenhagen, Denmark, May 28-31 2002.

[304] P. Duygulu, M. Bastan, and D. Forsyth. Translating images to words for recognizing objects in large image and video collections. In Jean Ponce, Martial Hebert, Cordelia Schmid, and Andrew Zisserman, editors, *Toward Category-Level Object Recognition*, vol. 4170 of *Lecture Notes in Computer Science*, pages 258–276. Berlin Heidelberg, Springer–Verlag, 2006.

[305] J. P. Eakins and M. E. Graham. Content-based image retrieval. *Technical Report JTAP-039, JISC Technology Application Program, University of Northumbria, UK.*, (10):1–10, 2000.

[306] M. J. Egenhofer and R. Franzosa. Point-Set Topological Spatial Relations. *Int. Journal of Geographic Information Systems*, 5(2):161–174, March 1991.

[307] C. Eickhoff and A. P. De Vries. How crowdsourcable is your task? In *Workshop on Crowdsourcing for Search and Data Mining*, pages 11–14, Hong Kong, February 9 2011.

[308] H. Ekanayake. P300 and emotiv epoc: Does emotiv epoc capture real EEG?

[309] R. El Kaliouby and P. Robinson. Real-time inference of complex mental states from facial expressions and head gestures. In *In Proc. Intl Conf. Computer Vision & Pattern Recognition*, 2004.

[310] M. Elhadad, S. Feiner, K. McKeown, and D. Seligmann. Generating customized text and graphics in the COMET explanation testbed. In *Proc. of the 23rd conf. on Winter simulation; Phoenix, AZ, USA*, pages 1058–1065. IEEE, 1991.

[311] Encyclopaedia Britannica, Inc., Chicago, USA. 2005 Ultimate Reference Suite DVD, 2005.

[312] P. G. B. Enser, C. J. Sandom, and P. H. Lewis. Surveying the reality of semantic image retrieval. In *Proceedings of the 8th international conference on Visual Information and Information Systems*, VISUAL'05, pages 177–188, Berlin, Heidelberg, Springer–Verlag, 2006.

[313] E. L. G. Escovar, C. A. Yaguinima, and M. Biajic. Using Fuzzy Ontologies to Extend Semantically Similar Data Mining. In *XXI Simposio*

*Brasileiro de Banco de Dados (SBBD), 2006, Florianopolis (SC)*, pages 16–30, 2006.

[314] E. T. Esfahani and V. Sundararajan. Using brain–computer interfaces to detect human satisfaction in human–robot interaction. *Int. J. Human. Robot.*, 8(01):87–101, March 2011.

[315] O. Etzion. Event Processing - past, present and future. In *Proceedings of VLDB*, vol. 3, pages 1651–1652, Singapore, 2010.

[316] M. Everingham, L. Van Gool, C. K. I. Williams, J. Winn, and A. Zisserman. The Pascal visual object classes challenge. *International Journal of Computer Vision*, 88(2):303–338, June 2010.

[317] F. M. Facca and P. L. Lanzi. Mining interesting knowledge from weblogs: a survey. *Data Knowl. Eng.*, 53(3):225–241, June 2005.

[318] G. P. Faconti and T. Rist, editors. *Proc. of ECAI-96 Workshop Towards a Standard Reference Model for Intelligent Multimedia Presentation Systems; Budapest, Hungary*, 1996.

[319] K. Falkovych and F. Nack. Context aware guidance for multimedia authoring: harmonizing domain and discourse knowledge. *Multimedia Syst.*, 11(3):226–235, 2006.

[320] K. Falkovych, F. Nack, J. van Ossenbruggen, and L. Rutledge. Sample: Towards a framework for system-supported multimedia authoring. In Yi-Ping Phoebe Chen, editor, *10th International Multimedia Modeling Conference; Brisbane, Australia*, pages 362–362. IEEE Computer Society, 2004.

[321] K. Falkovych, J. Werner, and F. Nack. Semantic-based support for the semi-automatic construction of multimedia presentations. In *1st Int. Workshop on Interaction Design and the Semantic Web*, May 2004.

[322] C. Faloutsos, R. Barber, M. Flickner, J. Hafner, W. Niblack, D. Petkovic, and W. Equitz. Efficient and Effective Querying by Image Content. *Journal of Intelligent Information Systems*, 3(3/4):231–262, 1994.

[323] J. Fan, Y. Gao, H. Luo, and R. Jain. Mining multilevel image semantics via hierarchical classification. *Multimedia, IEEE Transactions on*, 10(2):167–187, 2008.

[324] R.-E. Fan, P.-H. Chen, and C.-J. Lin. Working set selection using second order information for training support vector machines. *Journal of Machine Learning Research*, 6:1889–1918, December 2005.

[325] A. Fathi, Y. Li, and J. M. Rehg. Learning to recognize daily actions using gaze. In *European Conference on Computer Vision (ECCV'12), October 7–13, Florence, Italy*, 2012.

[326] F. Fauzi, J.-L. Hong, and M. Belkhatir. Webpage segmentation for extracting images and their surrounding contextual information. In *Proceedings of the ACM International Conference on Multimedia*, pages 649–652, Vancouver, Canada, October 19–24 2009.

[327] M. Fayzullin and V. S. Subrahmanian. An algebra for powerpoint sources. *Multimedia Tools Appl.*, 24(3):273–301, 2004.

[328] S. K. Feiner and K. R. McKeown. Automating the generation of coordinated multimedia explanations. In Maybury [655], Chapter 5, pages 117–138.

[329] H. Feng, R. Shi, and T.-S. Chua. A bootstrapping framework for annotating and retrieving www images. In *Proceedings of the 12th Annual ACM International Conference on Multimedia*, pages 960–967, New York, NY, USA, October 10–16 2004.

[330] H. Feng, R. Shi, and T.S. Chua. A bootstrapping framework for annotating and retrieving www images. In *Proceedings of the 12th Annual ACM International Conference on Multimedia*, MULTIMEDIA '04, pages 960–967, New York, NY, USA, ACM, 2004.

[331] S. L. Feng, R. Manmatha, and V. Lavrenko. Multiple bernoulli relevance models for image and video annotation. In *Proceedings of the IEEE Computer Society Conference on Computer Vision and Pattern Recognition*, pages 1002–1009, Washington, DC, USA, June 27–July 2, 2004.

[332] R. Fergus, P. Perona, and A. Zisserman. Object class recognition by unsupervised scale-invariant learning. In *Computer Vision and Pattern Recognition, 2003. Proceedings. 2003 IEEE Computer Society Conference on*, vol. 2, pages II–264 – II–271 vol.2, June 2003.

[333] J. Fink and A. Kobsa. User Modeling for Personalized City Tours. *Artificial Intelligence Review*, 18(1):33–74, 2002.

[334] J. Fink, A. Kobsa, and J. Schreck. Personalized hypermedia information through adaptive and adaptable system features: User modeling, privacy and security issues. In A. Mullery, M. Besson, M. Campolargo, R. Gobbi, and R. Reed, editors, *Intelligence in Services and Networks: Technology for Cooperative Competition*, pages 459–467. Springer–Verlag, 1997.

[335] J. G. Fiscus, J. Ajot, and J. S. Garofolo. The rich transcription 2007 meeting recognition evaluation. In *Multimodal Technologies for Perception of Humans*, pages 3–34, Baltimore, MD, Springer, May 2008.

[336] R. Ribon Fletcher, K. Dobson, M. S Goodwin, H. Eydgahi, O. Wilder-Smith, D. Fernholz, Y. Kuboyama, E. B. Hedman, M.-Z. Poh, and R. W Picard. icalm: Wearable sensor and network architecture for wirelessly communicating and logging autonomic activity. *Information Technology in Biomedicine, IEEE Transactions on*, 14(2):215–223, 2010.

[337] A. Fleury, M. Vacher, and N. Noury. SVM-based multimodal classification of activities of daily living in health smart homes: sensors, algorithms, and first experimental results. *Information Technology in Biomedicine, IEEE Transactions on*, 14(2):274–283, 2010.

[338] S. Fodeh, B. Punch, and P. N. Tan. On ontology-driven document clustering using core semantic features. *Knowledge and Information Systems*, 28(2):395–421, August 2011.

[339] S. Forough and J. Reza. A comparative study of context modeling approaches and applying in an infrastructure. *Canadian Journal on Data Information and Knowledge Engineering*, 3(1):1–6, January 2012.

[340] M. Forte and W. Lopez Souza. Using ontologies and web services for content adaptation in ubiquitous computing. *Journal of Systems and Software*, 81(3):368–381, March 2008.

[341] Wireless Application Protocol Forum. WAG UAProf. Technical report, WAP Forum, October 2001.

[342] J. G. K. Foss and A. I. Cristea. The next generation authoring adaptive hypermedia: using and evaluating the MOT3.0 and PEAL tools. In *Proceedings of the 21st ACM Conference on Hypertext and Hypermedia*, HT '10, pages 83–92, New York, NY, USA, ACM, 2010.

[343] M. Foulonneau. Smart Semantic Content for the Future Internet. In Elena García Barriocanal, Zeynel Cebeci, Mehmet C. Okur, and Aydin Öztürk, editors, *Metadata and Semantic Research – Proceedings of the 5th International Conference (MTSR 2011)*, vol. 240 of *Communications in Computer and Information Science*, pages 145–154. Springer Berlin Heidelberg, Izmir, Turkey, October 2011.

[344] R. François, R. Thierry, and T. Joëlle. HuMaFace : Human to machine facial animation. In Thierry Dutoit, editor, *QPSR of the numediart research program*, volume 5, pages 1–5. UMONS/numediart, March 2012.

[345] C. A. Frantzidis, C. Bratsas, M. A. Klados, E. Konstantinidis, C. D. Lithari, A. B. Vivas, C. L. Papadelis, E. Kaldoudi, C. Pappas, and P. D. Bamidis. On the classification of emotional biosignals evoked while viewing affective pictures: an integrated data-mining-based approach for healthcare applications. *Information Technology in Biomedicine, IEEE Transactions on*, 14(2):309–318, 2010.

[346] C. A. Frantzidis, C. Bratsas, C. L. Papadelis, E. Konstantinidis, C. Pappas, and P. D. Bamidis. Toward emotion aware computing: an integrated approach using multichannel neurophysiological recordings and affective visual stimuli. *Information Technology in Biomedicine, IEEE Transactions on*, 14(3):589–597, 2010.

[347] C. Freksa. Temporal reasoning based on semi-intervals. *Artif. Intell.*, 54(1-2):199–227, 1992.

[348] O. Frykholm, A. Lantz, K. Groth, and A. Walldius. Medicine meets engineering in cooperative design of collaborative decision-supportive system. In *Porceedings of the 23rd International Symposium on Computer-Based Medical Systems (CBMS)*, pages 116–121, Perth, Australia, IEEE, May 2010.

[349] B. Fuhrt. Multimedia Systems: An Overview. *IEEE MultiMedia*, 1(1):47–59, 1994.

[350] H. Fujimoto, M. Etoh, A. Kinno, and Y. Akinaga. Web user profiling on proxy logs and its evaluation in personalization. In *Proceedings of the 13th Asia-Pacific Web Conference on Web Technologies and Applications*, APWeb'11, pages 107–118, Berlin, Heidelberg, Springer–Verlag, 2011.

[351] K. Fukunaga. *Introduction to Statistical Pattern Recognition*. Academic Press, 1990.

[352] J. Fürnkranz and G. Widmer. Incremental reduced error pruning. In *International Conference on Machine Learning*, pages 70–77, New Brunswick, NJ, USA, July 10–15 1994.

[353] C. Gaffney, D. Dagger, and V. Wade. A survey of soft skill simulation authoring tools. In *Proceedings of the Nineteenth ACM Conference on Hypertext and Hypermedia*, HT '08, pages 181–186, New York, NY, USA, ACM, 2008.

[354] V. Galant and M. Paprzycki. Information personalization in an Internet based travel support system. In *Proceedings of the International Conference on Business Information*, pages 191–202, Poznan, Poland, May 2002.

[355] M. Galley, K. R. McKeown, E. Fosler-Lussier, and H. Jing. Discourse segmentation of multi-party conversation. In Erhard Hinrichs and Dan Roth, editors, *Proceedings of the 41st Annual Meeting of the Association for Computational Linguistics*, pages 562–569, Sapporo, Japan, July 2003.

[356] T. A. Galyean. Narrative Guidance. In *Proceedings of the AAAI Spring Symposium Series, Symposium on Interactive Story Systems: Plot and Character*, pages 52–55, Stanford, CA, US, 1995.

[357] J. S. Garofolo, C. D. Laprun, M. Michel, V. M. Stanford, and E. Tabassi. The NIST meeting room pilot corpus. In *Proc. 4th Intl. Conf. on Language Resources and Evaluation (LREC)*, pages 1411–1414, Lisbon, Portugal, ELRA, May 2004.

[358] S. Gauch, M. Speretta, A. Chandramouli, and A. Micarelli. The adaptive web. chapter User profiles for personalized information access, pages 54–89. Springer–Verlag, Berlin, Heidelberg, 2007.

[359] J. Gausemeier, J. Fruend, C. Matysczok, B. Bruederlin, and D. Beier. Development of a real time image based object recognition method for mobile ar-devices. In *Proceedings of the 2nd International Conference on Computer Graphics, Virtual Reality, Visualisation and Interaction in Africa*, AFRIGRAPH '03, pages 133–139, New York, NY, USA, ACM, 2003.

[360] J. Geurts. Constraints for multimedia presentation generation. Master's thesis, University of Amsterdam, Amsterdam, The Netherlands, January 2002.

[361] J. Geurts, J. van Ossenbruggen, and L. Hardman. Application-specific constraints for multimedia presentation generation. In *Proc. of the Int. Conf. on Multimedia Modeling; Amsterdam, The Netherlands*, pages 247–266. IEEE, November 2001.

[362] T. Gevers and A. W. M. Smeulders. Pictoseek: combining color and shape invariant features for image retrieval. *Image Processing, IEEE Transactions on*, 9(1):102–119, 2000.

[363] H. Ghasemzadeh, R. Jafari, and B. Prabhakaran. A body sensor network with electromyogram and inertial sensors: multimodal interpretation of muscular activities. *Information Technology in Biomedicine, IEEE Transactions on*, 14(2):198–206, 2010.

[364] A. Ghoshal, P. Ircing, and S. Khudanpur. Hidden markov models for automatic annotation and content-based retrieval of images and video. In *Proceedings of the 28th Annual International ACM SIGIR Conference on Research and Development in Information Retrieval*, pages 544–551, Salvador, Brazil, August 15–19 2005.

[365] G. A. Giannakakis, N. N. Tsiaparas, M.-F.S. Xenikou, C. Papageorgiou, and K. S. Nikita. Wavelet entropy differentiations of event related potentials in dyslexia. In *BioInformatics and BioEngineering, 2008. BIBE 2008. 8th IEEE International Conference on*, pages 1–6. IEEE, 2008.

[366] S. J. Gibbs and D. C. Tsichritzis. *Multimedia Programming: Objects, Environments and Frameworks*. ACM Press Books. Addison-Wesley, 2nd edition, 1995.

[367] B. Girma, L. Brunie, and J.-M. Pierson. Planning-based multimedia adaptation services composition for pervasive computing. In *Proc. of the 2nd International Conference on Signal-Image Technology and Internet-Based Systems (SITIS'09), Marrakech, Morocco*, pages 326–331, November 2009.

[368] C. Gkonela and K. Chorianopoulos. Videoskip: event detection in social web videos with an implicit user heuristic. *Multimedia Tools and Applications*, pages 1–14, 2012.

[369] M. Gleicher, R. Heck, and M. Wallick. A framework for virtual videography. In *Proceedings of the 2nd International Symposium on Smart Graphics*, SMARTGRAPH '02, pages 9–16, New York, NY, USA, ACM, 2002.

[370] B. Goldberg, R. Sottilare, K. Brawner, and H. Holden. Predicting learner engagement during well-defined and ill-defined computer-based inter-cultural interactions. In *Affective Computing and Intelligent Interaction*, pages 538–547. Springer, October 2011.

[371] D. Goldberg, D. Nichols, B. M. Oki, and D. Terry. Using collaborative filtering to weave an information tapestry. *Communications of the ACM*, 35(12):70, 1992.

[372] A. Gomez-Perez, O. Corcho, and M. Fernandez-Lopez. *Ontological Engineering*. Springer–Verlag London Limited, 2004.

[373] R. C. Gonzalez and R. E. Woods. *Digital Image Processing*. Addison-Wesley Longman Publishing Co., Inc., Boston, MA, USA, 2nd edition, 2001.

[374] J. Gonzalez-Sanchez, M. E. Chavez-Echeagaray, R. Atkinson, and W. Burleson. Abe: An agent-based software architecture for a multimodal emotion recognition framework. In *Proc. of 9th Working IEEE/IFIP Conference on Software Architecture (WICSA 11)*, pages 187–193, Boulder, CO, USA, IEEE, June 2011.

[375] Google. Introducing the knowledge graph: things, not strings. http://googleblog.blogspot.gr/2012/05/introducing-knowledge-graph-things-not.html.

[376] R. Götze, D. Boles, and H. Eirund. Multimedia user interfaces. In M. Sonnenschein, editor, *Final Report Arbeitsgruppe Informatik- Systeme*, Chapter 6, pages 120–165. Carl von Ossietzky University, Oldenburg, Germany, November 1996.

[377] K. Grauman and T. Darrell. The pyramid match kernel: Discriminative classification with sets of image features. In *IEEE International Conference on Computer Vision (ICCV'05), October 17–20, Beijing, China*, 2005.

[378] J. D. Green and A. Arduini. Hippocampal activity in arousal. *Journal of Neurophysiology*, 17(6):533–557, November 1954.

[379] S. Greenberg, N. Marquardt, T. Ballendat, R. Diaz-Marino, and M. Wang. Proxemic interactions: the new ubicomp? *interactions*, 18(1):42–50, January 2011.

[380] C. Greiner and T. Rose. A Web Based Training System for Cardiac Surgery: The Role of Knowledge Management for Interlinking Information Items. In *Proc. The World Congress on the Internet in Medicine; London, UK*, November 1998.

[381] G. A. Grimnes, P. Edwards, and A. Preece. Instance based clustering of semantic web resources. In *Proceedings of the 5th European Semantic Web Conference on the Semantic Web: Research and Applications*, ESWC'08, pages 303–317, Berlin, Springer–Verlag, 2008.

[382] B. N. Grosof, I. Horrocks, R. Volz R., and S. Decker. Description logic programs: Combining logic programs with description logic. In *Proc. of the Twelfth International World Wide Web Conference (WWW 2003)*, pages 48–57. ACM, 2003.

[383] V. Grouès, Y. Naudet, and O. Kao. Combining Linguistic Values and Semantics to Represent User Preferences. In *Proceedings of the 6th International Workshop on Semantic Media Adaptation and Personalization (SMAP2011)*, SMAP '11, pages 27–32, Washington, DC, USA, IEEE Computer Society, 2011.

[384] V. Grouès, Y. Naudet, and O. Kao. Adaptation and Evaluation of a Semantic Similarity Measure for DBPedia: A First Experiment. In *Proceedings of the 7th International Workshop on Semantic and Social Media Adaptation and Personalization (SMAP2012)*, SMAP '12, Washington, DC, USA, December 3-4, 2012.

[385] H. Gu and Q. Ji. An automated face reader for fatigue detection. In *Automatic Face and Gesture Recognition, 2004. Proceedings. Sixth IEEE International Conference on*, pages 111–116. IEEE, 2004.

[386] V. N. Gudivada and V. V. Raghavan. Design and evaluation of algorithms for image retrieval by spatial similarity. *ACM Trans. Inf. Syst.*, 13(2):115 144, April 1995.

[387] H. Gunes and M. Piccardi. Automatic temporal segment detection and affect recognition from face and body display. *Systems, Man, and Cybernetics, Part B: Cybernetics, IEEE Transactions on*, 39(1):64–84, 2009.

[388] J. Gwizdka and M. J. Cole. Inferring cognitive states from multimodal measures in information science. In *Proc. of ICMI Workshop*, Alicante, Spain, November 2011.

[389] A. Haag, S. Goronzy, P. Schaich, and J. Williams. Emotion recognition using bio-sensors: First steps towards an automatic system. In *Affective Dialogue Systems*, pages 36–48. Springer, 2004.

[390] L. F. Haas. Hans Berger (1873–1941), Richard Caton (1842–1926), and electroencephalography. *Journal of Neurology, Neurosurgery & Psychiatry*, 74(1):9–9, January 2003.

[391] J. Hafner, H. S. Sawhney, W. Equitz, M. Flickner, and W. Niblack. Efficient color histogram indexing for quadratic form distance functions. *IEEE Transactions on Pattern Analysis and Machine Intelligence*, 17(7):729–736, July 1995.

[392] G. Hakvoort, H. Gürkök, D. Plass-Oude Bos, M. Obbink, and M. Poel. Measuring immersion and affect in a brain-computer interface game. In *Human-Computer Interaction–INTERACT 2011*, pages 115–128, Lisbon, Portugal, Springer, September 2011.

[393] H. Hamdi, P. Richard, A. Suteau, and M. Saleh. Virtual reality and affective computing techniques for face-to-face communication. In *Proc. of the International Conference on Computer Graphics Theory and Applications*, pages 357–360, Vilamoura, Algarve, Portugal, March 2011.

[394] D. C. Hammond. What is neurofeedback? *Journal of Neurotherapy*, 10(4):25, September 2006.

[395] A. Hanbury. A survey of methods for image annotation. *Journal of Visual Languages & Computing*, 19(5):617–627, October 2008.

[396] A. Hanjalic, R. Lienhart, W.-Y. Ma, and J. R. Smith. The holy grail of multimedia information retrieval: So close or yet so far away? *Proceedings of the IEEE*, 96(4):541–547, 2008.

[397] L. Hardman. *Modeling and Authoring Hypermedia Documents*. PhD thesis, University of Amsterdam, Amsterdam, The Netherlands, March 1998.

[398] L. Hardman, D. C. A. Bulterman, and G. van Rossum. The Amsterdam hypermedia model: adding time and context to the Dexter model. *Communications of the ACM*, 37(2):50–62, February 1994.

[399] L. Hardman, G. van Rossum, J. Jansen, and S. Mullender. CMIFed: a transportable hypermedia authoring system. In *MULTIMEDIA*, pages 471–472. ACM, 1994.

[400] C. Harris and M. Stephens. A combined corner and edge detector. In *In Proc. of Fourth Alvey Vision Conference*, pages 147–151, 1988.

[401] C. Harrison, B. Amento, and L. Stead. IEPG: an ego-centric electronic program guide and recommendation interface. In *Proceedings of the 1st International Conference on Designing Interactive User Experiences for TV and Video*, UXTV '08, pages 23–26, New York, NY, USA, ACM, 2008.

[402] M. E. Hasselmo and H. Eichenbaum. Hippocampal mechanisms for the context-dependent retrieval of episodes. *Neural Networks*, 18(9):1172–1190, December 2005.

[403] L. He, E. Sanocki, A. Gupta, and J. Grudin. Auto-summarization of audio-video presentations. In *Proc of ACM International Conference on Multimedia*, pages 489–498, 1999.

[404] X. He, D. Cai, J.-R. Wen, W.-Y. Ma, and H.-J. Zhang. Clustering and searching www images using link and page layout analysis. *ACM Transactions on Multimedia Computing, Communications, and Applications*, 3(2), May 2007.

[405] M. A. Hearst. TextTiling: segmenting text into multi-paragraph subtopic passages. *Computational Linguistics*, 23(1):33–64, 1997.

[406] M. A. Hearst. *Search User Interfaces*. Cambridge University Press, New York, NY, USA, 1st edition, 2009.

[407] D. Heckmann, T. Schwartz, B. Brandherm, M. Schmitz, and M. von Wilamowitz-Moellendorff. Gumo–the general user model ontology. In *User Modeling 2005, 10th International Conference, UM 2005, Edinburgh, Scotland, UK, July 24-29, 2005. Proceedings*, vol. 3538, pages 428–432. Springer, 2005.

[408] R. S. Heller, C. D. Martin, N. Haneef, and S. Gievska-Krliu. Using a theoretical multimedia taxonomy framework. *J. Educ. Resour. Comput.*, 1(1es):6, 2001.

[409] S. Helmer and D. G. Lowe. Object class recognition with many local features. In *Computer Vision and Pattern Recognition Workshop, 2004. CVPRW '04. Conference on*, page 187, June 2004.

[410] W. R. Hendee and P. N. Wells. *The Perception of Visual Information, 2e.* Springer–Verlag New York, Inc., Secaucus, NJ, USA, 2nd edition, 1997.

[411] R. G. Herrtwich and R. Steinmetz. Towards Integrated Multimedia Systems: Why and How. In J. L. Encarnação, editor, *Jahrestagung Gesellschaft für Informatik*, volume 293 of *Informatik-Fachberichte*, pages 327–342. Springer–Verlag, 1991.

[412] K. Hirata and T. Kato. Query by visual example – content based image retrieval. In *Proceedings of the 3rd International Conference on Extending Database Technology: Advances in Database Technology,* EDBT '92, pages 56–71, London, UK, Springer–Verlag, 1992.

[413] P. S. Hiremath and J. Pujari. Content based image retrieval based on color, texture and shape features using image and its complement, 2007.

[414] P. S. Hiremath and J. Pujari. Content based image retrieval using color, texture and shape features. In *Advanced Computing and Communications, 2007. ADCOM 2007. International Conference on,* pages 780–784, 2007.

[415] Y. M. Hironobu, H. Takahashi, and R.H Oka. Image-to-word transformation based on dividing and vector quantizing images with words. In *in Boltzmann machines, Neural Networks,* pages 405–409, 1999.

[416] N. Hirzalla, B. Falchuk, and A. Karmouch. A Temporal Model for Interactive Multimedia Scenarios. *IEEE Multimedia,* 2(3):24–31, 1995.

[417] R. Hjelsvold, S. Vdaygiri, and Y. Léauté. Web-based personalization and management of interactive video. In *Proc. of International Conference on World Wide Web,* pages 129–139, 2001.

[418] J. R. Hobbs and F. Pan. Time Ontology. `http://www.w3.org/TR/owl-time/`.

[419] D. Hoiem, R. Sukthankar, H. Schneiderman, and L. Huston. Object-based image retrieval using the statistical structure of images. In *Computer Vision and Pattern Recognition, 2004. CVPR 2004. Proceedings of the 2004 IEEE Computer Society Conference on,* vol. 2, pages II–490 – II–497 Vol. 2, June-2 July 2004.

[420] D. Hoiem, R. Sukthankar, H. Schneiderman, and L. Huston. Object-based image retrieval using the statistical structure of images. In *Proc. IEEE Conference on Computer Vision and Pattern Recognition,* pages 490–497, 2004.

[421] L. Hollink, G. Schreiber, J. Wielemaker, and B. Wielinga. Semantic annotation of image collections. In *In Workshop on Knowledge Markup and Semantic Annotation, KCAP03, 2003. Available athttp://www.cs.vu.nl/guus,* pages 0–3, 2003.

[422] M. Holub and M. Bielikova. Estimation of user interest in visited web page. In *Proceedings of the 19th International Conference on World Wide Web,* WWW '10, pages 1111–1112, New York, NY, USA, ACM, 2010.

[423] T. P. Hong and C. Y. Lee. Induction of fuzzy rules and membership functions from training examples. *Fuzzy Sets Syst.*, 84(1):33–47, November 1996.

[424] M. Honkala, P. Cesar, and P. Vuorimaa. A Device Independent XML User Agent for Multimedia Terminals. In *Proc. of the IEEE 6th Int. Symposium on Multimedia Software Engineering; Miami, FL, USA*, pages 116–123. IEEE, December 2004.

[425] D. C. Hood and M. A. Finkelstein. Sensitivity to light. In K. R. Boff, L. Kaufman, and J. P. Thomas, editors, *Handbook of perception and human performance, Volume 1: Sensory processes and perception*, Chapter 5, pages 5.1–5.66. John Wiley & Sons, New York, NY, 1986.

[426] T. Horikoshi and H. KaSahara. 3-D shape indexing language. In *Computers and Communications, 1990. Conference Proceedings., Ninth Annual International Phoenix Conference on*, pages 493–499, 1990.

[427] J. Horkoff, A. Borgida, J. Mylopoulos, D. Barone, L. Jiang, E. Yu, and D. Amyot. Making data meaningful: The business intelligence model and its formal semantics in description logics. In Robert Meersman, Herve Panetto, Tharam Dillon, Stefanie Rinderle-Ma, Peter Dadam, Xiaofang Zhou, Siani Pearson, Alois Ferscha, Sonia Bergamaschi, and Isabel F. Cruz, editors, *On the Move to Meaningful Internet Systems: OTM 2012, Rome, Italy*, volume 7566 of *Lecture Notes in Computer Science*, pages 700–717. Springer Berlin Heidelberg, 2012.

[428] K. Hornsby. Retrieving event-based semantics from images. In *Multimedia Software Engineering, 2004. Proceedings. IEEE Sixth International Symposium on*, pages 529–536, 2004.

[429] T. Y. Hou, A. Hsu, P. Liu, and M. Y. Chiu. Content-based indexing technique using relative geometry features. pages 59–68, 1992.

[430] J. Howe. *Crowdsourcing: Why the Power of the Crowd is Driving the Future of Business*. Crown Business, 2008.

[431] C. C. Hsu, W. W. Chu, and R. K. Taira. A knowledge-based approach for retrieving images by content. *Knowledge and Data Engineering, IEEE Transactions on*, 8(4):522–532, 1996.

[432] P.-Y. Hsueh and J. D. Moore. Automatic decision detection in meeting speech. In Andrei Popescu-Belis, Steve Renals, and Hervé Bourlard, editors, *Machine Learning for Multimodal Interaction (MLMI '07)*, vol. 4892 of *Lecture Notes in Computer Science*, pages 168–179. Springer, Brno, Czech Republic, June 2007.

[433] P.-Y. Hsueh and J. D. Moore. Combining multiple knowledge sources for dialogue segmentation in multimedia archives. In *Proceedings of*

*the 45th Annual Meeting of the ACL*, pages 1016–1023, Prague, Czech Republic, June 2007. Association for Computational Linguistics.

[434] C.-R. Huang, P.-C. Chung, K.-W. Lin, and S.-C. Tseng. Wheelchair detection using cascaded decision tree. *Information Technology in Biomedicine, IEEE Transactions on*, 14(2):292–300, 2010.

[435] H.-M. Huang, U. Rauch, and S.-S. Liaw. Investigating learners attitudes toward virtual reality learning environments: Based on a constructivist approach. *Computers & Education*, 55(3):1171–1182, 2010.

[436] Data Hub. Data Hub LOD Datasets: `http://www4.wiwiss.fu-berlin.de/lodcloud/ckan/validator/`.

[437] S. Hubbard and K. Hornsby. Modeling Alternative Sequences of Events in Dynamic Geographic Domains. *Transactions in GIS*, 15:557–575, Oktober 2011.

[438] J. R. Hughes. Gamma, fast, and ultrafast waves of the brain: their relationships with epilepsy and behavior. *Epilepsy & Behavior*, 13(1):25–31, November 2008.

[439] M. J. Huiskes and M. S. Lew. The MIR flickr retrieval evaluation. In *Multimedia Information Retrieval, Proceedings of the 1st ACM International Conference on...*, pages 39–43, 2008.

[440] M. J. Huiskes, B. Thomee, and M. S. Lew. New trends and ideas in visual concept detection: the MIR flickr retrieval evaluation initiative. In *Multimedia Information Retrieval, Proceedings of the International Conference on...*, pages 527–536, 2010.

[441] D. Hyde. *Vagueness, Logic and Ontology*. Ashgate New Critical Thinking in Philosophy, 2008.

[442] S. Hyoseop, L. Minsoo, and K. Eun. Personalized digital TV content recommendation with integration of user behavior profiling and multimodal content rating. *IEEE Transactions on, Consumer Electronics*, 55(3):1417–1423, August 2009.

[443] E. Hyvonen, S. Saarela, and K. Viljanen. Intelligent image retrieval and browsing using semantic web techniques – a case study. In *International Sepia Conference at the Finnish Museum of Photography*, 2003.

[444] T. Ignatova and I. Bruder. Utilizing a Multimedia UML Framework for an Image Database Application. In J. Akoka, S. W. Liddle, I.-Y. Song, M. Bertolotto, I. Comyn-Wattiau, S. S.-S. Cherfi, W.-J. van den Heuvel, B. Thalheim, M. Kolp, P. Bresciani, J. Trujillo, C. Kop, and H. C. Mayr, editors, *ER (Workshops); Klagenfurt, Austria*, vol. 3770 of *Lecture Notes in Computer Science*, pages 23–32. Springer–Verlag, 2005.

[445] J. Indulska, R. Robinson, A. Rakotonirainy, and K. Henricksen. Experiences in using CC/PP in context-aware systems. In *Proc. of the 4th International Conference on Mobile Data Management (MDM'03), Melbourne, Australia*, pages 247–261, January 2003.

[446] P. S. Inventado, R. Legaspi, T. D. Bui, and M. Suarez. Predicting student's appraisal of feedback in an ITS using previous affective states and continuous affect labels from EEG data. In *Proceedings of the 18th International Conference on Computers in Education*, pages 71–75, Putrajaya, Malaysia, November–December 2010.

[447] B. Ionescu, C. Vertan, P. Lambert, and A. Benoit. A color-action perceptual approach to the classification of animated movies. In *ACM International Conference on Multimedia Retrieval (ICMR'11), April 18–20, Trento, Italy*, 2011.

[448] ISO/IEC (Int. Organization for Standardization/Int. Electrotechnical Commission). Information technology – Coding of audio-visual objects – Part 20: Lightweight Application Scene Representation (LASeR) and Simple Aggregation Format (SAF), June 2006.

[449] ISO/IEC (Int. Organization for Standardization/Int. Electrotechnical Commission) JTC1/SC29/WG11. LASeR and SAF editor's study: ISO/IEC 14496-20 Study; Poznan, Poland, July 2005.

[450] L. Itti. Quantifying the contribution of low-level saliency to human eye movements in dynamic scenes. *Visual Cognition*, 12(6):1093–1123, 2005.

[451] L. Itti and C. Koch. Computational modelling of visual attention. *Nature Review Neuroscience*, 2(3):194–203, 2001.

[452] International Telecommunication Union ITU. Methodology for the subjective assessment of the quality of television pictures. Recommendation BT.500-11, International Telecommunication Union ITU, 2002.

[453] C. E. Jacobs, A. Finkelstein, and D. H. Salesin. Fast multiresolution image querying. In *Proceedings of the 22nd Annual Conference on Computer Graphics and Interactive Techniques*, SIGGRAPH '95, pages 277–286, New York, NY, USA, 1995. ACM.

[454] V. Jacobson, D. K. Smetters, J. D. Thornton, M. F. Plass, N. H. Briggs, and R. L. Braynard. Networking named content. In *CoNEXT '09: Proceedings of the 5th international conference on Emerging networking experiments and technologies*, pages 1–12, Rome, Italy, ACM, December 1–4 2009.

[455] A. Jaimes and S. F. Chang. A conceptual framework for indexing visual information at multiple levels. In *Proc of SPIE Internet Imaging*, pages 2–15, 2000.

[456] A. Jaimes, K. Omura, T. Nagamine, and K. Hirata. Memory cues for meeting video retrieval. In *Proceedings of the the 1st ACM Workshop on Continuous Archival and Retrieval of Personal Experiences*, CARPE'04, pages 74–85, New York, NY, USA, ACM, October 2004.

[457] R. Jaimes, M. Christel, S. Gilles, R. Sarukkai, and W.-Y. Ma. Multimedia information retrieval: What is it, and why isn't anyone using it? In *Proceedings of the 7th ACM SIGMM International Workshop on Multimedia Information Retrieval*, pages 3–8, Singapore, November 10–11, 2005.

[458] A. K. Jain and A. Vailaya. Image retrieval using color and shape. *Pattern Recognition*, 29:1233–1244, 1996.

[459] P. Jain, P. Hitzler, P. Z Yeh, K. Verma, and A. P. Sheth. Linked data is merely more data. *Linked Data Meets Artificial Intelligence*, pages 82–86, 2010.

[460] R. Jain. NSF workshop on visual information management systems. *SIGMOD Rec.*, 22(3):57–75, September 1993.

[461] R. Jain. Multimedia information retrieval: watershed events. In *Proceeding of the 1st ACM International Conference on Multimedia Information Retrieval*, pages 229–236, Vancouver, BC, Canada, October 2008.

[462] M. Jakobsson, A. Juels, and J. Ratkiewicz. Privacy-Preserving History Mining for Web Browsers. In *Web 2.0 Security and Privacy*, Oakland, CA, May 2008. IEEE, IEEE.

[463] A. Janin, D. Baron, J. Edwards, D. Ellis, D. Gelbart, N. Morgan, B. Peskin, T. Pfau, E. Shriberg, A. Stolcke, and C. Wooters. The ICSI meeting corpus. In *IEEE International Conference on Acoustics, Speech, and Signal Processing, 2003. Proceedings. (ICASSP '03)*, vol. 1, pages 364–367, Hong Kong, China, April 2003.

[464] D. Jannach and K. Leopold. Knowledge-based multimedia adaptation for ubiquitous multimedia consumption. *Journal of Network and Computer Applications*, 30(3):958–982, August 2007.

[465] D. Jannach, K. Leopold, C. Timmerer, and H. Hellwagner. A knowledge-based framework for multimedia adaptation. *Applied Intelligence*, 24(2):109–125, 2006.

[466] J. Jansen and D. C. A. Bulterman. SMIL state: an architecture and implementation for adaptive time-based web applications. *Multimedia Tools and Applications*, 43:203–224, 2009.

[467] C. Jennett, A. L. Cox, P. Cairns, S. Dhoparee, A. Epps, T. Tijs, and A. Walton. Measuring and defining the experience of immersion in games. *International Journal of Human-Computer Studies*, 66(9):641–661, December 2008.

[468] J. Jeon, V. Lavrenko, and R. Manmatha. Automatic image annotation and retrieval using cross-media relevance models. In *Proceedings of the 26th Annual International ACM SIGIR Conference on Research and Development in Informaion Retrieval*, pages 119–126, Toronto, ON, Canada, July 28–August 1 2003.

[469] L. Jiang, J. Hou, Z. Chen, and D. Zhang. Automatic image annotation based on decision tree machine learning. In *Proceedings of International Conference on Cyber-Enabled Distributed Computing and Knowledge Discovery*, pages 170–175, Zhangjiajie, China, October 10–12 2009.

[470] W. Jiang, S.-F. Chang, and A.C. Loui. Context-based concept fusion with boosted conditional random fields. In *Acoustics, Speech and Signal Processing, 2007. ICASSP 2007. IEEE International Conference on*, vol. 1, pages I-949–I-952, April 2007.

[471] C. Jiannong, X. Na, A. T. S. Chan, F. Yulin, and J. Beihong. Service adaptation using fuzzy theory in context-aware mobile computing middleware. In *Proc. of the 11th IEEE International Conference on Embedded and Real-Time Computing Systems and Applications, 2005.*, pages 496–501, August 2005.

[472] E. Jiménez-Ruiz, B. Cuenca Grau, Y. Zhou, and I. Horrocks. Large-scale interactive ontology matching: Algorithms and implementation. In *ECAI*, pages 444–449, 2012.

[473] F. Jing, M. Li, H. J. Zhang, and B. Zhang. An effective region-based image retrieval framework. In *ACM International Conference on Multimedia, December 1-6, Juan les Pins, France*, 2002.

[474] F. Jing, M. Li, H. J. Zhang, and B. Zhang. Relevance feedback in region-based image retrieval. *Circuits and Systems for Video Technology, IEEE Transactions on*, 14(5):672–681, 2004.

[475] T. Joachims, L. Granka, B. Pang, H. Hembrooke, and Gay G. Accurately interpreting clickthrough data as implicit feedback. In *Proceedings of the 28th Annual International ACM SIGIR Conference*, pages 154–161, Salvador, Brazil, August 15–19, 2005.

[476] M. Johnson, G. J. Brostow, J. Shotton, O. Arandjelovic, V. Kwatra, and R. Cipolla. Semantic photo synthesis. *Computer Graphics Forum*, 25, 2006.

[477] Z. Johnson. SpatialKey: Insanely good geovisualization.

[478] Z. Johnson, M. Harrower, E. McGlynn, R. Roth, D. Sickle, and A. Woodruff. Development of an online visualization tool for the mapping and analysis of asthma exacerbations in space and time. In *Proceedings of the NACIS 2007. St. Louis, MO.*, 2007.

[479] C. Jorgensen. Image attributes in describing tasks: an investigation. In *Information Processing and Management*, pages 161–174, 1998.

[480] M. Jourdan and F. Bes. A new step towards multimedia documents generation. In *Int. Conf. on Media Futures*, pages 25–28, May 2001.

[481] M. Jourdan, N. Layaïda, and C. Roisin. Authoring Techniques for Temporal Scenarios of Multimedia Documents. In *Handbook of Internet and Multimedia Systems and Applications*, Chapter 8, pages 179–200. CRC press, IEEE press, 1999.

[482] M. Jourdan, N. Layaïda, C. Roisin, L. Sabry-Ismaïl, and L. Tardif. Madeus, an Authoring Environment for Interactive Multimedia Documents. In *MULTIMEDIA*, pages 267–272. ACM, 1998.

[483] M. Jourdan, N. Layaïda, and L. Sabry-Ismaïl. Authoring Environment for Interactive Multimedia Documents. In *Int. Conf. on Computer Communications*, pages 19–26, November 1997.

[484] W. Ju, B. A. Lee, and S. R. Klemmer. Range: Exploring implicit interaction through electronic whiteboard design. In *Proceedings of the 2008 ACM Conference on Computer Supported Cooperative Work*, CSCW '08, pages 17–26, New York, NY, USA, ACM, 2008.

[485] W. Junjian and X. Shujun. Fatigue detecting system. Master's thesis, Linnaeus University, 2011.

[486] I. Jurisica, J. Mylopoulos, and E. Yu. Ontologies for knowledge management: An information systems perspective. *Knowledge and Information Systems*, 6(4):380–401, July 2004.

[487] B. Kane and S. Luz. Information sharing at multidisciplinary medical team meetings. *Group Decision and Negotiation*, 20:437–464, 2011. Springer.

[488] B. Kane, S. Luz, and S. Jing. Capturing multimodal interaction at medical meetings in a hospital setting: Opportunities and challenges. In *Proceedings of Multimodal Corpora: Advances in Capturing, Coding and Analyzing Multimodality*, pages 140–145, Malta, June 2010. LREC.

[489] V. Kantere, I. Kiringa, and J. Mylopoulos. Supporting Distributed Event-Condition-Action Rules in a Multidatabase Environment. *World Scientific*, 16(3):467–506, September–December 2007.

[490] L. M. Kaplan, R. Murenzi, and K. R. Namuduri. Fast texture database retrieval using extended fractal features. pages 162–173, 1997.

[491] A. Kapoor, W. Burleson, and R. W. Picard. Automatic prediction of frustration. *International Journal of Human-Computer Studies*, 65(8):724–736, 2007.

[492] A. Kapoor and R. W. Picard. Multimodal affect recognition in learning environments. In *Proceedings of the 13th Annual ACM International Conference on Multimedia*, pages 677–682. ACM, 2005.

[493] G. Karagiannidis. *Telecommunication Systems*. Tziola Publications, 2010.

[494] S. Karaman, J. Benois-Pineau, R. Mégret, and A. Bugeau. Multi-layer local graph words for object recognition. In *Advances in Multimedia Modeling, January 4–6, Klagenfurt, Austria*, 2012.

[495] M. Karg, K. Kuhnlenz, and M. Buss. Recognition of affect based on gait patterns. *Systems, Man, and Cybernetics, Part B: Cybernetics, IEEE Transactions on*, 40(4):1050–1061, 2010.

[496] I. Karonis. Pattern recognition through PCA in odor detections system. Diploma thesis, Electrical and Computer Engineering, National Technical University of Athens, Greece, 2008.

[497] K. Karpouzis, G. Caridakis, L. Kessous, N. Amir, A. Raouzaiou, L. Malatesta, and S. Kollias. Modeling naturalistic affective states via facial, vocal, and bodily expressions recognition. In *Artifical Intelligence for Human Computing*, pages 91–112. Springer, 2007.

[498] I. Karydis, M. Avlonitis, and S. Sioutas. Collective intelligence in video user's activity. In *Artificial Intelligence Applications and Innovations (2)*, pages 490–499, 2012.

[499] J. O. Katz. Personal construct theory and the emotions: An interpretation in terms of primitive constructs. *British Journal of Psychology*, 75(3):315–327, 1984.

[500] S. D. Katz. *Film Directing Shot by Shot: Visualizing from Concept to Screen*. Michael Wiese Productions Series. Focal Press, London, UK, July 1991.

[501] G. Kazai, J. Kamps, M. Koolen, and N. Milic-Frayling. Crowdsourcing for book search evaluation: Impact of quality on comparative system

ranking. In *Proceedings of the 34th Annual International ACM SIGIR Conference on Research and Development in Information Retrieval,* pages 205–214, Beijing, China, July 24–28, 2011.

[502] R. Kazman, R. Al-Halimi, W. Hunt, and M. Mantey. Four paradigms for indexing video conferences. *IEEE Multimedia,* 3(1):63–73, 1996.

[503] Y. Ke and R. Sutkthankar. PCA-sift: A more distinctive representation for local image descriptors. In *Proceedings of IEEE International Conference on Computer Vision and Pattern Recognition,* vol. 2, pages II–506–II–513, Washington, DC, USA, June 27–July 2, 2004.

[504] S. S. Keerthi, S. K. Shevade, C. Bhattacharyya, and K. R. K. Murthy. Improvements to Platt's SMO algorithm for SVM classifier design. *Neural Computation,* 13(3):637–649, March 2001.

[505] D. Kelleher and S. Luz. Automatic hypertext keyphrase detection. In L. P. Kaelbling and A. Saffiotti, editors, *Proceedings of the Nineteenth International Joint Conference on Artificial Intelligence,* pages 1608–1609, Edinburgh, Scotland, IJCAI, July 2005.

[506] D. Kelly. Methods for evaluating interactive information retrieval systems with users. *Found. Trends Inf. Retr.,* 3(1–2):1–224, 2009.

[507] D. Kelly and J. Teevan. Implicit feedback for inferring user preference: a bibliography. *SIGIR Forum,* 37(2):18–28, September 2003.

[508] G. A. Kelly. *The Psychology of Personal Constructs.* Norton, 1955.

[509] F. H. Khan, M. Y. Javed, S. Bashir, A. Khan, and M. S. H. Khiyal. QoS Based Dynamic Web Services Composition & Execution. *International Journal of Computer Science and Information Security,* 7(2):147–152, February 2010.

[510] M. L. Kherfi and D. Ziou. Image collection organization and its application to indexing, browsing, summarization, and semantic retrieval. *Multimedia, IEEE Transactions on,* 9(4):893–900, 2007.

[511] M. L. Kherfi, D. Ziou, and A. Bernardi. Image retrieval from the World Wide Web: Issues, techniques, and systems. *ACM Comput. Surv.,* 36(1):35–67, March 2004.

[512] J. Kim, H. Kim, and K. Park. Towards optimal navigation through video content on interactive TV. *Interact. Comput.,* 18(4):723–746, 2006.

[513] S. Kim and J. Kwon. Effective context-aware recommendation on the semantic web. *International Journal of Computer Science and Network Security (IJCSNS),* 7(8):154, 2007.

[514] Y.-M. Kim, S.-Y. Koo, J. G. Lim, and D.-S. Kwon. A robust online touch pattern recognition for dynamic human-robot interaction. *Consumer Electronics, IEEE Transactions on*, 56(3):1979–1987, 2010.

[515] B. B. Kimia, J. Chan, D. Bertrand, S. Coe, Z. Roadhouse, and H. Tek. A shock-based approach for indexing of image databases using shape. In *Proceedings of the SPIE's Multimedia Storage and Archiving Systems II*, vol. 3229, pages 288–302, Dallas, Texas, November 1997.

[516] P. KiNam, J. Hyesung, L. Taemin, J. Soonyoung, and L. Heui-Seok. Automatic extraction of user's search intention from web search logs. *Multimedia Tools Appl.*, 61(1):145–162, 2012.

[517] R. King, N. Popitsch, and U. Westermann. Metis: a flexible foundation for the unified management of multimedia assets. *Multimedia Tools and Applications*, 33:325–349, 2007.

[518] A. Kinno, Y. Yonemoto, M. Morioka, and M. Etoh. Environment Adaptive XML Transformation and Its Application to Content Delivery. In *Proc. of the 2003 Symposium on Applications and the Internet; Washington, DC, USA*, pages 31–38. IEEE, 2003.

[519] S. Kisilevich, D. A. Keim, N. Andrienko, and G. Andrienko. Towards acquisition of semantics of places and events by multi-perspective analysis of geotagged photo collections. In Antoni Moore and Igor Drecki, editors, *Geospatial Visualisation*, Lecture Notes in Geoinformation and Cartography, pages 211–233. Springer Berlin Heidelberg, January 2013.

[520] L. Kjelldahl. Introduction. In L. Kjelldahl, editor, *Multimedia: Systems, Interaction and Applications*, Chapter 1, pages 3–5. Springer–Verlag, April 1991.

[521] W. Klas, C. Greiner, and R. Friedl. Cardio-OP: Gallery of Cardiac Surgery. In *IEEE Int. Conf. on Multimedia Computing and Systems; Florence, Italy*, pages 1092–1095, July 1999.

[522] T. Kleemann and A. Sinner. User profiles and matchmaking on mobile phones. *Declarative Programming for Knowledge Management*, pages 135–147, Berlin, Heidelberg, Springer–Verlag, 2006.

[523] T. Kliegr. UTA–NM: explaining stated preferences with additive non-monotonic utility functions. In *Preference Learning (PL-09) ECML/PKDD-09 Workshop*.

[524] G. Klir and B. Yuan. *Fuzzy Sets and Fuzzy Logic, Theory and Applications*. Prentice Hall, 1995.

[525] G. Klyne, F. Reynolds, C. Woodrow, O. Hidetaka, J. Hjelm, M. H. Butler, and L. Tran. Composite capability/preference profiles (CC/PP): Structure and vocabularies 1.0. Recommendation, W3C, 2004.

[526] D. E. Knuth. *The TeXbook*. Addison-Wesley, 1986.

[527] E. Knutov, P. De Bra, and M. Pechenizkiy. AH 12 years later: a comprehensive survey of adaptive hypermedia methods and techniques. *New Review of Hypermedia and Multimedia*, 15(1):5–38, 2009.

[528] E. Knutov, P. De Bra, and M. Pechenizkiy. Provenance meets adaptive hypermedia. In *Proceedings of the 21st ACM Conference on Hypertext and Hypermedia*, HT '10, pages 93–98, New York, NY, USA, ACM, 2010.

[529] M. Koch. Global Identity Management to Boost Personalization. In P. Schubert and U. Leimstoll, editors, *Proc. Research Symposium on Emerging Electronic Markets; Basel, Switzerland*, pages 137–147. Institute for Business Economics (IAB), University of Applied Sciences, Basel, Switzerland, September 2002.

[530] J. F. Koegel Buford. Architectures and issues for distributed multimedia systems. In J. F. Koegel Buford, editor, *Multimedia systems*, SIGGRAPH series, chapter 3, pages 45–64. ACM, December 1994.

[531] J. F. Koegel Buford. Uses of multimedia information. In J. F. Koegel Buford, editor, *Multimedia systems*, SIGGRAPH series, Chapter 1, pages 1–25. ACM, December 1994.

[532] R. Koenen and F. Pereira. MPEG-7: A standardised description of audiovisual content. *Signal Processing: Image Communication*, 16(12):5 13, 2000.

[533] R. Kohavi and F. Provost. Glossary of terms. *Machine Learning*, 30(2-3):271–274, January 1998.

[534] M. Koskela, J. Laaksonen, and E. Oja. Using mpeg-7 descriptors in image retrieval with self-organizing maps. In *Pattern Recognition, 2002. Proceedings. 16th International Conference on*, vol. 2, pages 1049–1052, 2002.

[535] M. Koskela, A. F. Smeaton, and J. Laaksonen. Measuring concept similarities in multimedia ontologies: Analysis and evaluations. *Multimedia, IEEE Transactions on*, 9(5):912–922, 2007.

[536] M. Koskela, A. F. Smeaton, and J. Laaksonen. Measuring concept similarities in multimedia ontologies: Analysis and evaluations. *Multimedia, IEEE Transactions on*, 9(5):912–922, August 2007.

[537] S. B. Kotsiantis. Supervised machine learning: A review of classification techniques. *Informatica*, 31:249–268, October 2007.

[538] T. Koutepova, Y. Liu, X. Lan, and J. Jeong. Enhancing video games in real time with biofeedback data. In *ACM SIGGRAPH ASIA 2010 Posters*, page 56, Seoul, S. Korea, ACM, December 2010.

[539]  P. Kraemer, J. Benois-Pineau, and J.-P. Domenger. Scene Similarity Measure for Video Content Segmentation in the Framework of Rough Indexing Paradigm. In *International Workshop on Adaptive Multimedia Retrieval (AMR'04), August 22–27, Valencia, Spain*, 2004.

[540]  M. Krötzsch, F. Simancik, and I. Horrocks. A description logic primer. *CoRR*, abs/1201.4089, 2012.

[541]  D. J. Krusienski, E. W. Sellers, F. Cabestaing, S. Bayoudh, D. J. McFarland, T. M. Vaughan, and J. R. Wolpaw. A comparison of classification techniques for the p300 speller. *Journal of Neural Engineering*, 3(4):299, December 2006.

[542]  H.-H. Ku and C.-M. Huang. Web2ohs: A web2. 0-based omnibearing homecare system. *Information Technology in Biomedicine, IEEE Transactions on*, 14(2):224–233, 2010.

[543]  J. Kuchař and I. Jelínek. Scoring Pageview Based on Learning Weight Function. *International Journal on Information Technologies and Security*, 2(4):19–28, November 2010.

[544]  J. Kuchař and T. Kliegr. Gain: Analysis of implicit feedback on semantically annotated content. In *Proceedings of The 7th Workshop on Intelligent and Knowledge Oriented Technologies (WIKT 2012)*, pages 75–78. Bratislava : Nakladatestvo STU, 2012.

[545]  F. Kuijk, R. L. Guimares, P. Cesar, and D. C. A. Bulterman. From photos to memories: A user-centric authoring tool for telling stories with your photos. In Petros Daras and Oscar Mayora Ibarra, editors, *User Centric Media*, vol. 40 of *Lecture Notes of the Institute for Computer Sciences, Social Informatics and Telecommunications Engineering*, pages 13–20. Springer Berlin Heidelberg, 2010.

[546]  L. Kuncheva, T. Christy, I. Pierce, and S. Mansoor. Multi-modal biometric emotion recognition using classifier ensembles. In *Modern Approaches in Applied Intelligence, Proc of IEA/AIE*, pages 317–326. Springer, June–July 2011.

[547]  Y.-M. Kuo, J.-S. Lee, and P.-C. Chung. A visual context-awareness-based sleeping-respiration measurement system. *Information Technology in Biomedicine, IEEE Transactions on*, 14(2):255–265, 2010.

[548]  H. Kwasnicka and M. Paradowski. Machine learning methods in automatic image annotation. In *Advances in Machine Learning II*, vol. 263 of *Studies in Computational Intelligence*, pages 387–411. 2010.

[549]  S. Laborie, J. Euzenat, and N. Layaïda. Semantic adaptation of multimedia documents. *Multimedia Tools and Applications*, 55(3):379–398, December 2011.

[550] J.-P. Lachaux, A. Lutz, D. Rudrauf, D. Cosmelli, M. Le Van Quyen, J. Martinerie, and F. Varela. Estimating the time-course of coherence between single-trial brain signals: an introduction to wavelet coherence. *Neurophysiologie Clinique/Clinical Neurophysiology*, 32(3):157–174, 2002.

[551] R. Laiola Guimarães, P. Cesar, D. C. A. Bulterman, V. Zsombori, and I. Kegel. Creating personalized memories from social events: community-based support for multi-camera recordings of school concerts. In *Proceedings of the 19th ACM International Conference on Multimedia*, MM '11, pages 303–312, New York, NY, USA, ACM, 2011.

[552] M. Lamolle, J. Gomez, and E. Exposito. MODA : une architecture multimédia dirigée par les ontologies pour des systèmes multimédia en réseau. In *Proc. of the 4ème Conférence francophone sur les Architectures Logicielles (CAL'10), Pau, France*, pages 137–151, March 2010.

[553] M. Land, N. Mennie, and J. Rusted. The roles of vision and eye movements in the control of activities of daily living. *Perception*, 28:1311–1328, 1999.

[554] P. J. Lang, M. M. Bradley, and B. N. Cuthbert. International affective picture system (iaps): Technical manual and affective ratings, 1999.

[555] S. Laplace, M. Dalmau, and P. Roose. Kalinahia: Considering quality of service to design and execute distributed multimedia applications. In *Proc. of IEEE/IFIP International Conference on Network Management and Management Symposium, Salvador de Bahia, Brasil*, pages 951–954, April 2008.

[556] I. Laptev. On space-time interest points. *International Journal on Computer Vision*, 2:107–123, 2005.

[557] E. A. Larsen. *Classification of EEG Signals in a Brain-Computer Interface System*. PhD thesis, Norwegian University of Science and Technology, 2011.

[558] V. Lavrenko, S. L. Feng, and R. Manmatha. Statistical models for automatic video annotation and retrieval. In *Acoustics, Speech, and Signal Processing, Proceedings of the IEEE International Conference on*, vol. 3, pages 1044–1047. IEEE, 2004.

[559] V. Lavrenko, R. Manmatha, and J. Jeon. A model for learning the semantics of pictures. In *Proceedings of Advances in Neural Information Processing Systems*, Lake Tahoe, NV, USA, December 9–10, 2003.

[560] S. Lazebnik, C. Schmid, and J. Ponce. Beyond bags of features: Spatial pyramid matching for recognizing natural scene categories. In *Proceedings IEEE Computer Society Conference on Computer Vision and Pattern Recognition*, vol. 2, pages 2169–2178, New York, NY, USA, June 17–22, 2006.

[561] S. Lazebnik, C. Schmid, and J. Ponce. Beyond bags of features: Spatial pyramid matching for recognizing natural scene categories. In *IEEE Conference on Computer Vision and Pattern Recognition (CVPR'06)*, *June 17–22, New York*, NY, USA, 2006.

[562] O. Le Meur, P. Le Callet, and D. Barba. Predicting visual fixations on video based on low-level video features. *Vision Research*, 47(19):1057–1092, 2007.

[563] B. Le Saux and G. Amato. Image recognition for digital libraries. In *Proceedings of the 6th ACM SIGMM International Workshop on Multimedia Information Retrieval*, MIR '04, pages 91–98, New York, NY, USA, ACM, 2004.

[564] C. S. Lee, Z. W. Jian, and L. K. Huang. A fuzzy ontology and its application to news summarization. *IEEE Transactions on Systems, Man and Cybernetics (Part B)*, 35(5):859–880, 2005.

[565] T. Lee, and Y.-T. Park. An empirical study on effectiveness of temporal information as implicit ratings. *Expert Systems with Applications*, 36(2):13157–1321, March 2009.

[566] T. Lee, L. Sheng, T. Bozkaya, N. H. Balkir, Özsoyoglu. Z. M., and G. Özsoyoglu. Querying Multimedia Presentations Based on Content. *IEEE Trans. on Knowledge and Data Engineering*, 11(3):361–385, May/June 1999.

[567] T. B. Lee, J. Hendler, O. Lassila, et al. The semantic web. *Scientific American*, 284(5):34–43, 2001.

[568] Y. J. Lee, J. Ghosh, and K. Grauman. Discovering important people and objects for egocentric video summarization. In *IEEE Conference on Computer Vision and Pattern Recognition (CVPR'12), June 16–21, Providence, Rhode Island, USA*, 2012.

[569] I. Leftheriotis, C. Gkonela, and K. Chorianopoulos. Efficient video indexing on the web: A system that crowdsources user interactions with a video player. In *UCMedia*, pages 123–131, 2010.

[570] T. Lemlouma and N. Layaïda. NAC: A basic core for the adaptation and negotiation of multimedia services. In *Opera project, INRIA*, September 2001.

[571] T. Lemlouma and N. Layaïda. Content adaptation and generation principles for heterogeneous clients. In *Proc. of the W3C Workshop on Device Independent Authoring Techniques, St. Leon-Rot, Germany,* September 2002.

[572] T. Lemlouma and N. Layaïda. Adapted Content Delivery for Different Contexts. In *Symposium on Applications and the Internet; Orlando, FL, USA,* pages 190–197. IEEE, January 2003.

[573] R. Lenman. *The Oxford Companion to the Photograph.* Oxford Companions Series. Oxford University Press, Incorporated, 2008.

[574] J. Lessiter, J. Freeman, E. Keogh, and J. Davidoff. A cross-media presence questionnaire: The ITC-sense of presence inventory. *Presence: Teleoperators & Virtual Environments,* 10(3):282–297, June 2001.

[575] C. H. C. Leung, A. W. S. Chan, A. Milani, J. Liu, and Y. Li. Intelligent social media indexing and sharing using an adaptive indexing search engine. *ACM Trans. Intell. Syst. Technol.,* 3(3):47:1–47:27, May 2012.

[576] C. H. C. Leung and Y. Li. Comparison of different ontology-based query expansion algorithms for effective image retrieval. In Tai-hoon Kim, Hojjat Adeli, Carlos Ramos, and Byeong-Ho Kang, editors, *Signal Processing, Image Processing and Pattern Recognition,* vol. 260 of *Communications in Computer and Information Science,* pages 291–299. Springer Berlin Heidelberg, 2011.

[577] C. H. C. Leung and Y. Li. Cyc based query expansion framework for effective image retrieval. In *Image and Signal Processing (CISP), 2011 4th International Congress on,* vol. 3, pages 1353–1357, 2011.

[578] C. H. C. Leung and Y. Li. The correlation between semantic visual similarity and ontology-based concept similarity in effective web image search. In *Proceedings of the 14th International Conference on Web Technologies and Applications,* APWeb'12, pages 125–130, Berlin, Heidelberg, Springer–Verlag, 2012.

[579] C. H. C. Leung and Y. Li. Semantic image retrieval using collaborative indexing and filtering. In *Web Intelligence and Intelligent Agent Technology (WI-IAT), 2012 IEEE/WIC/ACM International Conferences on,* vol. 3, pages 261–264, 2012.

[580] K. Levacher, S. Lawless, and V. Wade. Slicepedia: providing customized reuse of open-web resources for adaptive hypermedia. In *Proceedings of the 23rd ACM Conference on Hypertext and Social Media,* HT '12, pages 23–32, New York, NY, USA, ACM, 2012.

[581] E. Levina and P. J. Bickel. Maximum likelihood estimation of intrinsic dimension. *Advances in Neural Information Processing Systems,* 17:777–784, December 2004.

[582] A. Levinshtein, A. Stere, K. N. Kutulakos, D. J. Fleet, S. J. Dickinson, and K. Siddiqi. Turbopixels: Fast superpixels using geometric flows. *IEEE Transactions on Pattern Analysis and Machine Intelligence*, 31(12):2290–2297, 2009.

[583] F. Li, Q. Dai, W. Xu, and G. Er. Multilabel neighborhood propagation for region-based image retrieval. *Multimedia, IEEE Transactions on*, 10(8):1592–1604, 2008.

[584] F. C. Li, A. Gupta, E. Sanocki, L.-W. He, and Y. Rui. Browsing digital video. In *Proc. of SIGCHI Conference on Human Factors in Computing Systems*, pages 169–176, 2000.

[585] F.-F. Li, R. Fergus, and P. Perona. A Bayesian approach to unsupervised one-shot learning of object categories. In *Computer Vision, 2003. Proceedings. Ninth IEEE International Conference on*, vol. 2, pages 1134–1141, October 2003.

[586] J. Li and J. Z. Wang. Automatic linguistic indexing of pictures by a statistical modeling approach. *IEEE Transactions on Pattern Analysis and Machine Intelligence*, 25(9):1075–1088, 2003.

[587] J. Li and J. Z. Wang. Automatic linguistic indexing of pictures by a statistical modeling approach. *Pattern Analysis and Machine Intelligence, IEEE Transactions on*, 25(9):1075–1088, 2003.

[588] J. Li and J. Z. Wang. Real-time computerized annotation of pictures. In *Proceedings of the ACM Multimedia Conference*, pages 911–920, Santa Barbara, CA, USA, October 23-27 2006.

[589] J. Li and J. Z. Wang. Real-time computerized annotation of pictures. *Pattern Analysis and Machine Intelligence, IEEE Transactions on*, 30(6):985–1002, 2008.

[590] J. Li, J. Z. Wang, and G. Wiederhold. IRM: Integrated region matching for image retrieval. In *Proceedings of the Eighth ACM International Conference on Multimedia*, MULTIMEDIA '00, pages 147–156, New York, NY, USA, ACM, 2000.

[591] M. Li, Y. Sun, and H. Sheng. Temporal relations in multimedia systems. *Computers & Graphics*, 21(3):315–320, 1997. Computer Graphics in China.

[592] T. Li, T. Mei, I.-S. Kweon, and X.-S. Hua. Contextual bag-of-words for visual categorization. *IEEE Transactions on Circuits and Systems for Video Technology*, 21(4):381–392, April 2011.

[593] X. Li, L. Chen, L. Zhang, F. Lin, and W.-Y. Ma. Image annotation by large-scale content-based image retrieval. In *In Proceedings of the 14th Annual ACM International Conference on Multimedia*, page 27, 2006.

[594] X. Li, B. Hu, T. Zhu, J. Yan, and F. Zheng. Towards affective learning with an eeg feedback approach. In *Proceedings of the First ACM International Workshop on Multimedia Technologies for Distance Learning*, pages 33–38, Beijing, China, ACM, October 2009.

[595] X. Li and Q. Ji. Active affective state detection and user assistance with dynamic Bayesian networks. *Systems, Man and Cybernetics, Part A: Systems and Humans, IEEE Transactions on*, 35(1):93–105, 2005.

[596] X. Li, L. Shou, G. Chen, T. Hu, and J. Dong. Modeling image data for effective indexing and retrieval in large general image databases. *Knowledge and Data Engineering, IEEE Transactions on*, 20(11):1566–1580, 2008.

[597] K. C. Liang and C.-C. J. Kuo. Implementation and performance evaluation of a progressive image retrieval system. pages 37–48, 1997.

[598] S. Lie. What is consciousness, states of consciousness – http://www.suzanneliephd.com/solarpl/whatiscon2.html.

[599] H. Lieberman, H. Liu, P. Singh, and B. Barry. Beating common sense into interactive applications. *AI Magazine*, 25:63–76, 2004.

[600] T. S. Lim, W.-Y. Loh, and W. Cohen. A comparison of prediction accuracy, complexity, and training time of thirty-three old and new classification algorithms. *Machine Learning*, 40:203–228, 2000.

[601] S. Lin. On paradox of fuzzy modeling: Supervised learning for rectifying fuzzy membership function. *Artif. Intell. Rev.*, 23(4):395–405, June 2005.

[602] W. H. Lin, R. Jin, and A. Hauptmann. Web image retrieval re-ranking with relevance model. In *Proceedings of the 2003 IEEE/WIC International Conference on Web Intelligence*, WI '03, pages 242–252, Washington, DC, USA, 2003. IEEE Computer Society.

[603] G. Linden, B. Smith, and J. York. Amazon.com recommendations: item-to-item collaborative filtering. *Internet Computing, IEEE*, 7(1):76 – 80, January/February 2003.

[604] C. L. Lisetti and F. Nasoz. Maui: a multimodal affective user interface. In *Proceedings of the Tenth ACM International Conference on Multimedia*, pages 161–170. ACM, 2002.

[605] S. Little, J. Geurts, and J. Hunter. Dynamic generation of intelligent multimedia presentations through semantic inferencing. In *Conf. on Research and Advanced Technology for Digital Libraries*. Springer–Verlag, September 2002.

[606] S. Little and L. Hardman. Cuypers Meets Users: Implementing a User Model Architecture for Multimedia Presentation Generation. In Y.-P. P. Chen, editor, *10th Int. Multimedia Modeling Conf.; Brisbane, Australia*, pages 364–364. IEEE, January 2004.

[607] T. D. C. Little and A. Ghafoor. Network considerations for distributed multimedia object composition and communication. *IEEE Network*, 4(6):32–40, 45–49, November 1990.

[608] T. D. C. Little and A. Ghafoor. Interval-Based Conceptual Models for Time-Dependent Multimedia Data. *IEEE Trans. on Knowledge and Data Engineering*, 5(4):551–563, 1993.

[609] C. Liu. *SmallTalk, Objects, and Design*, Chapter 11, pages 115–126. iUniverse, 2000.

[610] D. Liu and Tsuhan C. Content-free image retrieval using Bayesian product rule. In *Multimedia and Expo, 2006 IEEE International Conference on*, pages 89–92, 2006.

[611] F. Liu and R. W. Picard. Periodicity, directionality, and randomness: World features for image modeling and retrieval. *Pattern Analysis and Machine Intelligence, IEEE Transactions on*, 18(7):722–733, 1996.

[612] H. Liu, S. Jiang, Q. Huang, C. Xu, and W. Gao. Region-based visual attention analysis with its application in image browsing on small displays. In *Proceedings of the 15th International Conference on Multimedia*, MULTIMEDIA '07, pages 305–308, New York, NY, USA, ACM, 2007.

[613] H. Liu and H. Lieberman. Robust photo retrieval using world semantics. In *Proceedings of LREC2002 Workshop: Using Semantics for IR, Canary Islands*, pages 15–20, 2002.

[614] Y. Liu, E. Shriberg, A. Stolcke, D. Hillard, M. Ostendorf, and M. Harper. Enriching speech recognition with automatic detection of sentence boundaries and disfluencies. *IEEE Transactions on Audio, Speech, and Language Processing*, 14(5):1526–1540, September 2006.

[615] F. Long, H. Zhang, and D.D. Feng. Fundamentals of content-based image retrieval. In *Multimedia Information Retrieval and Management*, 2003.

[616] R. López-Cózar, Z. Callejas, M. Kroul, J. Nouza, and J. Silovský. Two-level fusion to improve emotion classification in spoken dialogue systems. In *Text, Speech and Dialogue*, pages 617–624. Springer, 2008.

[617] M. Lopez-Nores, J. Pazos-Arias, J. Garcia-Duque, Y. Blanco-Fernandez, M. Martin-Vicente, A. Fernandez-Vilas, M. Ramos-Cabrer, and A. Gil-Solla. MiSPOT: dynamic product placement for Digital TV

through MPEG-4 processing and semantic reasoning. *Knowledge and Information Systems*, 22(1):101–128, January 2010.

[618] E. Loupias, N. Sebe, and N. Sebe. Wavelet-based salient points: Applications to image, 2000.

[619] D. G. Lowe. Distinctive image features from scale invariant keypoints. *International Journal of Computer Vision*, 60(2):91–110, November 2004.

[620] Y. Lu, C. Hu, X. Zhu, H. Zhang, and Q. Yang. A unified framework for semantics and feature based relevance feedback in image retrieval systems. In *Proceedings of the Eighth ACM International Conference on Multimedia*, MULTIMEDIA '00, pages 31–37, New York, NY, USA, ACM, 2000.

[621] Y. Lu, N. Sebe, R. Hytnen, and Q. Tian. Personalization in multimedia retrieval: A survey. *Multimedia Tools and Applications*, 51:247–277, 2011.

[622] D. Luckham. *The Power of Events: An Introduction to Complex Event Processing in Distributed Enterprise Systems.* Addison Wesley, London, UK, May 2002.

[623] O. Ludwig, D. Delgado, V. Goncalves, and U. Nunes. Trainable classifier-fusion schemes: An application to pedestrian detection. In *Proceedings of 12th International IEEE Conference on Intelligent Transportation Systems*, pages 432–437, St.Louis, MO, USA, October 4–7 2009.

[624] J. Luo and A. Savakis. Indoor vs outdoor classification of consumer photographs using low-level and semantic features. In *Image Processing, 2001. Proceedings. 2001 International Conference on*, vol.2, pages 745–748, October 2001.

[625] S. Luz. Interleave factor and multimedia information visualisation. In H. Sharp, P. Chalk, J. LePeuple, and J. Rosbottom, editors, *Proceedings of Human Computer Interaction 2002*, volume 2, pages 142–146, London, September 2002.

[626] S. Luz. Locating case discussion segments in recorded medical team meetings. In *Proceedings of the ACM Multimedia Workshop on Searching Spontaneous Conversational Speech (SSCS'09)*, pages 21–30, Beijing, China, ACM Press, October 2009.

[627] S. Luz. The non-verbal structure of patient case discussions in multidisciplinary medical team meetings. *ACM Transactions on Information Systems*, 30(3):article 17, August 2012.

[628] S. Luz and M. Masoodian. A mobile system for non-linear access to time-based data. In *Proceedings of Advanced Visual Interfaces AVI'04*, pages 454–457. ACM Press, May 2004.

[629] S. Luz and M. Masoodian. A model for meeting content storage and retrieval. In Y.-P. P. Chen, editor, *11th International Conference on Multi-Media Modeling (MMM 2005)*, pages 392–398, Melbourne, Australia, IEEE Signal Processing Society, January 2005.

[630] S. Luz and D. M. Roy. Meeting browser: A system for visualising and accessing audio in multicast meetings. In *Proceedings of the International Workshop on Multimedia Signal Processing*, pages 489–494, Copenhagen, Denmark, IEEE Signal Processing Society, September 1999.

[631] M. Maloof and R. Michalski. Selecting examples for partial memory learning. *Machine Learning*, 41(1):27–52, October 2000.

[632] M. Pazzani and D. Billsus. Learning and revising user profiles: the identification of interesting web sites. *Machine Learning*, 27(3):313–331, June 1997.

[633] H. Ma, H. Yang, M. R. Lyu, and I. King. Sorec: social recommendation using probabilistic matrix factorization. In *Proceedings of the 17th ACM Conference on Information and Knowledge Management*, CIKM '08, pages 931–940, New York, NY, USA, ACM, 2008.

[634] W. Y. Ma and B. S. Manjunath. Netra: a toolbox for navigating large image databases. In *Image Processing, 1997. Proceedings., International Conference on*, vol. 1, pages 568–571 vol.1, October 1997.

[635] W. Y. Ma and B. S. Manjunath. A texture thesaurus for browsing large aerial photographs. *J. Am. Soc. Inf. Sci.*, 49(7):633–648, May 1998.

[636] W. Y. Ma and B. S. Manjunath. Netra: A toolbox for navigating large image databases. In *Multimedia Systems*, pages 568–571, 1999.

[637] C. Macdonald and I. Ounis. Usefulness of quality clickthrough data for training. In *Proceedings of the 2009 Workshop on Web Search Click Data*, pages 75–79, Barcelona, Spain, February 9 2009.

[638] A. Maedche and V. Zacharias. Clustering ontology-based metadata in the semantic web. In *Proceedings of the 6th European Conference on Principles of Data Mining and Knowledge Discovery*, PKDD '02, pages 348–360, London, UK, Springer–Verlag, 2002.

[639] T. P. Mailis, G. Stoilos, and G. B. Stamou. Expressive reasoning with horn rules and fuzzy description logics. *Knowledge and Information Systems*, 25(1):105–136, 2010.

[640] J. Mairal, F. Bach, J. Ponce, and G. Sapiro. Online learning for matrix factorization and sparse coding. *Journal of Machine Learning Research*, 11:19–60, 2010.

[641] A. Makadia, V. Pavlovic, and S. Kumar. A new baseline for image annotation. In *Proceedings of European Conference on Computer Vision*, pages 316–329, Marseille, France, October 12 18, 2008.

[642] U. Malinowska, P. J. Durka, J. Żygierewicz, W. Szelenberger, and A. Wakarow. Explicit parameterization of sleep EEG transients. *Computers in Biology and Medicine*, 37(4):534–541, 2007.

[643] E. T. Mampusti, J. S. Ng, J. J. I. Quinto, G. L. Teng, M. T. C. Suarez, and R. S. Trogo. Measuring academic affective states of students via brainwave signals. In *Knowledge and Systems Engineering (KSE), 2011 Third International Conference on*, pages 226–231, Hanoi, Vietnam, IEEE, October 2011.

[644] B. S. Manjunath and W. Y. Ma. Texture features for browsing and retrieval of image data. *Pattern Analysis and Machine Intelligence, IEEE Transactions on*, 18(8):837–842, 1996.

[645] B. S. Manjunath, J. R. Ohm, V. V. Vasudevan, and A. Yamada. Colour and texture descriptors. *IEEE Transactions on Circuits and Systems for Video Technology*, 11(6):703–715, 2001.

[646] B. S. Manjunath, P. Salembier, and T. Sikora. *Introduction to MPEG 7: Multimedia Content Description Language*. Ed. Wiley, 2002.

[647] S. Marat, T. Ho Phuoc, L. Granjon, N. Guyader, D. Pellerin, and A. Guérin-Dugué. Modelling spatio-temporal saliency to predict gaze direction for short videos. *International Journal on Computer Vision*, 82(3):231–243, 2009.

[648] D. L. Martin, M. Paolucci, S. A. McIlraith, M. H. Burstein, D. V. McDermott, D. L. McGuinness, B. Parsia, T. R. Payne, M. Sabou, M. Solanki, N. Srinivasan, and J. P. Sycara. Bringing semantics to Web Services: The OWL-S approach. *Lecture Notes in Computer Science*, 3387(1):26–42, July 2005.

[649] A. B. B. Martinez, J. J. Pazos Arias, A. F. Vilas, J. G. Duque, and M. L. Nores. What's on TV tonight? An efficient and effective personalized recommender system of TV programs. In *Consumer Electronics, 2009. ICCE '09. IEEE Transactions on*, vol. 55, pages 286–294, February 2009.

[650] J. M. Martinez, R. Koenen, and F. Pereira. MPEG-7: the generic multimedia content description standard, part 1. *MultiMedia, IEEE*, 9(2):78–87, 2002.

[651] L. Martinez, M. J. Barranco, L. G. Perez, M. Espinilla, and F. Siles. A knowledge based recommender system with multigranular linguistic information. *International Journal of Computational Intelligence Systems*, 1(3):225–236, 2008.

[652] C. L. Marvel and J. E. Desmond. The contributions of cerebro-cerebellar circuitry to executive verbal working memory. *Cortex*, 46(7):880–895, 2010.

[653] J. V. Mascelli. *The five C's of cinematography: motion picture filming techniques*. Silman-James Press, Los Angeles, CA, US, June 1998.

[654] T. Massey, G. Marfia, M. Potkonjak, and M. Sarrafzadeh. Experimental analysis of a mobile health system for mood disorders. *Information Technology in Biomedicine, IEEE Transactions on*, 14(2):241–247, 2010.

[655] M. T. Maybury, editor. *Intelligent Multimedia Interfaces*. American Association for Artificial Intelligence, Menlo Park, CA, USA, 1993.

[656] I. McCowan, D. Gatica-Perez, S. Bengio, G. Lathoud, M. Barnard, and D. Zhang. Automatic analysis of multimodal group actions in meetings. *IEEE Transactions on Pattern Analysis and Machine Intelligence*, 27(3):305–317, March 2005.

[657] V. McGee and B. McLaughlin. Distinctions Without a Difference. *Southern Journal of Philosophy*, 33, 1994.

[658] K. McKeown, J. Robin, and M. Tanenblatt. Tailoring lexical choice to the user's vocabulary in multimedia explanation generation. In *Proc. of the 31st Conf. on Association for Computational Linguistics; Columbus, OH, USA*, pages 226–234, 1993.

[659] C. Meghini, F. Sebastiani, and U. Straccia. A model of multimedia information retrieval. *J. ACM*, 48(5):909–970, 2001.

[660] A. Mehrabian. Pleasure-arousal-dominance: A general framework for describing and measuring individual differences in temperament. *Current Psychology*, 14(4):261–292, December 1996.

[661] R. Mehrotra and J.E. Gary. Similar-shape retrieval in shape data management. *Computer*, 28(9):57–62, 1995.

[662] J. Melton and A. Eisenberg. SQL multimedia and application packages (SQL/MM). *SIGMOD Rec.*, 30:97–102, December 2001.

[663] R. Mendoza and M. A. Williams. Ontology based object categorisation for robots. In *Proceedings of the 2005 Australasian Ontology Workshop - Volume 58*, AOW '05, pages 61–67, Darlinghurst, Australia, Australian Computer Society, Inc, 2005.

[664] Merriam-Webster, Inc., USA. Definition of multimedia, 2006. `http://www.m-w.com/dictionary/multimedia`, last accessed: 23/1/2013.

[665] Merriam-Webster, Inc., USA. Definition of personalize, 2006. `http://www.m-w.com/dictionary/personalization`, last accessed: 23/1/2013.

[666] T. O. Meservy, M. L. Jensen, J. Kruse, J. K. Burgoon, J. F. Nunamaker Jr, D. P. Twitchell, G. Tsechpenakis, and D. N. Metaxas. Deception detection through automatic, unobtrusive analysis of nonverbal behavior. *Intelligent Systems, IEEE*, 20(5):36–43, 2005.

[667] V. Mezaris, I. Kompatsiaris, and M. G. Strintzis. Region-based image retrieval using an object ontology and relevance feedback. In *Eurasip Journal on Applied Signal Processing*, pages 886–901, 2004.

[668] E. Michaletou. *Customization speech's signal for recognition speaker's emotion*. PhD thesis, Electrical Engineering and Computer Technologies, University of Patras, Greece, 2008.

[669] A. Michlmayr, F. Rosenberg, P. Leitner, and S. Dustdar. Advanced event processing and notifications in service runtime environments. In *Proceedings of the Second International Conference on Distributed Event-Based Systems*, DEBS '08, pages 115–125, New York, NY, USA, ACM, July 2008.

[670] Microsoft. Ambient intelligence at microsoft. `http://www.engadget.com/2011/05/03/microsofts-home-of-the-future-lulls-teens-to-sleep-with-tweets/`.

[671] Microsoft. From real humans to avatars in real-time. `http://www.microsoft.com/presspass/features/2012/jan12/01-03Future.mspx`.

[672] Microsoft. Microsoft's kinect., 2010. `http://www.xbox.com/kinect`.

[673] S. E. Middleton, N. R. Shadbolt, and D. C. De Roure. Ontological user profiling in recommender systems. *ACM Trans. Inf. Syst.*, 22(1):54–88, January 2004.

[674] S. E. Middleton, N. R. Shadbolt, and D. C. De Roure. Ontological user profiling in recommender systems. *ACM Transactions on Information Systems (TOIS)*, 22(1):54–88, 2004.

[675] K. Mikolajczyk and C. Schmid. Scale and affine invariant interest point detectors. *Int. J. Comput. Vision*, 60(1):63–86, October 2004.

[676] K. Mikolajczyk and C. Schmid. A performance evaluation of local descriptors. *IEEE Transactions on Pattern Analysis and Machine Intelligence*, 27(10):1615–1630, October 2005.

[677] B. N. Miller, I. Albert, S. K. Lam, J. A. Konstan, and J. Riedl. Movielens unplugged: experiences with an occasionally connected recommender system. In *Proceedings of the 8th International Conference on Intelligent User Interfaces*, IUI '03, pages 263–266, New York, NY, USA, ACM, 2003.

[678] F. P. Miller, A. F. Vandome, and M. B. John. *Independent Component Analysis*. VDM Publishing, Saarbrcken, Germany, December 2010.

[679] G. A. Miller and W. G. Charles. Contextual correlates of semantic similarity. *Language & Cognitive Processes*, 6(1):1–28, 1991.

[680] G. A. Miller et al. Wordnet: a lexical database for english. *Communications of the ACM*, 38(11):39–41, 1995.

[681] R. Mizoguchi. Ontology engineering environments. *International Handbooks on Information Systems, Handbook on Ontologies*, pages 275–295, 2004.

[682] J.-C. Moissinac. Automatic discovery and composition of multimedia adaptation services. In *Proc. of the 4th International Conferences on Advances in Multimedia (MMEDIA), Chamonix, France*, pages 155–160, April 2012.

[683] F. Mokhtarian and R. Suomela. Robust image corner detection through curvature scale space. *IEEE Transactions on Pattern Analysis Machine Intelligence*, 20(12):1376–1381, 1998.

[684] A. G. Money and H. W. Agius. Video summarisation: A conceptual framework and survey of the state of the art. *J. Visual Communication and Image Representation*, pages 121–143, 2008.

[685] T. P. Moran, L. Palen, S. Harrison, P. Chiu, D. Kimber, S. Minneman, W. van Melle, and P. Zellweger. "I'll get that off the audio": A case study of salvaging multimedia meeting records. In *Proceedings of ACM CHI 97 Conference on Human Factors in Computing Systems*, vol. 1, pages 202–209, San Jose, CA, ACM, May 1997.

[686] Y. Mori, H Takahashi, and R. Oka. Image-to-word transformation based on dividing and vector quantizing images with words. In *Proceedings of International Workshop on Multimedia Intelligent Storage and Retrieval Management*, Orlando, FL, USA, October 30, 1999.

[687] A. Mosca. A review essay on Antonio Damasio's the feeling of what happens: Body and emotion in the making of consciousness. *PSYCHE*, 6:10, 2000.

[688] M. D. Mulvenna, S. S. Anand, and A. G. Büchner. Personalization on the Net using Web mining: introduction. *Communications of the ACM*, 43(8):122–125, 2000.

[689] H. Murase and S. Nayar. Visual learning and recognition of 3-D objects from appearance. *International Journal of Computer Vision*, 14(1):5–24, January 1995.

[690] P. Mylonas. Understanding how visual context influences multimedia content analysis problems. In Ilias Maglogiannis, Vassilis Plagianakos, and Ioannis Vlahavas, editors, *Artificial Intelligence: Theories and Applications*, volume 7297 of *Lecture Notes in Computer Science*, pages 361–368. Springer, May 2012.

[691] P. Mylonas, E. Spyrou, Y. Avrithis, and S. Kollias. Using visual context and region semantics for high-level concept detection. *IEEE Transactions on Multimedia*, 11(2):229–243, February 2009.

[692] P. Mylonas, E. Spyrou, Y. Avrithis, and S. Kollias. Using visual context and region semantics for high-level concept detection. *Multimedia, IEEE Transactions on*, 11(2):229–243, 2009.

[693] P. Mylonas, D. Vallet, P. Castells, M. Fernandez, and Y. Avrithis. Personalized information retrieval based on context and ontological knowledge. *The Knowledge Engineering Review*, 23(01):73–100, 2008.

[694] P. Mylonas and M. Wallace. Using ontologies and fuzzy relations in multimedia personalization. In *Proceedings of the First International Workshop on Semantic Media Adaptation and Personalization (SMAP)*, pages 146–150. IEEE Computer Society, 2006.

[695] G. R. Naik, D. Kant Kumar, et al. Twin SVM for gesture classification using the surface electromyogram. *Information Technology in Biomedicine, IEEE Transactions on*, 14(2):301–308, 2010.

[696] A. Natsev, A. Haubold, J. Tešić, L. Xie, and R. Yan. Semantic concept-based query expansion and re-ranking for multimedia retrieval. In *Proceedings of the 15th International Conference on Multimedia*, MULTIMEDIA '07, pages 991–1000, New York, NY, USA, ACM, 2007.

[697] A. Natsev, R. Rastogi, and K. Shim. Walrus: A similarity retrieval algorithm for image databases. pages 395–406, 1999.

[698] A. Natsev, R. Rastogi, and K. Shim. Walrus: a similarity retrieval algorithm for image databases. *Knowledge and Data Engineering, IEEE Transactions on*, 16(3):301–316, March 2004.

[699] Y. Naudet. Reconciling context, observations and sensors in ontologies for pervasive computing. In *Proceedings of the 6th International Workshop on Semantic Media Adaptation and Personalization (SMAP2011)*, 2011.

[700] Y. Naudet, A. Aghasaryan, S. Mignon, Y. Toms, and C. Senot. Ontology-Based Profiling and Recommendations for Mobile TV. In M. Wallace et al., editors, *Semantics in Adaptive and Personalized Services*, vol. 279/2010 of *Studies in Computational Intelligence*, pages 23–48. Springer Berlin / Heidelberg, 2010.

[701] Y. Naudet, G. Arnould, S. Mignon, M. Foulonneau, K. Devooght, and D. Nicolas. A mixed CCN-MAS architecture to handle context-awareness in hybrid networks. In Cunningham P. and Cunningham M., editors, *Proc. of Future Network and Mobile Summit 2011*. IIMC International Information Management Corporation, 2011. ISBN: 978-1-905824-23-6.

[702] Y. Naudet, V. Groués, and M. Foulonneau. Introduction to Fuzzy-Ontological Context-Aware Recommendations in Mobile Environments. In *Proc. of 1st International Workshop on Adaptation, Personalization and REcommendation in the Social-semantic Web (APRESW 2010), 7th Extended Semantic Web Conference (ESWC 2010)*, Heraklion, Greece, CEUR, May 31, 2010.

[703] Y. Naudet, S. Mignon, L. Lecaque, C. Hazotte, and V. Groués. Ontology-based Matchmaking Approach for Context-aware Recommendations. In *Proc. of the Fourth International Conference on Automated Solutions for CrossMedia Content and Multi-Channel Distribution (AXMEDIS 2008)*, Florence, Italy, November 2008.

[704] Y. Naudet, L. Schwartz, S. Mignon, and M. Foulonneau. Applications of user and context-aware recommendations using ontologies. In *Proc. of the 22th Conference Internationale Francophone sur I'Interaction Homme-Machine (IHM'10)*, IHM '10, pages 165–172, New York, NY, USA, ACM, Sept. 2010.

[705] NAZOU. NAZOU project Website: http://nazou.fiit.stuba.sk/.

[706] C. W. Niblack, R. Barber, W. Equitz, M. D. Flickner, E. H. Glasman, D. Petkovic, P. Yanker, C. Faloutsos, and G. Taubin. QBIC project: querying images by content, using color, texture, and shape. In *Proceedings of SPIE Storage and Retrieval for Image and Video Databases*, pages 173–187, San Jose, CA, USA, January 31–February 5, 1993.

[707] W. Niblack, R. Barber, W. H. R. Equitz, E. Glasman, D. Petovic, P. Yanker, C. Faloutsos, and G. Taubin. *The QBIC Project: Querying Images by Content Using Color, Texture, and Shape*. Computer science. IBM Research Division, 1993.

[708] E. Niedermeyer and F. H. L. Da Silva. *Electroencephalography: basic principles, clinical applications, and related fields*. Lippincott Williams & Wilkins, Philadelphia, USA, November 2004.

[709] J. Nielsen. Personalization is Over-Rated, October 1998. http://www.nngroup.com/articles/personalization-is-over-rated/, last accessed: 23/1/2013.

[710] A. Nijholt, D. P.-O. Bos, and B. Reuderink. Turning shortcomings into challenges: Brain computer interfaces for games. *Entertainment Computing*, 1(2):85–94, 2009.

[711] D. Nister and H. Stewenius. Scalable recognition with a vocabulary tree. In *IEEE Conference on Computer Vision and Pattern Recognition (CVPR'06), June 17–22, New York, NY, USA*, 2006.

[712] M. Nixon and A. S. Aguado. *Feature Extraction & Image Processing*. Academic Press, second edition, 2008.

[713] NoTube. Project Website: http://www.notube.tv.

[714] D. W. Oard and J. Kim. Implicit feedback for recommender system. Technical report, Massachusetts Institute of Technology, Department of Electrical Engineering and Computer, 1998.

[715] Open Geospatial Consortium (OGC). OpenGIS KML Encoding Standard (OGC KML), url=http://www.opengeospatial.org/standards/kml/.

[716] Open Geospatial Consortium (OGC). Web Map Service, url=http://www.opengeospatial.org/standards/wms/.

[717] H. Ohmura, T. Kitasuka, and M. Aritsugi. Web browsing behavior recording system. In *Proceedings of the 15th International Conference on Knowledge-Based and Intelligent Information and Engineering Systems - Volume Part IV*, KES'11, pages 53–62, Berlin, Heidelberg, 2011. Springer–Verlag.

[718] D. Ostwald, C. Porcaro, and A. P. Bagshaw. Voxel-wise information theoretic eeg-fmri feature integration. *Neuroimage*, 55(3):1270–1286, 2011.

[719] T. Otsuka and J. Ohya. Spotting segments displaying facial expression from image sequences using hmm. In *Automatic Face and Gesture Recognition, 1998. Proceedings. Third IEEE International Conference on*, pages 442–447. IEEE, 1998.

[720] P. Ganesan, H. Garcia-Molina and J. Widom. Exploiting Hierarchical Domain Structure to Compute Similarity. *ACM Transactions on Information Systems*, 21(1):64–93, January 2003.

[721] P. Mell and T. Grance. The NIST definition of cloud computing. http://csrc.nist.gov/publications/nistpubs/800-145/SP800-145.pdf, 2011.

[722] C. Palmisano, A. Tuzhilin, and M. Gorgoglione. Using context to improve predictive modeling of customers in personalization applications. *Knowledge and Data Engineering, IEEE Transactions on*, 20(11):1535 –1549, November 2008.

[723] H. Pan, S. E. Levinson, T. S. Huang, and Z.-P. Liang. A fused hidden Markov model with application to bimodal speech processing. *Signal Processing, IEEE Transactions on*, 52(3):573–581, 2004.

[724] J. Y. Pan, H. J. Yang, P. Duygulu, and C. Faloutsos. Automatic image captioning. In *International Conference on Multimedia and Expo (ICME)*, 2004.

[725] E. Panofsky. *Meaning in the Visual Arts: Papers in and on Art History*. Doubleday, 1967.

[726] A. Panteli, C. Tsinaraki, L. Ragia, F. Kazasis, and S. Christodoulakis. MObile Multimedia Event Capturing and Visualization (MOME). In *Proceedings of FTRA Conference Multimedia and Ubiquitous Engineering (MUE 2011)*, pages 101–106, Loutraki, Greece, 2011.

[727] A. Pantelopoulos and N. G Bourbakis. A survey on wearable sensor-based systems for health monitoring and prognosis. *Systems, Man, and Cybernetics, Part C: Applications and Reviews, IEEE Transactions on*, 40(1):1–12, 2010.

[728] D. Papadias and T. Sellis. Qualitative Representation of Spatial Knowledge in Two-Dimensional Space. *The VLDB Journal: The International Journal on Very Large Data Bases*, 3(4):479–516, October 1994.

[729] D. Papadias, T. Sellis, Y. Theodoridis, and M. J. Egenhofer. Topological relations in the world of minimum bounding rectangles: a study with r-trees. In *SIGMOD*, pages 92–103. ACM, 1995.

[730] D. Papadias, Y. Theodoridis, T. Sellis, and M. J. Egenhofer. Topological Relations in the World of Minimum Bounding Rectangles: A Study with R-Trees. In *Proc. of the ACM SIGMOD Conf. on Management of Data; San Jose, CA, USA*, pages 92–103, March 1995.

[731] S. Papadopoulos, C. S. Zigkolis, S. Kapiris, Y. Kompatsiaris, and A. Vakali. City exploration by use of spatio-temporal analysis and clustering of user contributed photos. In *Proceedings of the 1st ACM International Conference on Multimedia Retrieval*, ICMR '11, pages 65:1–65:2, New York, NY, USA, ACM, 2011.

[732] B. Parkinson and M. Lea. Investigating personal constructs of emotions. *British Journal of Psychology*, 82(1):73–86, 1991.

[733] D. Parry. A fuzzy ontology for medical document retrieval. In *Proceedings of the Second Workshop on Australasian Information Security, Data Mining and Web Intelligence, and Software Internationalisation - Volume 32 (Dunedin, New Zealand)*. J. Hogan, P. Montague, M. Purvis, and C. Steketee, Eds. *ACM International Conference Proceeding Series, vol. 54. Australian Computer Society, Darlinghurst, Australia, 121–126*, 2004.

[734] A. Passant. Measuring Semantic Distance on Linking Data and Using it for Resources Recommendations. In *2010 AAAI Spring Symposium Series*, 2010.

[735] A. Pentland, R. W. Picard, and S. Sclaroff. Photobook: Content-based manipulation of image databases, 1995.

[736] A. Pentland, R. W. Picard, and S. Sclaroff. Photobook: Content-based manipulation of image databases. *International Journal of Computer Vision*, 18(3):233–254, June 1996.

[737] F. Pereira. The MPEG-21 Standard: Why an Open Multimedia Framework? In D. Shepherd, J. Finney, L. Mathy, and N. Race, editors, *Interactive Distributed Multimedia Systems*, volume 2158 of *Lecture Notes in Computer Science*, pages 219–220. Springer Berlin Heidelberg, September 2001.

[738] F. Perronnin and C. Dance. Fisher kernels on visual vocabularies for image categorization. In *IEEE Conference on Computer Vision and Pattern Recognition (CVPR'07), June 18–23, Minneapolis, Minnesota, USA*, 2007.

[739] M. Perry, M. Janik, C. Ramakrishnan, C. Ibanez, B. Arpinar, and A. Sheth. Peer-to-Peer discovery of semantic associations. In *2nd International Workshop on Peer-to-Peer Knowledge Management*, pages 1–12, San Diego, USA, July 2005.

[740] M. Petersen, C. Stahlhut, A. Stopczynski, J. Larsen, and L. Hansen. Smartphones get emotional: mind reading images and reconstructing the neural sources. In *Affective Computing and Intelligent Interaction*, pages 578–587, Memphis, Tennessee, USA, Springer, October 2011.

[741] P. C. Petrantonakis and L. J. Hadjileontiadis. Emotion recognition from EEG using higher order crossings. *Information Technology in Biomedicine, IEEE Transactions on*, 14(2):186–197, 2010.

[742] D. Petrovska-Delacrêtaz, G. Chollet, and B. Dorizzi. *Guide to biometric reference systems and performance evaluation*. Springer, 2009.

[743] T. N. Phyu. Survey of classification techniques in data mining. In *Proceedings of the International Multiconference of Engineers and Computer Scientists*, volume 1, pages 18–20, Hong Kong, China, March 2009.

[744] R. W. Picard. *Affective computing.* MIT press, 1995.

[745] R. W. Picard. Light-years from lena: video and image libraries of the future. In *Proceedings of International Conference on Image Processing*, pages 310–313, Washington, DC, USA, October 23–26 1995.

[746] R. W. Picard, T. P. Minka, et al. Vision texture for annotation. *Multimedia systems*, 3(1):3–14, 1995.

[747] M. Piccardi and Ó. Pérez. Hidden Markov models with kernel density estimation of emission probabilities and their use in activity recognition. In *Computer Vision and Pattern Recognition, 2007. CVPR'07. IEEE Conference on*, pages 1–8. IEEE, 2007.

[748] H. Pirsiavash and D. Ramanan. Detecting activities of daily living in first-person camera views. In *IEEE Conference on Computer Vision and Pattern Recognition (CVPR'12), June 16–21, Providence, Rhode Island, USA*, 2012.

[749] K. B. Plaban, S. Sudeshna, and B. Anupam. Ontology based user modeling for personalized information access. *IJCSA*, 7(1):1–22, 2010.

[750] D. Plas, M. Verheijen, H. Zwaal, and M. Hutschemaekers. Manipulating context information with SWRL. Report of A-MUSE project, 2006.

[751] J. C. Platt. *Advances in kernel methods.* Chapter Fast training of support vector machines using sequential minimal optimization, pages 185–208. MIT Press, Cambridge, MA, USA, 1999.

[752] R. Plutchik. The nature of emotions. *American Scientist*, 89(4):344–350, 2001.

[753] M. Pomplun, H. Ritter, and B. Velichkovsky. Disambiguating complex visual information: Towards communication of personal views of a scene. *Perception*, 25:931–948, 1995.

[754] P. Poolman, R. M. Frank, P. Luu, S. M. Pederson, and D. M. Tucker. A single-trial analytic framework for EEG analysis and its application to target detection and classification. *Neuroimage*, 42(2):787–798, 2008.

[755] A. Popescu-Belis, D. Lalanne, and H. Bourlard. Finding information in multimedia meeting records. *MultiMedia, IEEE*, 19(2):48–57, February 2012.

[756] C. Porcel and E. Herrera-Viedma. A fuzzy linguistic recommender system to disseminate the own academic resources in universities. In *Proceedings of the 2009 IEEE/WIC/ACM International Joint Conference on Web Intelligence and Intelligent Agent Technology - Volume 03*, WI-IAT '09, pages 179–182, Washington, DC, USA, 2009. IEEE Computer Society.

[757] C. Porcel, A. G. López-Herrera, and E. Herrera-Viedma. A recommender system for research resources based on fuzzy linguistic modeling. *Expert Systems with Applications*, 36(3):5173–5183, 2009.

[758] S. Lo-Presti, D. Bert, and A. Duda. TAO: Temporal algebraic operators for modeling multimedia presentations. *J. Network and Computer Applications*, 25(4):319–342, 2002.

[759] N. U. Qazi, M. Woo, and A. Ghafoor. A synchronization and communication model for distributed multimedia objects. In *MULTIMEDIA*, pages 147–155, 1993.

[760] G.-J. Qi, X.-S. Hua, Y. Rui, J. Tang, T. Mei, and H.-J. Zhang. Correlative multi-label video annotation. In *Multimedia, Proceedings of the 15th International Conference on*, pages 17–26, 2007.

[761] J. R. Quinlan. Discovering rules by induction from large collections of examples. In D. Michie, editor, *Expert Systems in the Micro-Electronic Age*, pages 168–201. Edinburgh University Press, Edinburgh, 1979.

[762] J. R. Quinlan. *C4.5: Programs for machine learning*. Morgan Kaufmann, 1993.

[763] C. H. C. Leung and R. C. F. Wong. A knowledge framework for histogram-based image retrieval. In *Proceeding of International Conference on Imaging Engineering*, pages 904–909, 2009.

[764] R. L. Eubank. *Nonparametric Regression and Spline Smoothing*. CRC Press, 1999.

[765] M. D. Rabin and M. J. Burns. Multimedia authoring tools. In *Conf. Companion on Human Factors in Computing Systems; Vancouver, British Columbia, Canada*, pages 380–381. ACM, 1996.

[766] L. R. Rabiner. A tutorial on hidden markov models and selected applications in speech recognition. *Proceedings of the IEEE*, 77(2):257–286, 1989.

[767] A. Rabinovich, A. Vedaldi, C. Galleguillos, E. Wiewiora, and S. Belongie. Objects in context. In *Computer Vision, Proceedings of the IEEE 11th International Conference on*, pages 1–8, 2007.

[768] R. Rada, H. Mili, E. Bicknell, and M. Blettner. Development and application of a metric on semantic nets. *IEEE Transactions on Systems, Man, and Cybernetics*, 19(1):17–30, January 1989.

[769] A. Rafaeli and I. Vilnai-Yavetz. Instrumentality, aesthetics and symbolism of physical artifacts as triggers of emotion. *Theoretical Issues in Ergonomics Science*, 5(1):91–112, 2004.

[770] A. Rahimi and B. Recht. Clustering with normalized cuts is clustering with a hyperplane. In *Statistical Learning in Computer Vision workshop, ECCV04*, volume 56, Prague, Czech Republic, May 2004.

[771] C. Ramakrishnan, W. Milnor, M. Perry, and A. Sheth. Discovering informative connection subgraphs in multi-relational graphs. *SIGKDD Explorations Newsletter*, 7(2):56–63, December 2005.

[772] J. M. Ramirez-Cortes, V. Alarcon-Aquino, G. Rosas-Cholula, P. Gomez-Gil, and J. Escamilla-Ambrosio. P-300 rhythm detection using anfis algorithm and wavelet feature extraction in EEG signals. In *Proceedings of the World Congress on Engineering and Computer Science*, volume 1, pages 963–968, San Francisco, USA, October 2010.

[773] N. Rasiwasia and N. Vasconcelos. Holistic context modeling using semantic co-occurrences. In *Computer Vision and Pattern Recognition, Proceedings of the IEEE Conference on*, pages 1889–1895, June 2009.

[774] S. Ravela and R. Manmatha. On computing global similarity in images. In *Applications of Computer Vision, 1998. WACV '98. Proceedings., Fourth IEEE Workshop on*, pages 82–87, 1998.

[775] S. Ravela and R. Manmatha. Retrieving images by appearance. In *In Proceedings of the International Conference on Computer Vision*, pages 608–613, 1998.

[776] G. Rebolledo-Mendez and S. De Freitas. Attention modeling using inputs from a brain computer interface and user-generated data in second life. In *The Tenth International Conference on Multimodal Interfaces (ICMI 2008)*, Crete, Greece, October 2008.

[777] X. Ren and M. Philipose. Egocentric recognition of handled objects: Benchmark and analysis. In *Computer Vision and Pattern Recognition Workshop (CVPR'09), June 20–25, Miami, Florida, USA*, 2009.

[778] S. Renals. Recognition and understanding of meetings. In *Human Language Technologies: Conference of the North American Chapter of the Association of Computational Linguistics, Proceedings*, pages 1–9, Los Angeles, CA, ACL Press, June 2010.

[779] S. Renals, T. Hain, and H. Bourlard. Recognition and understanding of meetings: The AMI and AMIDA projects. In *Proc. IEEE Workshop on Automatic Speech Recognition and Understanding (ASRU '07)*, pages 238–247, Kyoto, Japan, IEEE, December 2007.

[780] P. Resnick, N. Iacovou, M. Suchak, P. Bergstrom, and J. Riedl. Grouplens: an open architecture for collaborative filtering of netnews. In *Proceedings of the 1994 ACM Conference on Computer Supported Cooperative Work*, CSCW '94, pages 175–186, New York, NY, USA, ACM, 1994.

[781] P. Resnick and H. R. Varian. Recommender systems. *Communications of the ACM*, 40(3):56–58, 1997.

[782] E. Rich. User modeling via stereotypes. *International Journal of Cognitive Science*, 3(1):329–354, January 1979.

[783] D. Riecken. Personalized communication networks. *Communications of the ACM*, 43(8):41–42, 2000.

[784] G. Rizzo and R. Troncy. Nerd: evaluating named entity recognition tools in the web of data. In *Workshop on Web Scale Knowledge Extraction (WEKEX11), Bonn, Germany*, pages 1–16, 2011.

[785] J. Robert and Howlett L. C. J. *Radial basis function networks 2: New advances in design*. Physica-Verlag, 2001.

[786] M. A. Rodriguez and M. J. Egenhofer. Determining semantic similarity among entity classes from different ontologies. *Knowledge and Data Engineering, IEEE Transactions on*, 15(2):442–456, 2003.

[787] T. Rohlfing, D. B. Russakoff, and C. R. Maurer, Jr., Performance-based classifier combination in atlas-based image segmentation using expectation-maximization parameter estimation. *Medical Imaging, IEEE Transactions on*, 23(8):983–994, 2004.

[788] A. Rosenthal. *Writing, Directing, and Producing Documentary Film*. Southern Illinois University Press, Illinois, IL, US, June 2007.

[789] R. Roth and K. Ross. Extending the Google Maps API for Event Animation Mashups. *Cartographic Perspectives*, 64:21–40, Fall 2009.

[790] R. Roth, K. Ross, B. Finch, W. Luo, and A. MacEachren. A user-centered approach for designing and developing spatiotemporal crime analysis tools. In *Proceedings of GIScience*, Zurich, Switzerland, September 2010.

[791] N. Roussopoulos, C. Faloutsos, and T. Sellis. An efficient pictorial database system for PSQL. *Software Engineering, IEEE Transactions on*, 14(5):639–650, 1988.

[792] D. M. Roy and S. Luz. Audio meeting history tool: Interactive graphical user-support for virtual audio meetings. In *Proceedings of the ESCA Tutorial and Research Workshop (ETRW) on Accessing Information in Spoken Audio*, pages 107–110, Cambridge, Cambridge University, April 1999.

[793] Y. Rubner, L. Guibas, and C. Tomasi. The earth movers distance, multi-dimensional scaling, and color-based image retrieval. In *in Proceedings of the ARPA Image Understanding Workshop*, pages 661–668, 1997.

[794] L. Ruen Shan, F. Ibrahim, and M. Moghavvemi. Assessment of steady-state visual evoked potential for brain computer communication. In *3rd Kuala Lumpur International Conference on Biomedical Engineering 2006*, pages 352–354, Kuala Lumpur, Malaysia, Springer, December 2007.

[795] B. C. Russell, A. Torralba, K. P. Murphy, and W. T. Freeman. Labelme: a database and web-based tool for image annotation. *International Journal of Computer Vision*, 77(1):157–173, 2008.

[796] S. J. Russell and P. Norvig. *Artificial Intelligence: A Modern Approach*. Prentice Hall, 2nd edition, 2003.

[797] L. Rutledge, B. Bailey, J. van Ossenbruggen, L. Hardman, and J. Geurts. Generating presentation constraints from rhetorical structure. In *Hypertext and Hypermedia*, pages 19–28. ACM, 2000.

[798] L. Rutledge, L. Hardman, J. van Ossenbruggen, and D. C. A. Bulterman. Implementing Adaptability in the Standard Reference Model for Intelligent Multimedia Presentation Systems. In *Proc. of the 1998 Conf. on MultiMedia Modeling*, pages 12–20. IEEE, 1998.

[799] L. Rutledge, J. van Ossenbruggen, L. Hardman, and D. C. A. Bulterman. Practical application of existing hypermedia standards and tools. In *Proc. of the 3rd ACM Conf. on Digital libraries; Pittsburgh, PA, USA*, pages 191–199. ACM, 1998.

[800] S. Berkovsky, M. Coombe, J. Freyne and D. Bhandari. Recommender algorithms in activity motivating games. In *Proceedings of the 4th ACM Conference on Recommender Systems*, pages 175–182, Barcelona, Spain, September 2010.

[801] S. Mukherjea and B. Bamba. BioPatentMiner: an information retrieval system for biomedical patents. In *Proceedings of 30th International Conference on Very Large Data Bases*, pages 1066–1077, Toronto, Canada, September 2004.

[802] S. M. Sadjadi and P. K. McKinley. A survey of adaptive middleware. *Michigan State University Report MSU-CSE-03-35*, 2003.

[803] H. Sahbi, J. Y. Audibert, J. Rabarisoa, and R. Keriven. Robust matching and recognition using context-dependent kernels. In *ACM International Conference on Machine Learning (ICML'08), June 5–9, Helsinki, Finland*, 2008.

[804] P. Salembier and T. Sikora. *Introduction to MPEG-7: Multimedia Content Description Interface*. John Wiley & Sons, Inc., New York, NY, USA, 2002.

[805] P. Salembier and J.R. Smith. MPEG-7 multimedia description schemes. *Circuits and Systems for Video Technology, IEEE Transactions on*, 11(6):748 –759, June 2001.

[806] M. Salmela and V. E. Mayer. *Emotions, ethics, and authenticity*, vol. 5. John Benjamins, 2009.

[807] A. Salomaa. *Computation and Automata*. Cambridge University Press, 1989.

[808] Sananews. Samsung integrates cameras directly in their TVs:. http:// www.sananews.net/english/2012/01/samsung-will-include-int egrated -camera-and-microphone-in-smart-tv-2-0/.

[809] D. Sanchez, D. Isern, and M. Millan. Content annotation for the semantic web: an automatic web-based approach. *Knowledge and Information Systems*, 27(3):393–418, June 2011.

[810] P. Sandhaus, S. Thieme, and S. Boll. Processes of photo book production. *Multimedia Syst.*, 14(6):351–357, 2008.

[811] A. Sano, R. W. Picard, S. E. Goldman, B. A. Malow Rana el Kaliouby, and R. Stickgold. Analysis of autonomic sleep patterns. Research Projects. available from http://www.media.mit.edu/research/ groups/affective-computing.

[812] J. B. Schafer, J. A. Konstan, and J. Riedl. E-commerce recommendation applications. *Data mining and knowledge discovery*, 5(1):115–153, 2001.

[813] Schema.org. Website: http://schema.org/.

[814] A. Scherp. Software development process model and methodology for virtual laboratories. In *Proc. of the 20th IASTED Int. Multi-Conf. Applied Informatics; Innsbruck, Austria*, pages 47–52. IASTED, February 2002.

[815] A. Scherp. *A Component Framework for Personalized Multimedia Applications*. PhD thesis, University of Oldenburg, Germany, 2006. http: //www.ansgarscherp.net/dissertation, last accessed: 20/1/2013.

[816] A. Scherp and S. Boll. Context-driven smart authoring of multimedia content with xSMART. In *Proc. of the 13th Annual ACM Int. Conf. on Multimedia; Hilton, Singapore*, pages 802–803. ACM, 2005.

[817] A. Scherp and S. Boll. *Managing Multimedia Semantics*, Chapter MM4U - A framework for creating personalized multimedia content, page 246. Idea Publishing, 2005.

[818] A. Scherp and S. Boll. Paving the Last Mile for Multi-Channel Multimedia Presentation Generation. In Y.-P. P. Chen, editor, *Proc. of the 11th Int. Conf. on Multimedia Modeling; Melbourne, Australia*, pages 190–197. IEEE, January 2005.

[819] A. Scherp and S. Boll. Paving the last mile for multi-channel multimedia presentation generation. In *Multimedia Modeling*, pages 190–197. IEEE, 2005.

[820] B. Schiele and J. L. Crowley. Probabilistic object recognition using multidimensional receptive field histograms. In *Proceedings of the 13th International Conference on Pattern Recognition*, volume 2, pages 50–54, Vienna, Austria, August 25–29 1996.

[821] C. Schmid and R. Mohr. Local grayvalue invariants for image retrieval. *IEEE Transactions on Pattern Analysis and Machine Intelligence*, 19(5):530–535, May 1997.

[822] L. A. Schmidt and L. J. Trainor. Frontal brain electrical activity (EEG) distinguishes valence and intensity of musical emotions. *Cognition & Emotion*, 15(4):487–500, September 2001.

[823] P. Schmitz, J. Yu, and P. Santangeli. Timed Interactive Multimedia Extensions for HTML (HTML+TIME) – Extending SMIL into the Web Browser, September 1998. http://www.w3.org/TR/1998/NOTE-HTMLplusTIME-19980918, last accessed: 23/1/2013.

[824] B. Schopman, D. Palmisano, R. Siebes, C. van Aart, M. Minno, V. Malaisé, and L. Aroyo. Linking TV and Web using semantics, a notube application. In *Future Television: Integrating the Social and Semantic Web (workshop at EuroITV 2010)*, 2010.

[825] G. Schreiber, H. Akkermans, A. Anjewierden, R. Hoog, N. Shadbolt, W. Velde, and B.J. Wielinga. *Knowledge Engineering and Management: The CommonKADS Methodology*. MIT Press, 2nd edition, 2002.

[826] M. A. Schreibman. *The Indie Producers Handbook: Creative Producing from A to Z*. Crown Publishing Group, New York, NY, US, August 2012.

[827] A. Schroeder, M. van der Zwaag, and M. Hammer. A middleware architecture for human-centred pervasive adaptive applications. In *Self-Adaptive and Self-Organizing Systems Workshops, 2008. SASOW 2008. Second IEEE International Conference on*, pages 138–143. IEEE, 2008.

[828] C. Schuldt, I. Laptev, and B. Caputo. Recognizing human actions: A local SVM approach. In *International Conference on Pattern Recognition (ICPR'04), August 23–26, Cambridge, UK*, pages 32–36, 2004.

[829] R. Schulmeister. Intelligent tutorial systems. In *Hypermedia Learning Systems: Theory - Didactics - Design*, Chapter 6. Oldenbourg, Munich, Germany, Hamburg, Germany, 2nd edition, 1997.

[830] I. Schwab, J Kobsa, and I. Koychev. Learning about users from observation. In *Proceedings of AAAI Spring Symposium: Adaptive User Interface*, pages 241–247, Palo Alto, USA, March 2000.

[831] T. Serre, L. Wolf, S. Bileschi, M. Riesenhuber, and T. Poggio. Robust object recognition with cortex-like mechanisms. *Pattern Analysis and Machine Intelligence, IEEE Transactions on*, 29(3):411–426, 2007.

[832] M. Sewell. Ensemble learning, 2008.

[833] H. A. Shaban, M. Abou El-Nasr, and R. M. Buehrer. Toward a highly accurate ambulatory system for clinical gait analysis via UWB radios. *Information Technology in Biomedicine, IEEE Transactions on*, 14(2):284–291, 2010.

[834] D. A. Shamma, R. Shaw, P. L. Shafton, and Y. Liu. Watch what i watch: using community activity to understand content. In *Proc. of International Workshop on Multimedia Information Retrieval*, pages 275–284, 2007.

[835] D. A. Shamma, J. Yew, L. Kennedy, and E. F. Churchill. Viral actions: Predicting video view counts using synchronous sharing behaviors. In *Proc. of ICWSM*, 2011.

[836] G. Shani, A. Meisles, Y. Gleyzer, L. Rokach, and D. Ben-Shimon. A stereotypes-based hybrid recommender system for media items. In *Workshop on Intelligent Techniques for Web Personalization*, pages 201–209, Pasadena, USA, July 2007.

[837] S. Shapiro. *Vagueness in Context*. Oxford University Press, 2006.

[838] U. Shardanand and P. Maes. Social information filtering: algorithms for automating word of mouth. In *Proceedings of the SIGCHI Conference on Human Factors in Computing Systems*, CHI '95, pages 210–217, New York, NY, USA, ACM Press/Addison-Wesley Publishing Co., 1995.

[839] J. Shi and J. Malik. Normalized cuts and image segmentation. *IEEE Trans. Pattern Anal. Mach. Intell.*, 22(8):888–905, August 2000.

[840] J. Shi and J. Malik. Normalized cuts and image segmentation. *Pattern Analysis and Machine Intelligence, IEEE Transactions on*, 22(8):888–905, August 2000.

[841] L. C. Shi and B. L. Lu. Dynamic clustering for vigilance analysis based on EEG. In *Engineering in Medicine and Biology Society, 2008. EMBS 2008. 30th Annual International Conference of the IEEE*, pages 54–57, Vancouver, Canada, IEEE, August 2008.

[842] E. Shriberg, A. Stolcke, D. Hakkani-Tür, and G. Tür. Prosody-based automatic segmentation of speech into sentences and topics. *Speech communication*, 32(1-2):127–154, September 2000.

[843] H. Y. Shum, M. Hebert, and K. Ikeuchi. On 3D shape similarity. In *Computer Vision and Pattern Recognition, 1996. Proceedings CVPR '96, 1996 IEEE Computer Society Conference on*, pages 526–531, 1996.

[844] A. Sieg, B. Mobasher, and R. Burke. Improving the effectiveness of collaborative recommendation with ontology-based user profiles. In *Proceedings of the 1st International Workshop on Information Heterogeneity and Fusion in Recommender Systems*, HetRec '10, pages 39–46, 2010.

[845] A. Sieg, B. Mobasher, and R. D. Burke. Learning ontology-based user profiles: A semantic approach to personalized web search. *IEEE Intelligent Informatics Bulletin*, 8(1):7–18, 2007.

[846] H. A. Simon. Motivational and emotional controls of cognition. *Psychological review*, 74(1):29, 1967.

[847] N. Simou and S. Kollias. Fire : A fuzzy reasoning engine for impecise knowledge. In *K-Space PhD Students Workshop, Berlin, Germany, 14 September 2007*, 2007.

[848] R. Sitaram, A. Caria, and N. Birbaumer. Hemodynamic brain-computer interfaces for communication and rehabilitation. *Neural networks*, 22(9):1320–1328, 2009.

[849] J. Sivic and A. Zisserman. Video google: A text retrieval approach to object matching in videos. In *Proceedings of IEEE International Conference on Computer Vision*, pages 1470–1477, Nice, France, October 13–16 2003.

[850] J. Sivic and A. Zisserman. Video google: A text retrieval approach to object matching in videos. In *IEEE International Conference on Computer Vision (ICCV'03), October 14–17, Nice, France*, 2003.

[851] J. Sivic and A. Zisserman. Efficient visual search for objects in videos. *Proceedings of the IEEE*, 96(4):548–566, April 2008.

[852] A. Sloman. *Prolegomena to a theory of communication and affect.* Springer, 1992.

[853] A. Sloman. Review of affective computing. *AI magazine*, 20(1):127–133, 1999.

[854] A. Sloman and M. Croucher. Why robots will have emotions. In *Proceedings of 7th International Joint Conference on Artificial Intelligence*, 1981.

[855] A. F. Smeaton. Indexing, browsing, and searching of digital video and digital audio information. In M. Agosti, F. Crestani, and G. Pasi, editors, *3rd European Summer School on Information Retrieval*, vol. 1980 of *Lecture Notes in Computer Science*, pages 93–110, Varenna, Italy, September 2001.

[856] A. W. M. Smeulders, M. Worring, S. Santini, A. Gupta, and R. Jain. Content-based image retrieval at the end of the early years. *IEEE Transactions on Pattern Analysis and Machine Intelligence*, 22(2):1349–1380, December 2000.

[857] A. W. M. Smeulders, M. Worring, S. Santini, A. Gupta, and R. Jain. Content-based image retrieval at the end of the early years. *IEEE Trans. on Pattern Analysis and Machine Intelligence*, 22(12):1349–1380, 2000.

[858] J. R. Smith. Universal multimedia access. *Proc. SPIE, Multimedia Systems and Applications*, 4209(III):21–32, March 2001.

[859] J. R. Smith and S.-F. Chang. Visualseek: a fully automated content-based image query system. In *Proceedings of the Fourth ACM International Conference on Multimedia*, MULTIMEDIA '96, pages 87–98, New York, NY, USA, ACM, 1996.

[860] J. R. Smith and S. F. Chang. Querying by color regions using the visualseek content-based visual query system. In *Intelligent Multimedia Information Retrieval*, pages 23–41. AAAI Press, 1997.

[861] J. R. Smith and C.-S. Li. Image classification and querying using composite region templates. *Computer Vision and Image Understanding*, 75(1):165–174, 1999.

[862] R. Smith, R. Pawlicki, I. Kókai, J. Finger, and T. Vetter. Navigating in a shape space of registered models. *IEEE Transactions on Visualization and Computer Graphics*, 13(6):1552–1559, November 2007.

[863] D. Smits and P. De Bra. Gale: a highly extensible adaptive hypermedia engine. In *Proceedings of the 22nd ACM Conference on Hypertext and Hypermedia*, HT '11, pages 63–72, New York, NY, USA, ACM, 2011.

[864] C. G. M. Snoek and M. Worring. Multimodal video indexing: A review of the state-of-the-art. *Multimedia Tools and Applications*, 25(1):5–35, January 2005.

[865] C. G. M. Snoek and M. Worring. Concept-based video retrieval. *Found. Trends Inf. Retr.*, 2(4):215–322, April 2009.

[866] SoftKinetic. Softkinetic company. http://www.softkinetic.com/.

[867] M. Song, M. You, N. Li, and C. Chen. A robust multimodal approach for emotion recognition. *Neurocomputing*, 71(10):1913–1920, 2008.

[868] S. Songbo, M. Hassnaa, and A. Hossam. A survey on personalized tv and ngn services through context-awareness. *ACM Comput. Surv.*, 44(1):4:1–4:18, January 2012.

[869] M. Spring. EEG or the consequences of sleigh riding – https://secure.flickr.com/photos/25405741@n00/3190760941. Flickr upload. Creative Commons BY-NC-SA 2.0 License, https://creativecommons.org/licenses/by-nc-sa/2.0/.

[870] Special Issue of Comput. Stand. Interfaces on Standard Reference Model for Intelligent Multimedia Presentation Systems, vol. 18, no. 6-7, December 1997.

[871] T. Starner, B. Schiele, and A. Pentland. Visual contextual awareness in wearable computing. In *International Symposium on Wearable Computers (ISWC'98), October 19–20, Pittsburgh, Pennsylvania, USA*, 1998.

[872] N. Stash, A. Cristea, and P. De Bra. Authoring of Learning Styles in Adaptive Hypermedia. In *World Wide Web Conf., Education Track; New York, NY, USA*, pages 114–123, ACM, 2004.

[873] N. Stash and P. De Bra. Building Adaptive Presentations with AHA! 2.0. In *Proc. of the PEG Conf.; Saint Petersburg, Russia*, June 2003.

[874] K. Stefanidis, I. Ntoutsi, K. Nørvåg, and H.-P. Kriegel. A framework for time-aware recommendations. In *Database and Expert Systems Applications*, pages 329–344. Springer, 2012.

[875] B. Steichen, A. O'Connor, and V. Wade. Personalisation in the wild: providing personalisation across semantic, social and open-web resources. In *Proceedings of the 22nd ACM Conference on Hypertext and Hypermedia*, HT '11, pages 73–82, New York, NY, USA, ACM, 2011.

[876] R. Steinmetz and K. Nahrstedt. *Multimedia: Computing, Communications and Applications*. Prentice Hall, 1995.

[877] G. Stoilos, G. Stamou, J. Z. Pan, N. Simou, and V. Tzouvaras. Reasoning with the Fuzzy Description Logic f-SHIN: Theory, Practice and Applications. In *In Paulo CG Costa, Claudia d'Amato et al. (eds) Uncertainty Reasoning for the Semantic Web I*, 2008.

[878] G. Stoilos, U. Straccia, G. Stamou, and J. Z. Pan. General concept inclusions in fuzzy description logics. In *17th European Conference on Artificial Intelligence (ECAI 06), Riva del Garda, Italy, 2006*, 2006.

[879] A. Stolcke, K. Ries, N. Coccaro, E. Shriberg, R. Bates, D. Jurafsky, P. Taylor, R. Martin, C. Van Ess-Dykema, and M. Meteer. Dialogue act modeling for automatic tagging and recognition of conversational speech. *Computational Linguistics*, 26(3):339–373, September 2000.

[880] K. Stolze. SQL/MM Spatial – the standard to manage spatial data in a relational database system. In *BTW*, vol. 26 of *LNI*, pages 247–264. Gesellschaft für Informatik e.V., 2003.

[881] U. Straccia. Reasoning within Fuzzy Description Logics. *Journal of Artificial Intelligence Research*, 14, 2001.

[882] N. Streitz, T. Prante, C. Röcker, D. Van Alphen, C. Magerkurth, R. Stenzel, and D. Plewe. *Ambient displays and mobile devices for the creation of social architectural spaces*, Chapter 16, pages 387–409. Springer Netherlands, 2003.

[883] M. Stricker, A. Dimai, and E. Dimai. Color indexing with weak spatial constraints. In *Proc. SPIE Storage and Retrieval for Image and Video Databases*, pages 29–40, 1996.

[884] M. Stricker and M. Orengo. Similarity of color images. pages 381–392, 1995.

[885] F. M. Suchanek, G. Kasneci, and G. Weikum. Yago: a core of semantic knowledge. In *Proceedings of the 16th International Conference on World Wide Web*, pages 697–706. ACM, 2007.

[886] SumTotal Systems, Inc., USA. ToolBook, 2006. `http://www.toolbook.com/`, last accessed: 23/1/2013.

[887] Y. Sun, S. Shimada, and M. Morimoto. Visual pattern discovery using web images. In *Proceedings of the 8th ACM International Workshop on Multimedia Information Retrieval*, MIR '06, pages 127–136, New York, NY, USA, ACM, 2006.

[888] W. Sunayama, A. Nagata, and M. Yachida. Image clustering system on www using web texts. In *Hybrid Intelligent Systems, 2004. HIS '04. Fourth International Conference on*, pages 230–235, 2004.

[889] A. Sutcliffe. *Multimedia and virtual reality: designing multisensory user interfaces*. Psychology Press, 2004.

[890] Scalable Vector Graphics (SVG) Full 1.2 Specification, April 2005. `http://www.w3.org/TR/2005/WD-SVG12-20050413/`, last accessed: 22/1/2013.

[891] M. J. Swain and D. H. Ballard. Color indexing. *International Journal of Computer Vision*, 7(1):11–32, November 1991.

[892] T. Syeda-Mahmood and D. Ponceleon. Learning video browsing behavior and its application in the generation of video previews. In *Proc. of ACM International Conference on Multimedia*, pages 119–128, 2001.

[893] Y. Takahashi, N. Nitta, and N. Babaguchi. Video summarization for large sports video archives. *Multimedia and Expo, IEEE International Conference on*, 0:1170–1173, 2005.

[894] T. Taleb, D. Bottazzi, and N. Nasser. A novel middleware solution to improve ubiquitous healthcare systems aided by affective information. *Information Technology in Biomedicine, IEEE Transactions on*, 14(2):335–349, 2010.

[895] A. M. Tam and C. H. C. Leung. Semantic content retrieval and structured annotation: Beyond keywords. *ISO/IEC JTC1/SC29/WG11 MPEG00/M5738.*, March 2000.

[896] A. M. Tam and C. H. C. Leung. Structured natural-language descriptions for semantic content retrieval of visual materials. *J. Am. Soc. Inf. Sci. Technol.*, 52(11):930–937, September 2001.

[897] H. Tamura, S. Mori, and T. Yamawaki. Textural features corresponding to visual perception. *Systems, Man and Cybernetics, IEEE Transactions on*, 8(6):460–473, 1978.

[898] H. Tamura and N. Yokoya. Image database systems: A survey. *Pattern Recognition*, 17(1):29–43, 1984.

[899] J. Tang, X.-S. Hua, M. Wang, Z. Gu, G.-J. Qi, and X. Wu. Correlative linear neighborhood propagation for video annotation. *Systems, Man, and Cybernetics, IEEE Transactions on*, 39(2):409–416, April 2009.

[900] J. Tang, S. Yan, R. Hong, G.-J. Qi, and T.-S. Chua. Inferring semantic concepts from community-contributed images and noisy tags. In *Multimedia, Proceedings of the 17th ACM international conference on*, MM '09, pages 223–232, 2009.

[901] R. S. Tannenbaum. *Theoretical Foundations of Multimedia*. W. H. Freeman & Co., New York, NY, USA, 1998.

[902] N. Tarantilis, C. Tsinaraki, F. Kazasis, N. Gioldasis, and S. Christodoulakis. Evisuge: Event visualization on google earth. In *Sixth International Workshop on Semantic Media Adaptation and Personalization (SMAP), Vigo, Spain*, pages 68–73, December 2011.

[903] B. W. Tatler. The central fixation bias in scene viewing: Selecting an optimal viewing position independently of motor biases and image feature distributions. *Journal of Vision*, 7(14):1–17, 2007.

[904] D. M. J. Tax and R. P. W. Duin. Using two-class classifiers for multi-class classification. In *Proceedings of the of 16th International Conference of Pattern Recognition*, pages 124–127, Quebec, Canada, August 11–15, 2002.

[905] TechCrunch. Intel and softkinetic develop interfaces for interactive adds:. http://techcrunch.com/2012/01/30/softkinetic-and-intel-partner-for-minority-report-style-ads/.

[906] S. Thakur and A. Hanson. A 3D Visualization of Multiple Time Series on Maps. In *2010 14th International Conference Information Visualisation*, pages 336–343, Los Alamitos, CA, USA, 2010. IEEE Computer Society.

[907] The Economist. A world of hits. *The Economist*, November 2009.

[908] Z. Theodosiou, C. Kasapi, and N. Tsapatsoulis. Semantic gap between people: An experimental investigation based on image annotation. In *Seventh International Workshop on Semantic and Social Media Adaptation and Personalization (SMAP)*, pages 73–77, Luxembourg, December 3–4, 2012.

[909] Z. Theodosiou, A. Kounoudes, N. Tsapatsoulis, and M. Milis. Mulvat: A video annotation tool based on xml-dictionaries and shot clustering. In *Proceedings of the 19th International Conference on Artificial Neural Networks*, pages 913–922, limassol, Cyprus, September 14–15, 2009.

[910] Z. Theodosiou and N. Tsapatsoulis. Modelling crowdsourcing originated keywords within the athletics domain. In L. Iliadis, I. Maglogiannis, and H. Papadopoulos, editors, *Artificial Intelligence Applications and Innovations*, vol. 381 of *IFIP Advances in Information and Communication Technology*, pages 404–413. Springer Berlin Heidelberg, September 2012.

[911] Q. T. Tho, S. C. Hui, A. C. M. Fong, and T. H. Cao. Automatic fuzzy ontology generation for semantic web. *IEEE Transactions on Knowledge and Data Engineering*, 18(6):824–856, 2006.

[912] C. Thomas and A. Sheth. On the Expressiveness of the Languages for the Semantic Web – Making a Case for A Little More. *Fuzzy Logic and the Semantic Web*, 2006.

[913] Q. Tian, Y. Wu, and T. S. Huang. Combine user defined region-of-interest and spatial layout for image retrieval. In *Image Processing, 2000. Proceedings. 2000 International Conference on*, vol.3, pages 746–749 2000.

[914] S. Tirthapura, D. Sharvit, P. Klein, and B. B. Kimia. Indexing based on edit-distance matching of shape graphs. In *Multimedia Storage And Archiving Systems III*, pages 25–36, 1998.

[915] E. Tola, V. Lepetit, and P. Fua. Daisy: An efficient dense descriptor applied to wide-baseline stereo. *IEEE Transactions on Pattern Analysis and Machine Intelligence*, 32(5):815–830, May 2010.

[916] C. Town. Ontology-driven Bayesian networks for dynamic scene understanding. In *Computer Vision and Pattern Recognition Workshop, 2004. CVPRW '04. Conference on*, page 116, June 2004.

[917] N. Tractinsky. Tools over solutions? comments on interacting with computers special issue on affective computing. *Interacting with Computers*, 16(4):751–757, 2004.

[918] S. E. Tranter and D. A. Reynolds. An overview of automatic speaker diarization systems. *IEEE Transactions on Audio, Speech, and Language Processing*, 14(5):1557–1565, September 2006.

[919] A. M. Treisman and G. Gelade. A feature-integration theory of attention. *Cognitive Psychology*, 12(1):97–136, 1980.

[920] G. Tryfou, Z. Theodosiou, and N. Tsapatsoulis. Web image context extraction based on semantic representation of web page visual segments. In *Proceedings of the 7th Int. Workshop on Semantic and Social Media Adaptation and Personalization*, pages 63–67, Luxembourg, December 3–4, 2012.

[921] N. Tsapatsoulis and Z. Theodosiou. Object classification using the mpeg-7 visual descriptors: An experimental evaluation using state of the art data classifiers. In C. Alippi, M. Polycarpou, C. Panayiotou, and G. Ellinas, editors, *Artificial Neural Networks ICANN 2009*, vol. 5769/2009 of *Lecture Notes in Computer Science*, pages 905–912. Springer Berlin Heidelberg, September 2009.

[922] D. Tsatsou, F. Menemenis, I. Kompatsiaris, and P. C. Davis. A semantic framework for personalized ad recommendation based on advanced textual analysis. In *Proceedings of the Third ACM Conference on Recommender Systems*, RecSys '09, pages 217–220, New York, NY, USA, ACM, 2009.

[923] D. Tsatsou, S. Papadopoulos, I. Kompatsiaris, and P. C. Davis. Distributed technologies for personalized advertisement delivery. *Online Multimedia Advertising: Techniques and Technologies*, page 233, 2011.

[924] G. Tselentis, J. Domingue, A. Galis, A. Gavras, and D. Hausheer. *Towards the Future Internet: A European Research Perspective*. IOS Press, Amsterdam, The Netherlands, The Netherlands, 2009.

[925] T. Tsikrika, C. Diou, A. P De Vries, and A. Delopoulos. Image annotation using clickthrough data. In *Proceedings of the 8th International Conference on Image and Video Retrieval*, pages 1–8, Santorini, Greece, July 8–10, 2009.

[926] C. Tsinaraki and S. Christodoulakis. Domain Knowledge Representation in Semantic MPEG-7 Descriptions. In M. Angelides and H. Agius, editors, *The Handbook of MPEG applications: Standards in Practice*, pages 293–316. Wiley Publishers, West Sussex, UK, November 2010.

[927] C. Tsinaraki, P. Polydoros, and S. Christodoulakis. Interoperability Support between MPEG-7/21 and OWL in DS-MIRF. *Knowledge and Data Engineering, IEEE Transactions on*, 19(2):219–232, February 2007.

[928] T. Tsunoda and M. Hoshino. Automatic metadata expansion and indirect collaborative filtering for tv program recommendation system. *Multimedia Tools Appl.*, 36(1-2):37–54, January 2008.

[929] M. Tuceryan and A. K. Jain. Texture analysis. *Handbook of Pattern Recognition and Computer Vision*, 276, 1993.

[930] Tumult, Inc. Hype, 2013. `http://tumult.com/hype/`, last accessed: 20/1/2013.

[931] G. Tur, A. Stolcke, L. Voss, S. Peters, D. Hakkani-Tur, J. Dowding, B. Favre, R. Fernndez, M. Frampton, and M. Frandsen. The CALO meeting assistant system. *Audio, Speech, and Language Processing, IEEE Transactions on*, 18(6):1601–1611, August 2010.

[932] M. Turk and A. Pentland. Eigenfaces for recognition. *Journal of Cognitive Neuroscience*, 3(1):71–86, 1991.

[933] M. A. Turk and A. P. Pentland. Face recognition using eigenfaces. In *Proceedings of the IEEE Computer Society Conference on Computer Vision and Pattern Recognition*, pages 586–591, Maui, HI, USA, June 3–6, 1991.

[934] R. M. Turner. Context-mediated behavior for intelligent agents. *International Journal of Human-Computer Studies, special issue on Using Context in Applications*, 48(3):307–330, March 1998.

[935] T. Tuytelaars and K. Mikolajczyk. Local invariant feature detectors: A survey. *Computer Graphics and Vision*, 3(3):177–280, January 2008.

[936] A. Tuzhilin. Personalization: The state of the art and future directions. In *Business Computing*, volume 3 of *Handbooks in Information Systems*, pages 3–43. Emerald Group Publishing Limited, Bingley UK, 2009.

[937] Unidex, Inc., USA. Universal Turing Machine in XSLT, March 2001. http://unidex.com/turing/utm.htm, last accessed: 23/1/2013.

[938] J. Urban and J. M. Jose. Evaluating a workspaces usefulness for image retrieval. *Multimedia Systems*, 12(4-5):355–373, 2007.

[939] M. Uschold and M. King. Towards a Methodology for Building Ontologies. *IJCAI-95 Workshop Basic Ontological Issues in Knowledge Sharing*, pages 6.1–6.10, 1995.

[940] A. Vailaya, M. A. T. Figueiredo, A. K. Jain, and H. J. Zhang. Image classification for content-based indexing. *Image Processing, IEEE Transactions on*, 10(1):117–130, 2001.

[941] A. Vailaya, A. Jain, and H.J. Zhang. On image classification: City images vs. landscapes. *Pattern Recognition*, 31(12):1921–1935, 1998.

[942] D. Vallet, I. Cantador, M. Fernández, and P. Castells. A multipurpose ontology-based approach for personalized content filtering and retrieval. In *Proc. of the First International Workshop on Semantic Media Adaptation and Personalization (SMAP2006)*, pages 19–24. IEEE, 2006.

[943] D. Vallet, M. Fernandez, P. Castells, P. Mylonas, and Y. Avrithis. A contextual personalization approach based on ontological knowledge. In *International Workshop on Context and Ontologies (C&O 2006) at the 17th European Conference on Artificial Intelligence (ECAI 2006)*. Citeseer, 2006.

[944] G. C. Van den Eijkel, M. M. Lankhorst, H. Van Kranenburg, and M. E. Bijlsma. Personalization of next-generation mobile services. In *Wireless World Research forum; Munich, Germany*, March 2001.

[945] P. van Emde Boas. Machine models and simulation. In *Handbook of Theoretical Computer Science, Volume A: Algorithms and Complexity (A)*, pages 1–66. Elsevier, 1990.

[946] J. van Gemert, C. Veenman, A. Smeulders, and J-M. Geusebroek. Visual word ambiguity. *IEEE Transactions on Pattern Analysis Machine Intelligence*, 32:1271–1283, 2010.

[947] J. van Ossenbruggen, J. Geurts, F. Cornelissen, L. Hardman, and L. Rutledge. Towards second and third generation web-based multimedia. In *WWW*, pages 479–488, 2001.

[948] J. van Ossenbruggen, L. Hardman, J. Geurts, and L. Rutledge. Towards a multimedia formatting vocabulary. In *Proc. of the 12th Int. Conf. on World Wide Web; Budapest, Hungary*, pages 384–393. ACM, 2003.

[949] J. R. van Ossenbruggen, F. J. Cornelissen, J. P. T. M. Guerts, L. W. Rutledge, and H. L. Hardman. Cuypers: a semi-automatic hypermedia presentation system. Technical Report INS-R0025, CWI, The Netherlands, December 2000.

[950] G. van Rossum, J. Jansen, S. Mullender, and D. C. A. Bulterman. CMIFed: a presentation environment for portable hypermedia documents. In *MULTIMEDIA*, pages 183–188. ACM, 1993.

[951] E. Vanmarcke. *Random fields, analysis and synthesis*. MIT Press, 1983.

[952] V. Vapnik. *The Nature of Statistical Learning Theory*. Springer–Verlag, 1995.

[953] A. V. Vasilakos and C. Lisetti. Guest editorial special section on affective and pervasive computing for healthcare. *Information Technology in Biomedicine, IEEE Transactions on*, 14(2):183–185, 2010.

[954] Vicon. Vicon motion capture systems. http://www.vicon.com/products/viconmx.html.

[955] E. Vig, M. Dorr, and D. Cox. Space-variant descriptor sampling for action recognition based on saliency and eye movements. In *European Conference on Computer Vision (ECCV'12), October 7–13, Florence, Italy*, 2012.

[956] A. Vinciarelli, M. Pantic, and H. Bourlard. Social signal processing: Survey of an emerging domain. *Image Vision Comput.*, 27(12):1743–1759, November 2009.

[957] M. Virvou, A. Savvopoulos, G.-A. Tsihrintzis, and D. N. Sotiropoulos. Constructing stereotypes for an adaptive e-shop using AIN-based clustering. *Lecture Notes in Computer Science*, 4431:837–845, April 2007.

[958] J. Vlaxavas, P. Kefalas, N. Vasileiadis, F. Kokkoras, and H. Sakellariou. *Artifical Intelligence*. Giourdas Publications, 2006.

[959] D. Vogel and R. Balakrishnan. Interactive public ambient displays: transitioning from implicit to explicit, public to personal, interaction with multiple users. In *Proceedings of the 17th Annual ACM Symposium on User Interface Software and Technology*, UIST '04, pages 137–146, New York, NY, USA, 2004. ACM.

[960] J. Vogel and B. Schiele. Semantic modeling of natural scenes for content-based image retrieval. *International Journal of Computer Vision*, 72(2):133–157, April 2007.

[961] J. Vogel, A. Schwaninger, C. Wallraven, and H. H. Bülthoff. Categorization of natural scenes: Local versus global information and the role of color. *ACM Trans. Appl. Percept.*, 4(3), November 2007.

[962] J. Volz, C. Bizer, M. Gaedke, and G. Kobilarov. Silk a link discovery framework for the web of data. In *Proc. of the 2nd Workshop about Linked Data on the Web (LDOW2009)*, vol, 538, Madrid, Spain, CEUR-WS, April 2009.

[963] L. von Bertalanffy. *General System Theory: Foundations, Development, Applications (Revised Edition)*. George Braziller Inc., revised edition, May 1969.

[964] A. Vourvopoulos and F. Liarokapis. Brain-controlled nxt robot: Teleoperating a robot through brain electrical activity. In *Games and Virtual Worlds for Serious Applications (VS-GAMES), 2011 Third International Conference on*, pages 140–143, Athens, Greece, IEEE, May 2011.

[965] E. Vozalis and K. G. Margaritis. Analysis of recommender systems' algorithms. In *Proceedings of the 6th Hellenic-European Conference on Computer Mathematics and Its Applications (HERCMA 2003)*, pages 732–745, Athens, Greece, September 25–27 2003. Athens: LEA.

[966] M. Škrabal, R.and Šimůnek, S. Vojíř, A. Hazucha, T. Marek, D. Chudán, and T. Kliegr. Association rule mining following the web search paradigm. In *Proceedings of the 2012 European Conference on Machine Learning and Knowledge Discovery in Databases - Volume Part II*, ECML PKDD'12, pages 808–811, Berlin, Heidelberg, Springer–Verlag, 2012.

[967] World-Wide Web Consortium (W3C). Scalable Vector Graphics (SVG) Full 1.2 Specification, url=http://www.w3.org/TR/SVG12/.

[968] W3C OWLWorking Group. OWL 2 Web Ontology Language: Document Overview, 2009, http://www.w3.org/TR/owl2-overview.

[969] W3C (World Wide Web Consortium). Cascading Style Sheets, 2006. http://www.w3.org/Style/CSS/, last accessed: 23/1/2013.

[970] W3C (World Wide Web Consortium). HTML5, December 2012. http://www.w3.org/TR/html5/, last accessed: 20/1/2013.

[971] T. Wahl and K. Rothermel. Representing Time in Multimedia Systems. In *Proc. IEEE Int. Conf. on Multimedia Computing and Systems; Boston, MA, USA*, pages 538–543, May 1994.

[972] A. Waibel, M. Bett, M. Finke, and R. Stiefelhagen. Meeting browser: Tracking and summarizing meetings. In D. E. M. Penrose, editor,

*Proceedings of the Broadcast News Transcription and Understanding Workshop*, pages 281–286, Lansdowne, VA, Morgan Kaufmann, February 1998.

[973] A. Waibel, M. Brett, F. Metze, K. Ries, T. Schaaf, T. Schultz, H. Soltau, H. Yu, and K. Zechner. Advances in automatic meeting record creation and access. In *Proceedings of the IEEE International Conference on Acoustics, Speech and Signal Processing*, vol. 1, pages 597–600, Pittsburgh, PA, IEEE Press, May 2001.

[974] M. Wallace, Ph. Mylonas, G. Akrivas, Y. Avrithis, and S. Kollias. Automatic thematic categorization of multimedia documents using ontological information and fuzzy algebra. *Studies in Fuzziness and Soft Computing, Soft Computing in Ontologies and Semantic Web*, 204, 2006.

[975] C. Wang, L. Zhang, and H. J. Zhang. Learning to reduce the semantic gap in web image retrieval and annotation. In *Proceedings of the 31st Annual International ACM SIGIR Conference on Research and Development in Information Retrieval*, SIGIR '08, pages 355–362, New York, NY, USA, ACM, 2008.

[976] H. H. Wang, D. Mohamad, and N. A. Ismail. Approaches, challenges and future direction of image retrieval. *CoRR*, abs/1006.4568, 2010.

[977] J. Wang and Thiesson. M. Image and video segmentation by anisotropic kernel mean shift. In Toms Pajdla and Ji Matas, editors, *Computer Vision - ECCV 2004*, volume 3022 of *Lecture Notes in Computer Science*, pages 238–249. Springer Berlin Heidelberg, 2004.

[978] J. Z. Wang, J. Li, R. M. Gray, and G. Wiederhold. Unsupervised multiresolution segmentation for images with low depth of field. *Pattern Analysis and Machine Intelligence, IEEE Transactions on...*, 23(1):85–90, January 2001.

[979] J. Z. Wang, J. Li, and G. Wiederhold. Simplicity: Semantics-sensitive integrated matching for picture libraries. *Pattern Analysis and Machine Intelligence, IEEE Transactions on...*, 23(9):947–963, 2001.

[980] P. Wang. Recommendation based on personal preference. *Series In Machine Perception And Artificial Intelligence*, vol. 58: pages 101–116, 2004.

[981] X.-J. Wang, L. Zhang, F. Jing, and W.-Y. Ma. Image annotation using search and mining technologies. In *Proceedings of the 15th International Conference on World Wide Web*, WWW '06, pages 1045–1046, New York, NY, USA, ACM, 2006.

[982] N. Weissenberg, A. Voisard, and R. Gartmann. Using ontologies in personalized mobile applications. In *Proceedings of the 12th Annual ACM*

*International Workshop on Geographic Information Systems*, GIS '04, pages 2–11, New York, NY, USA, ACM, 2004.

[983] P. Welinder and P. Perona. Online crowdsourcing: rating annotators and obtaining cost effective labels. In *Proceedings of IEEE Conference on Computer Vision and Pattern Recognition*, pages 25–32, San Francisco, CA, USA, June 13–18, 2010.

[984] P. Wellner, M. Flynn, S. Tucker, and S. Whittaker. A meeting browser evaluation test. In *CHI '05 Extended abstracts on Human factors in computing systems*, pages 2021–2024, New York, NY, ACM Press, May 2005.

[985] I. Q. Whishaw and C. H. Vanderwolf. Hippocampal EEG and behavior: Change in amplitude and frequency of RSA (theta rhythm) associated with spontaneous and learned movement patterns in rats and cats1. *Behavioral biology*, 8(4):461–484, June 1973.

[986] S. Whittaker, R. Laban, and S. Tucker. Analysing meeting records: An ethnographic study and technological implications. In Steve Renals and Samy Bengio, editors, *Machine Learning for Multimodal Interaction*, vol. 3869 of *Lecture Notes in Computer Science*, pages 101–113. Springer, Bethesda, MD, May 2006.

[987] Wikipedia. Animated mapping, url=http://en.wikipedia.org/wiki/Animated_mapping. Wikipedia, The Free Encyclopedia.

[988] Wikipedia. Geographic Information System, url=http://en.wikipedia.org/wiki/Geographic_information_system. Wikipedia, The Free Encyclopedia.

[989] Wikipedia. Geovisualization, url=http://en.wikipedia.org/wiki/Geovisualization. Wikipedia, The Free Encyclopedia.

[990] Wikipedia. Historical Geographic Information System, url=http://en.wikipedia.org/wiki/Historical_geographic_information_system. Wikipedia, The Free Encyclopedia.

[991] R. C. F. Wong. Automatic semantic image annotation and retrieval. *Thesis for the degree of Doctor of Philosophy.*, Hong Kong Baptist University, August 2010.

[992] R. C. F. Wong and C. H. C. Leung. Automatic semantic annotation of real-world web images. *IEEE Trans. Pattern Anal. Mach. Intell.*, 30(11):1933–1944, November 2008.

[993] R. C. F. Wong and C. H. C. Leung. Knowledge-based expansion for image indexing. In *Proceedings of the International Computer Symposium ICS 2008*, pages 161–165, 2008.

[994] R. C. F. Wong and C. H. C. Leung. Incorporating concept ontology into multi-level image indexing. In *Proceedings of the First International Conference on Internet Multimedia Computing and Service*, ICIMCS '09, pages 90–96, New York, NY, USA, ACM, 2009.

[995] M. Woo, N. U. Qazi, and A. Ghafoor. A synchronization framework for communication of pre-orchestrated multimedia information. *Network, IEEE*, 8(1):52–61, January–February 1994.

[996] D. Wooding. Eye movements of large populations: Ii. deriving regions of interest, coverage, and similarity using fixation maps. *Behavior Research Methods*, 34:518–528, 2002.

[997] C.-H. Wu and W.-B. Liang. Emotion recognition of affective speech based on multiple classifiers using acoustic-prosodic information and semantic labels. *Affective Computing, IEEE Transactions on*, 2(1):10–21, 2011.

[998] D. Wu, C. G. Courtney, B. J. Lance, S. S. Narayanan, M. E. Dawson, K. S. Oie, and T. D. Parsons. Optimal arousal identification and classification for affective computing using physiological signals: virtual reality stroop task. *Affective Computing, IEEE Transactions on*, 1(2):109–118, 2010.

[999] H. Wu and P. De Bra. Link-Independent Navigation Support in Web-Based Adaptive Hypermedia. In *11th Int. World Wide Web Conf.; Honolulu, HI, USA*, May 2002.

[1000] H. Wu, E. de Kort, and P. De Bra. Design issues for general-purpose adaptive hypermedia systems. In *Proc. of the 12th ACM Conf. on Hypertext and Hypermedia; Aarhus, Denmark*, pages 141–150. ACM, 2001.

[1001] Z. Wu and M. Palmer. Verb semantics and lexical selection. In *Proceedings of 32nd Annual Meeting of the Association for Computational Linguistics*, pages 133–138, New Mexico State University, New Mexico, June 1994.

[1002] J. Xiao, X. Zhang, P. Cheatle, Y. Gao, and C. B. Atkins. Mixed-initiative photo collage authoring. In *Proceedings of the 16th ACM International Conference on Multimedia*, MM '08, pages 509–518, New York, NY, USA, ACM, 2008.

[1003] L. Xie, Y. Zhao, and Z. Zhu. A unified system for web personal image retrieval. In *Intelligent Information Hiding and Multimedia Signal Processing, 2008. IIHMSP '08 International Conference on*, pages 787–790, 2008.

[1004] Z. Xingquan, W. Xindong, A. K. Elmagarmid, F. Zhe, and W. Lide. Video data mining: semantic indexing and event detection from the association perspective. *Knowledge and Data Engineering, IEEE Transactions on*, 17(5):665–677, May 2005.

[1005] C. Xu and J. L. Prince. Snakes, shapes, and gradient vector flow. *Image Processing, IEEE Transactions on*, 7(3):359–369, 1998.

[1006] G. Xu, Y. Zhang, and L. Li. Web mining and recommendation systems. In *Web Mining and Social Networking*, vol. 6 of *Web Information Systems Engineering and Internet Technologies Book Series*, pages 169–188. Springer US, 2011.

[1007] P. Xuwei and Z. Li. A service-oriented middleware architecture for building context-aware personalized information service. In *in Proc. of Int. Symp. On Intelligent Ubiquitous Computing and Education*, pages 457–460. IEEE Computer Society, 2009.

[1008] Y. Blanco-Fernandez, M. Lopez-Nores, J. Pazos-Arias, A. Gil-Solla and M. Ramos-Cabrer. Personalizing e-commerce by semantics-enhanced strategies and time-aware recommendations. In *Proceedings of 3rd International Workshop on Semantic Media Adaptation and Personalization*, pages 193–198, Prague, Czech Republich, December 2008.

[1009] Y. Blanco-Fernandez, M. Lopez-Nores, J.J. Pazos-Arias and J. Garcia-Duque. An improvement for semantics-based recommender systems grounded on attaching temporal information to ontologies and user profiles. *Engineering Applications of Artificial Intelligence*, 24(8):1385–1397, December 2011.

[1010] R. Yan and A. Hauptmann. A review of text and image retrieval approaches for broadcast news video. *Information Retrieval*, 10:445–484, 2007.

[1011] T. A. Yanar and Z. Akyürek. Fuzzy model tuning using simulated annealing. *Expert Systems and Applications*, 38(7):8159–8169, July 2011.

[1012] C. C. Yang. Content-based image retrieval: A comparison between query by example and image browsing map approaches. *J. Information Science*, 30(3):254–267, 2004.

[1013] J. Yang, Q. Li, L. Wenyin, and Y. Zhuang. Content-based retrieval of FlashTM movies: research issues, generic framework, and future directions. *Multimedia Tools Appl.*, 34(1):1–23, 2007.

[1014] J. Yang, J.-Y. Yang, D. Zhang, and J.-F. Lu. Feature fusion: Parallel strategy vs. serial strategy. *Pattern Recognition*, 33:1369–1381, June 2003.

[1015] J. Yang, K. Yu, Y. Gong, and T. Huang. Linear spatial pyramid matching using sparse coding for image classification. In *IEEE Conference on Computer Vision and Pattern Recognition (CVPR'09), June 20–25, Miami, Florida, USA*, 2009.

[1016] S.-H. Yang, B. Long, A. Smola, N. Sadagopan, Z. Zheng, and H. Zha. Like like alike: joint friendship and interest propagation in social networks. In *Proceedings of the 20th International Conference on World Wide Web*, WWW '11, pages 537–546, New York, NY, USA, ACM, 2011.

[1017] J. Yew and D. A. Shamma. Know your data: Understanding implicit usage versus explicit action in video content classification. In *Proc. of IS&T/SPIE Symp. on Electronic Imaging: Science & Technology*, pages 297–306, 2011.

[1018] J. Yew, D. A. Shamma, and E. F. Churchill. Knowing funny: genre perception and categorization in social video sharing. In *Proc. of Annual Conference on Human Factors in Computing Systems*, pages 297–306, 2011.

[1019] C. Yin and S.-Y. Kung. A hierarchical algorithm for image retrieval by sketch. In *IEEE First Workshop on Multimedia Signal Processing*, pages 564–569, 1997.

[1020] R. Yong, T. S. Huang, M. Ortega, and S. Mehrotra. Relevance feedback: a power tool for interactive content-based image retrieval. *Circuits and Systems for Video Technology, IEEE Transactions on*, 8(5):644–655, September 1998.

[1021] YouTube. The acceptance of free laptops, that have been given to secondary education students, 2012. http://www.youtube.com/watch?v=Z09ythJT9Wk.

[1022] YouTube. Protagonists tv series, 2012. http://www.youtube.com/watch?v=GOQfIXxbjlE.

[1023] YouTube. Soufle sokolatas - cooking lesson, 2012. http://www.youtube.com/watch?v0LzkYvtqlT5I.

[1024] YouTube. Upload videos longer than 15 minutes, 2013. http://support.google.com/youtube/bin/answer.py?hl=en-US&answer=71673&rd=1.

[1025] B. Yu, W.-Y. Ma, K. Nahrstedt, and H.-J. Zhang. Video summarization based on user log enhanced link analysis. In *Proc. of ACM International Conference on Multimedia*, pages 382–391, 2003.

[1026] Z. Yu, Z. Yu, X. Zhou, C. Becker, and Y. Nakamura. Tree-based mining for discovering patterns of human interaction in meetings. *IEEE Transactions on Knowledge and Data Engineering*, 24(4):759–768, April 2012.

[1027] L. A. Zadeh. Fuzzy logic= computing with words. *Fuzzy Systems, IEEE Transactions on*, 4(2):103–111, 1996.

[1028] L. A. Zadeh. From search engines to question-answering systems the need for new tools. In *12th IEEE International Conference on Fuzzy Systems*, volume 2, pages 1107–1109, 2003.

[1029] P. M. Zadeh and M. S. Moshkenani. Mining social network for semantic advertisement. In *Proceedings of the 2008 Third International Conference on Convergence and Hybrid Information Technology – Volume 01*, ICCIT '08, pages 611–618, Washington, DC, USA, IEEE Computer Society, 2008.

[1030] T. Zahariadis, P. Daras, J. Bouwen, N. Niebert, D. Griffin, F. Alvarez, and G. Camarillo. *Towards the Future Internet: A European Research Perspective*, chapter Towards a Content-Centric Internet, pages 227–234. IOS Press, Amsterdam, 2010.

[1031] K. Zechner and A. Waibel. DiaSumm: flexible summarization of spontaneous dialogues in unrestricted domains. In *Proceedings of the 18th Conference on Computational linguistics*, pages 968–974, Hong Kong, China, ACL, October 2000.

[1032] Z. Zeng, M. Pantic, G. I Roisman, and T. S Huang. A survey of affect recognition methods: Audio, visual, and spontaneous expressions. *Pattern Analysis and Machine Intelligence, IEEE Transactions on*, 31(1):39–58, 2009.

[1033] J. Zhai and A. Barreto. Stress detection in computer users based on digital signal processing of noninvasive physiological variables. In *Engineering in Medicine and Biology Society, 2006. EMBS'06. 28th Annual International Conference of the IEEE*, pages 1355–1358. IEEE, 2006.

[1034] J. Zhai, Y. Liang, J. Jiang, and Y. Yu. Fuzzy ontology models based on fuzzy linguistic variable for knowledge management and information retrieval. *Intelligent Information Processing IV*, pages 58–67, 2008.

[1035] D. Zhang, M. M. Islam, and G. Lu. A review on automatic image annotation techniques. *Pattern Recognition*, 45:346–362, January 2012.

[1036] H. Zhang, C. Low, and S. Smoliar. Video parsing and browsing using compressed data. *Multimedia Tools and Applications*, 1:89–111, March 1995.

[1037] M. Zhang and A. A. Sawchuk. Motion primitive-based human activity recognition using a bag-of-features approach. In *Proceedings of the 2nd ACM SIGHIT International Health Informatics Symposium*, pages 631–640, Miami, FL, USA, January 28–30 2012.

[1038] Q. Zhang and E. Izquierdo. Combining low-level features for semantic inference in image retrieval. *Journal on Advances in Signal Processing*, 2007, 2007.

[1039] R. Zhang and Z. Zhang. Effective image retrieval based on hidden concept discovery in image database. *Image Processing, IEEE Transactions on*, 16(2):562–572, 2007.

[1040] R. Zhang, Z. Zhang, M. Li, and H. J. Zhang. A probabilistic semantic model for image annotation and multi-modal image retrieval. *Multimedia Systems*, pages 27–33, August 2006.

[1041] X. Zhang, H. Ju, and J. Wang. *Electrochemical sensors, biosensors and their biomedical applications*. Elsevier, 2008.

[1042] Y. Zhang, Y. Chen, S.L. Bressler, and M. Ding. Response preparation and inhibition: the role of the cortical sensorimotor beta rhythm. *Neuroscience*, 156(1):238, September 2008.

[1043] R. Zhao and W. I. Grosky. Narrowing the semantic gap – improved text-based web document retrieval using visual features. *Multimedia, IEEE Transactions on*, 4(2):189–200, June 2002.

[1044] Y. Zhiwen, Z. XingShe, Z. Daqing, C. Chung-Yau, W. Xiaohang, and M. Ji. Supporting context-aware media recommendations for smart phones. *Pervasive Computing, IEEE*, 5(3):68 –75, July–September 2006.

[1045] Y. Zhou, D. Wilkinson, R. Schreiber, and R. Pan. Large-scale parallel collaborative filtering for the netflix prize. In R. Fleischer and J. Xu, editors, *Algorithmic Aspects in Information and Management, 4th International Conference, AAIM 2008, Shanghai, China, June 23-25, 2008. Proceedings*, volume 5034 of *LNCS*, pages 337–348. Springer, June 23–25 2008.

[1046] Z. H. Zhou and M. L. Zhang. Multi-instance multi-label learning with application to scene classification. In *Advances in Neural Information Processing Systems 19*, pages 1609–1616, December 2006.

[1047] S. Zhu and X. Tan. A novel automatic image annotation method based on multi-instance learning. *Procedia Engineering*, 15:3439–3444, 2011.

[1048] X. Zhu, X. Wu, A. K. Elmagarmid, Z. Feng, and L. Wu. Video data mining: Semantic indexing and event detection from the association perspective. *Knowledge and Data engineering, IEEE Transactions on*, 17(5):665–677, May 2005.

[1049] S. Zillner, U. Westermann, and W. Winiwarter. EMMA - a query algebra for enhanced multimedia meta objects. In *CoopIS/DOA/ODBASE*, pages 1030–1049. Springer, 2004.

[1050] J. Zimmerman, K. Kurapati, A. Buczak, D. Schafer, S. Gutta, and J. Martino. TV personalization system: design of a TV show recommender engine and interface. In *Personalized Digital Television: Targeting Programs to Individual Viewers*, pages 27–51. Kluwer Academics Publishers, 2004.

[1051] T. Zoller and J. M. Buhmann. Robust image segmentation using resampling and shape constraints. *IEEE Trans. Pattern Anal. Mach. Intell.*, 29(7):1147–1164, July 2007.

# Index